Gerhard Gottschalk

**Discover the World of
Microbes**

Related Titles

Dale, J. W., Park, S. F.

Molecular Genetics of Bacteria

2010
ISBN: 978-0-470-74184-9

Krämer, R., Jung, K. (eds.)

Bacterial Signaling

2010
ISBN: 978-3-527-32365-4

Feldmann, H.

Yeast

Molecular and Cell Biology

2010
ISBN: 978-3-527-32609-9

Wiegel, J., Maier, R., Adams, M. (eds.)

Incredible Anaerobes

From Physiology to Genomics to Fuels

ISBN: 978-1-57331-705-4

Wilson, M.

Bacteriology of Humans

An Ecological Perspective

2008
ISBN: 978-1-4051-6165-7

Gerhard Gottschalk

Discover the World of Microbes

Bacteria, Archaea, and Viruses

The Author

Prof. Dr. Gerhard Gottschalk
Institut für Mikrobiologie und
Genetik
Georg-August-Universität
Grisebachstr. 8
37077 Göttingen
Germany

Cover
The cover was designed by Anne Kemmling, Göttingen. Source of the micrographs (left to right: row 1, Anne Kemmling; row 2, Manfred Rohde, Braunschweig, fig. 8a (this book), Michael Hoppert, Göttingen; row 3, Michael Hoppert, Manfred Rohde, figures 74b and 75 (this book); row 4, Jim Hogle, Boston, Anne Kemmling.

■ **Limit of Liability/Disclaimer of Warranty:** While the publisher and author have used their best efforts in preparing this book, they make no representations or warranties with respect to the accuracy or completeness of the contents of this book and specifically disclaim any implied warranties of merchantability or fitness for a particular purpose. No warranty can be created or extended by sales representatives or written sales materials. The advice and strategies contained herein may not be suitable for your situation. You should consult with a professional where appropriate. Neither the publisher nor authors shall be liable for any loss of profit or any other commercial damages, including but not limited to special, incidental, consequential, or other damages.

Library of Congress Card No.: applied for

British Library Cataloguing-in-Publication Data
A catalogue record for this book is available from the British Library.

Bibliographic information published by the Deutsche Nationalbibliothek
The Deutsche Nationalbibliothek lists this publication in the Deutsche Nationalbibliografie; detailed bibliographic data are available on the Internet at <http://dnb.d-nb.de>.

© 2012 Wiley-VCH Verlag & Co. KGaA, Boschstr. 12, 69469 Weinheim, Germany

Wiley-Blackwell is an imprint of John Wiley & Sons, formed by the merger of Wiley's global Scientific, Technical, and Medical business with Blackwell Publishing.

All rights reserved (including those of translation into other languages). No part of this book may be reproduced in any form – by photoprinting, microfilm, or any other means – nor transmitted or translated into a machine language without written permission from the publishers. Registered names, trademarks, etc. used in this book, even when not specifically marked as such, are not to be considered unprotected by law.

Composition   Toppan Best-set Premedia Ltd., Hong Kong

Printing and Binding   Fabulous Printers Pte Ltd., Singapore

Printed in Singapore
Printed on acid-free paper

Print ISBN: 978-3-527-32845-1

Contents

Preface   IX
Prolog   XI

Part One   Reading Section   1

Chapter 1   Extremely small but incredibly active   3

Chapter 2   Bacteria are organisms like you and me   7

Chapter 3   My name is LUCA   15

Chapter 4   From the Big Bang to LUCA   23

Chapter 5 O_2 33

Chapter 6   Life in boiling water   39

Chapter 7   Life in the Dead Sea   45

Chapter 8   Bacteria and archaea are everywhere   53

Chapter 9   The power of photosynthesis, even in almost complete darkness   65

Chapter 10   Man and his microbes   73

Chapter 11   Without bacteria there is no protein   81

Chapter 12   Napoleon's victory gardens   87

Chapter 13   Alessandro Volta's and George Washington's combustible air   91

Chapter 14	Microbes as climate makers	*99*
Chapter 15	How a state was founded with the aid of *Clostridium acetobutylicum*	*105*
Chapter 16	Pulque, wine, and biofuel	*111*
Chapter 17	Energy conservation from renewable resources	*117*
Chapter 18	Cheese and vinegar	*121*
Chapter 19	The periodic table of bioelements	*127*
Chapter 20	Bacterial sex life	*133*
Chapter 21	Bacteria can also catch viruses	*145*
Chapter 22	Antibiotics: from microorganisms against microorganisms	*149*
Chapter 23	Plasmids and resistances	*159*
Chapter 24	*Agrobacterium tumefaciens*, a genetic engineer *par excellence*	*165*
Chapter 25	*Eco* R1 and PCR – molecular biology at its finest	*169*
Chapter 26	Interbacterial relationships	*177*
Chapter 27	From life as a nomad to life as an endosymbiont	*185*
Chapter 28	Bacteria as production factories	*191*
Chapter 29	Plants, animals, and humans as food resources for bacteria	*203*
Chapter 30	Viruses, chemicals causing epidemics?	*221*
Chapter 31	The "omics" era	*235*
Chapter 32	Incredible microbes	*245*

Epilog   *256*

Part Two   Study Guide   *257*

Overview to the Study Guide   *259*

Section 1   Microbial growth   *261*

Section 2	Molecules that make up microbes	267
Section 3	Evolution, from the RNA world to the tree of life	277
Section 4	Archaea	281
Section 5	Bacterial diversity	289
Section 6	Membranes and energy	297
Section 7	Carbon metabolism	311
Section 8	Regulation of microbial metabolism	325
Section 9	Genomes, genes, and gene transfer	333
Section 10	In-depth study of four special topics	337

Appendix A Selected literature 345
Appendix B Glossary 351
Appendix C Subject index of figures and tables 373
Credits 379
Index 381

Preface

Numerous discussions have repeatedly made it clear to me that microbes are a mystery to most people. After all, they are invisible. Microbes cause disease; they are involved in all kinds of novel production processes; they can easily be manipulated genetically – so it's best to keep your distance! Besides, it requires some effort to fathom the secrets of these organisms. However, it is well worth the time. Microbes are fascinating in their diversity, activities, and achievements. Considering their abundance and their involvement in the global cycles of carbon, nitrogen, and sulfur, it is no exaggeration to say that microbes rule our planet. The object of this book is to spark interest in these multifaceted organisms. It consists of two parts, a reading section and a study guide which has been added to the thirty-two essays to allow this book to serve as an introductory text. Formulas and equations have been kept to a minimum whenever possible.

This book was written to include a fictitious conversation with someone who has an interest in the subject without being an expert and who asks questions every now and then. This form has been chosen to keep the text down-to-earth. Statements of highly respected colleagues have lent authenticity and brilliance to many of the chapters, for which I am truly grateful. I am also honored that they have read the respective chapters and complemented them in such a persuasive and competent manner. They are: for Chapter 1: Frank Mayer, Stade (DE); Chapter 3: Ralph Wolfe, Urbana, IL (US); Chapter 4: Manfred Eigen, Goettingen (DE), Gerald Joyce, La Jolla, CA (US); Chapter 5: Joachim Reitner, Goettingen (DE); Chapter 6: Karl Stetter, Regensburg (DE), Gregory Zeikus, East Lansing, MI (US); Chapter 7: Aharon Oren, Jerusalem (IL), Colleen Cavanaugh, Cambridge, MA (US), Antje Boetius, Bremen (DE); Chapter 8: William Whitman, Athens, GA (US), Karl-Heinz Schleifer and Wolfgang Ludwig, Munich (DE), Dieter Oesterhelt, Martinsried (DE), Volker Mueller, Frankfurt/Main (DE), Andrew Benson, Santa Barbara, CA (US); Chapter 9: Joerg Overmann, Munich (DE), Jack C. Meeks, Davis, CA (US); Chapter 10: Holger Brueggemann, Berlin (DE), Michael Blaut, Potsdam-Rehbruecke (DE); Chapter 11: Oliver Einsle, Freiburg (DE), Alfred Puehler, Bielefeld (DE); Chapter 12: Gijs Kuenen, Delft (NL); Chapter 13: Douglas Eveleigh, New Brunswick, NJ (US), Rolf Thauer, Marburg (DE); Chapter 15: Peter Duerre, Ulm and Hubert Bahl, Rostock (DE), Michael Young, Aberystwyth (UK); Chapter 16: Douglas Clark, Berkeley, CA (US); Chapter 18: Hermann Sahm,

Juelich (DE), Karl Sanford, Palo Alto, CA (US); Chapter 19: Jan Andreesen, Goettingen (DE), Hans Guenter Schlegel, Goettingen (DE); Chapter 20: Timothy Palzkill, Houston, TX (US), Beate Averhoff, Frankfurt/Main (DE); Chapter 22: David Hopwood, Norwich (UK); Chapter 23: Julian Davies, Vancouver (CA); Chapter 25: Werner Arber, Basel (CH), Chapter 26: Peter Greenberg, Seattle, WA (US), Anne Kemmling, Goettingen (DE); Chapter 27: Eugene Rosenberg, Tel Aviv (IL); Chapter 28: Michael Rey, Davis, CA (US), Gregory Whited, Palo Alto, CA (US), Alexander Steinbuechel, Muenster (DE), Garabed Antranikian Hamburg (DE); Chapter 29: Stefan Kaufmann, Berlin (DE), Joerg Hacker, Berlin (DE), Werner Goebel, Munich (DE), Michael Gilmore, Cambridge, MA (US), Julia Vorholt, Zuerich (CH), Ulla Bonas, Halle (DE); Chapter 30: Eckard Wimmer, Stony Brook, NY (US), Karin Moelling, Zurich (CH), Stephen Gottschalk, Houston, TX (US), Patrick Forterre, Paris (France); Chapter 31: Claire Fraser-Liggett, Baltimore, MD (US), Michael Hecker, Greifswald (DE), Edward DeLong, Cambridge, MA (US), Rolf Daniel, Goettingen (DE); Chapter 32: Kenneth Nealson, San Francisco, CA (US), Friedrich Widdel, Bremen (DE), Douglas Nelson, Davis, CA (US), Michael McInerney, Norman, OK (US), and Koki Horikoshi, Tokyo (JP).

This book tells the story of microbes and the discoveries revolving around these organisms. Some of the scientists involved in these discoveries have been mentioned, whereby the list of names makes no claim to be complete. If in connection with such discoveries the work of some colleagues has not been acknowledged, I would explicitly like to request their understanding.

The compilation of this manuscript was supported by Daniela Dreykluft, for which I am truly grateful. I would especially like to point out the contribution of Dr. Anne Kemmling, who designed a great number of the drawings and illustrations. Dr. Petra Ehrenreich also deserves credit for some of the figures. I am indebted to Theodor Wolpers, Emeritus Professor of English Literature, for his contribution of Shakespeare quotes, Colleen Cavanaugh for discussions on the title of this book, Martin Keller for his advice on Chapter 14, and Eckard Wimmer for his advice on Chapter 30.

Special thanks go to Wiley-VCH, notably Anne du Guerny, Dr. Gregor Cicchetti, and Dr. Andreas Sendtko, for the constructive cooperation in a pleasant atmosphere.

This book would not be what it is without Dr. Lynne Rogers-Blaut. She grew up in Ohio (US), came to Germany, and obtained her PhD in my department, so she is qualified to convert my English into readable English and to discover any inaccuracies and inadequate explanations. Thank you, Lynne – it was a wonderful cooperation.

Last but not least, I thank my wife Ellen for her patience and her support.

Goettingen, 2011   *Gerhard Gottschalk*

Prolog

> *Bacteria: they are a real threat to mankind. They cause plague and cholera, and in previous centuries more people were killed by them than in wars. Still today we may suffer from tuberculosis, intestinal infections, bronchitis, and many other diseases.*
>
> *Bacteria contaminate water and food. No, I don't want to know more about these creatures. There is already a lot of information in the press and in package inserts of drugs.*

This is not fair towards the bacteria. I admit, they do cause diseases, but only a very small percentage of bacteria is responsible for this. Most microbial species on our planet are peaceful and extremely useful. Without them, life on earth, would not be possible. They also affect our climate and are irreplaceable in the manufacture of biotechnological products.

> *Aren't you exaggerating? I have read that bacteria help clean beaches and waters after tanker accidents. They are also used in waste water treatment. But I can't believe that life on earth, depends on them, and I am sure viruses are not useful at all. By the way, what are archaea?*

The discovery of archaea is an exciting story. It will be reported in Chapter 3. These microorganisms look more or less like bacteria, but they are fundamentally different. They represent a separate domain of life, and were already different from the bacterial domain three billion years ago.

Let's start with bacteria, with their smallness and their enormous activity.

Part One
Reading Section

> It is the greatest dream of a bacterial cell
> to become two bacterial cells
>
> *François Jacob*

Chapter 1
Extremely small but incredibly active

A visit to our Department of Microbiology was on the agenda of a high-ranking politician. How to impress him? We started with the smallness of bacteria but not in the usual way by stating that bacteria are approximately 1 µm long, so 1000 bacterial cells lined up end-to-end would measure just 1 mm. We tried a different way:

"Sir, this test tube contains nearly 6.5 billion bacterial cells in a spoonful of water. Thus, the number of bacteria nearly equals the number of human beings on our planet." He took the test tube, looked at it, and could hardly recognize the slight turbidity. One billion bacterial cells in one ml or 1000 billion cells in a liter are barely visible. Then we pulled out a photograph the size of a letter pad and said, "Here are two of these 6.5 billion cells (Figure 1)." The Minister was impressed with the smallness of bacteria, which makes them barely visible even in large numbers, and with the enormous power of the methods used to examine them, for example, electron microscopy.

Electron microscopy? I used a light microscope when I was at school, but what is the principle behind the electron microscope?

Let's have the expert Frank Mayer (Goettingen, Germany) tell us about this:

> "Well, the "light" required for electron microscopy is a beam of electrons. This one is invisible to our eyes, but the pictures produced can be made visible. Because of the shorter wavelength of electron beams, much smaller details of biological objects can be seen than by light microscopy. Even enzyme molecules can be made visible, for example, on photographic paper. The disadvantage of using electrons is that a vacuum is required. Therefore, water has to be removed from samples before they can be examined, and this may cause damage to the objects. But recent improvements in electron microscopy make it possible to avoid damage to the objects by removing water from the objects in the frozen state."

Isn't it fascinating that electron microscopy makes it possible to magnify objects 100 000 times? Even light microscopy is capable of enlarging objects 1000-fold.

Discover the World of Microbes: Bacteria, Archaea, and Viruses, First Edition. Gerhard Gottschalk.
© 2012 Wiley-VCH Verlag GmbH & Co. KGaA. Published 2012 by Wiley-VCH Verlag GmbH & Co. KGaA.

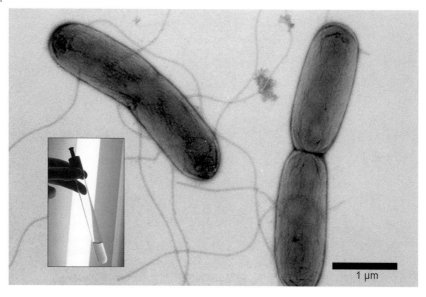

Figure 1 Test tube with a suspension of 6.5 billion bacteria, of which two are shown in an electron micrograph. The cell on the right has nearly completed cell division. The flagellae (long thread-like structures) provide motility to the cells. (Source: Frank Mayer and Anne Kemmling, Goettingen, Germany.)

This already impressed the plant physiologist Ferdinand Cohn (1828–1898), who wrote,

> "If one could inspect a man under a similar lens system, he would appear as big as Mont Blanc [in the Alps] or even Mount Chimborazo [in Ecuador]. But even under these colossal magnifications, the smallest bacteria look no larger than the periods and commas of a good print; little or nothing can be distinguished of the inner parts and of most of them their very existence would have remained unsuspected if it had not been for their countless numbers."

Ferdinand Cohn obviously exaggerated somewhat: a man two meters tall magnified 1000 times would be two thousand meters (6600 feet) tall, nearly half the elevation of Mont Blanc and one third that of Mount Chimborazo.

It is difficult to imagine that clear water can actually be highly contaminated, or that one cubic meter of air can contain one thousand microbial cells. Air, of course, is only slightly inhabited by microorganisms, but it is different when we look at our skin, which is densely populated by bacterial cells (see Chapter 10) with amazing biological activities. There are many sites in nature where they are able to multiply rapidly. *Escherichia coli* (*E. coli*, for short) resides in our intestine and is able to divide every 20 minutes! To put it casually, if one trillion bacterial cells in my intestine go with me to the movies, and if they manage to grow and divide optimally, then 16 trillion cells will leave the cinema with me 80 minutes later.

Good example, but why do bacteria multiply so astonishingly fast?

It's because bacteria have a high metabolic activity due to their high surface-to-volume ratio. Let me give you an example: If we put a cube of sugar into a glass of tea and, at the same time, the same amount of table sugar into a second glass, the table sugar will dissolve faster than the cube of sugar. Its surface-to-volume ratio is larger. A cube with an edge length of 1 cm has a surface-to-volume ratio of 6:1, between the total surface area of the sides, 6 cm^2, and the total volume, 1 cm^3. If we cut the cube into "bacteria-size" cubes with an edge length of 1 µm, we would end up with 100 million cubes with an overall surface area of 60 000 cm^2. The total volume would be the same but the surface-to-volume ratio would increase by a factor of 10 000.

That has its consequences. Compared to cells of higher organisms, bacteria have a much larger surface area at their disposal, allowing the faster import of nutrients and export of waste products. Therefore, cell constituents can be synthesized more rapidly, a prerequisite for the rapid multiplication of cells. That's why bacteria have the highest multiplication rates: some species have a record of around 12 minutes, so every 12 minutes two cells emerge from one. This, of course, cannot be generalized. There are also slow-growing bacteria that divide every 6 hours or even once every few days. Bacteria living in the "land of milk and honey" grow and divide rapidly, whereas the organisms in nutrient-deficient habitats such as oceans are much slower when it comes to cell division.

The ability of bacterial cells to divide every 20 minutes, or even every 12 minutes, is quite impressive. What does that mean for a bacterial population?

Let's look at a single bacterial cell multiplying every 20 minutes under optimal conditions. How many cells and how much cell mass would be produced after 48 hours? We have to do some simple calculations. One cell (2^0) would give rise to two cells (2^1) after 20 minutes; four cells (2^2) after 40 minutes; and eight cells (2^3) after 60 minutes. Three divisions per hour would make a total of 144 divisions in 48 hours, resulting in a total of 2^{144} cells. This number probably doesn't impress you. Let's do a few more calculations: Conversion into a common logarithm (144 × 0.3010), with 10 as a base, yields 10^{43} cells. The weight of one bacterial cell is around 10^{-12} g, so 10^{43} cells weigh 10^{31} g or 10^{25} tons. Our planet weighs 6×10^{21} tons, so after 48 hours the total bacterial mass would be nearly 1000 times that of our planet.

Very impressive, but certainly not realistic.

Of course not, but the calculation is correct. However, the assumption that cells would divide every 20 minutes for a period of 48 hours is incorrect. Nutrients would have become limited after a few hours, so growth would have slowed down and stopped eventually. Perhaps the situation can be compared to that of a large pumpkin, which after reaching a critical size will also stop growing because of shortage of nutrients and accumulation of metabolic byproducts.

I have learned something new. I would like to know how bacteria compare with higher organisms.

> Nature would not invest herself in such
> shadowing passion without some instruction
>
> William Shakespeare, Othello

Chapter 2
Bacteria are organisms like you and me

But what about archaea and viruses?

Archaea, to be introduced in Chapter 3, are living organisms like bacteria, but viruses aren't like them at all because several characteristic features are missing. Viruses look and often act like little golf balls, just lying around or flying through the air. They aren't able to do much by themselves. But as soon as they have entered a host cell, they start their devilish work. Viruses are able to cause epidemics, so they must somehow have life in them (see Chapter 30).

Bacterial and archaeal cells actually have much in common with plant and animal cells. Of course, you can't compare a single-celled organism such as our intestinal bacterium *Escherichia coli* with an oak tree or an elephant. Comparisons have to be made at eye level, for example, comparing an *E. coli* cell with a cell from an oak leaf or with a muscle cell from an elephant. Then, the features common to all cells will become apparent. Let's first look at the cell constituents.

All cells contain DNA (DeoxyriboNucleic Acid), but there is one qualitative difference. The DNA in plant and animal cells is localized in the nucleus, a compartment surrounded by a membrane. Plants and animals are therefore called eukaryotic organisms. A simple eukaryotic cell, the yeast cell, is depicted schematically in Figure 2a. Bacteria, on the other hand, are prokaryotic organisms whose DNA more or less floats in the cytoplasm (Figure 2b), which is a sort of gel. This intracellular space contains many proteins, nucleic acids, amino acids, vitamins, and salts.

All cells contain three types of RNA (RiboNucleic Acid). The ribosomal RNA, together with the ribosomal proteins, makes up the ribosomes, the protein synthesis factories of the cells. The second type is messenger RNA which transmits DNA-imprinted messages to the protein synthesis factory. Messenger RNA passes the instructions from the DNA to the ribosomes, where proteins are synthesized on the basis of these instructions. There are mechanisms to ensure that only those proteins are synthesized that are required under certain physiological conditions. Not all proteins encoded on the DNA are continuously needed. The third type of RNA, transfer RNA, is required for the alignment of amino acids to form proteins. Each cell contains at least 20 of these transfer RNAs, which are specific for the 20 amino acids present in proteins. According to the synthesis protocol of the

Discover the World of Microbes: Bacteria, Archaea, and Viruses, First Edition. Gerhard Gottschalk.
© 2012 Wiley-VCH Verlag GmbH & Co. KGaA. Published 2012 by Wiley-VCH Verlag GmbH & Co. KGaA.

Chapter 2 Bacteria are organisms like you and me

a)

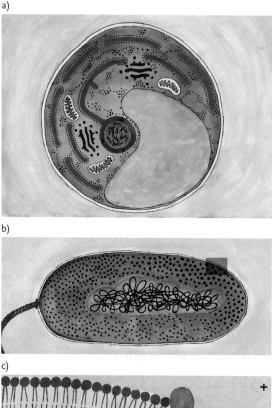

b)

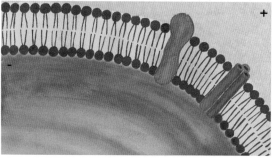

c)

Figure 2 The eukaryotic and the prokaryotic cell. (a) The eukaryotic cell contains a nucleus (center) surrounded by a membrane (with pores), a vacuole (light blue), the endoplasmatic reticulum (green), the Golgi-apparatus (purple), mitochondria (yellow/orange), ribosomes (black dots) and cytoplasm. The cell is surrounded by a cytoplasmic membrane and a cell wall. Diameter of the yeast cell depicted: 10 μm. (b) The prokaryotic cell contains a circular, coiled-up chromosome; ribosomes; and cytoplasm. The cell is also surrounded by a cytoplasmic membrane and a cell wall. A flagellum is depicted on the left (not present in all bacteria). Bacterial cells have an average length of 1 μm. (c) The cytoplasmic membrane consists of a phospholipid bilayer. In living organisms the membrane is charged, negative inside and positive outside. Proteins (red) are inserted into the membrane. (Watercolor and gouache: Anne Kemmling, Goettingen, Germany.)

messenger RNA, the transfer RNAs, each linked to a respective amino acid, are lined up in the prescribed order then the amino acids are connected.

In all cells the entire machinery discussed above is surrounded by the cytoplasmic membrane, which is negatively charged on the inside and positively charged on the outside (Figure 2c). The membrane contains checkpoints for the transport of materials into the cells. These transport processes are highly specific; for example, there are checkpoints that allow potassium ions to pass but not sodium ions. The interior of most bacteria is high in potassium ions but low in sodium ions. If we were somehow able to taste the interior of a bacterium from the ocean (intracellular volume around $1\,\mu m^3$, 1 cubic micrometer), it wouldn't taste salty. Without its charge, the cytoplasmic membrane would be unable to fulfill its functions to ensure that the composition of the cell's interior differs dramatically from the surrounding fluid. Inside the cell there are favorable conditions for cell division, irregardless of the conditions outside. The cytoplasmic membrane and its functions is one of the greatest miracles of evolution. How the membrane is charged will be described in Chapter 8.

Those are the cell constituents, but how does one cell become two cells?

To answer this question we have to look at the processes of life at a cellular level. Which processes are involved when two cells are formed from one? As already discussed, DNA is the carrier of genetic information needed to generate two *E. coli*-cells from one *E. coli*-cell. First of all, energy is required for the generation of a new cell. Here, the magic word is ATP, the abbreviation for adenosine-5′-triphosphate. ATP is the energy currency of all organisms on our planet. It powers processes such as thinking or muscle work, also growth, motility, and reproduction in bacteria. When ATP fulfills its role as an energy source, it is at the same time devaluated; it loses one phosphate residue, and adenosine-5′-diphosphate (ADP) is formed. This conversion is coupled with a release of chemical energy that can be invested in the energy-requiring reactions mentioned above.

Before a cell can divide into two cells, the chemical constituents of the cell have to be synthesized. It is as if a completely furnished house is to be converted into a completely furnished duplex. The "furniture" has to be assembled and set up or installed, so that two viable cells will have been formed from one. If we disregard the membrane, the cell wall, and any reserve material such as starch, the cell essentially has to deal with the synthesis of DNA, RNAs, and proteins, the three types of constituents already introduced. Before we go into protein synthesis, let's look at the role of proteins.

Most of the proteins of a cell are enzymes, except for proteins such as collagen in higher organisms (part of the supporting tissues) or capsular proteins in certain bacteria. Enzymes are also called biocatalysts, and their names usually end in "-ase," as in lipase or protease. The enzymes consist of 20 different building blocks, called the 20 natural amino acids (▶Study Guide). These amino acids are found in proteins, not only once but in multiple copies. The chain length of proteins is variable; proteins may consist of 100–300 building blocks. The amino acids have different chemical properties, so their chains fold to yield complicated structures to which metal ions, such as magnesium or ferrous (iron) ions are often bound. Every enzyme contains a catalytic center. This is the place of action, where the enzyme-catalyzed reactions take place. The diversity of enzymes is fantastic. Even our commensal *E. coli* is able to synthesize approximately four thousand different enzymes. They all have a specific function at defined sites of metabolism. For example, enzymes make the synthesis of DNA and RNA possible. Enzymes exhibit specificity, which means that the catalytic center of a particular enzyme has been designed to fit certain reaction partners. A DNA polymerase is capable of elongating DNA strands but it cannot cleave fat–that's the job of the lipases. It is important to note that enzymes dramatically increase reaction rates because they bring the reaction partners into optimal spatial positions. Without enzymes, even enzymes themselves would not exist. The interdependence of DNA, RNA, and protein (enzyme) synthesis will become clear when we look at these processes (Figure 3).

The genetic information of a bacterial cell is present in the form of a circular chromosome. It consists of double-stranded DNA, and the double strands are stabilized, as we say, by base pairing. This principle can be considered, without exaggerating, the secret of the conservation and transfer of genetic information in nature. The DNA consists of deoxyribose (a sugar), of phosphate to connect the deoxyribose molecules; and of four chemical compounds linked to deoxyribose. These are adenine, guanine, cytosine, and thymine, or A, G, C, and T, for short. The chemistry of these four compounds, commonly called the four bases, is such that two bases with a high affinity to each other tend to form pairs, referred to as base pairing (▶Study Guide). The two most-favored base pairs are AT and GC. With this information it is relatively easy to understand how two double-stranded chromosomes are formed from one double-stranded chromosome. In turn, this is a prerequisite for formation of two cells from one. The underlying process is called replication or identical reduplication. Obviously, the circular chromosome has to be replicated precisely, otherwise the genetic information in the two resulting bacterial cells will not be the same as in the original cell. The apparatus by which this is accomplished is commonly called the replication factory. Several enzymes are at work in this factory, DNA polymerase and helicase, to mention two. In the case of *E. coli*, the task is to exactly replicate the DNA ring, which is 4 938 975 base pairs long, and this in about 20 minutes. This length corresponds to the chromosome size of *E. coli* strain 536, a strain that causes urinary infections.

We will now attempt to grasp the principle of replication. The helicase manages to separate the double strand into single strands, initiating the replication at a distinct point. The single strands then enter the replication factory through different gates. If you understand the principle of base pairing, the next steps are easy to follow. The precursors of the four bases, dATP, dGTP, dCTP, and dTTP, are floating around in the cytoplasm and, of course, are also present in the factory. If a single strand with the sequence ATTCGGA becomes available, the precursors

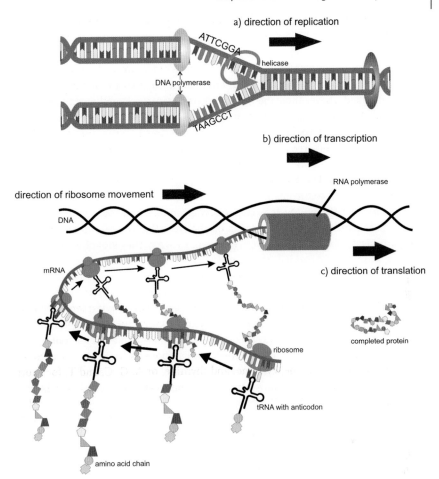

Figure 3 The three principal processes for the transmission and utilization of genetic information. (a) Replication: The enzyme helicase separates the double strand into its two single strands. The exposed bases (letters) are then subject to base pairing, and the second, complementary strands are formed by the action of DNA polymerase. (b) Transcription: RNA polymerase widens the double helix and transcribes the base sequence of one DNA strand into the complementary sequence of an RNA strand, messenger RNA (mRNA). The DNA strand subject to transcription is called the codogenic strand. (c) Translation: As soon as mRNA becomes available, it binds to ribosomes (blue) and protein synthesis is initiated. The ribosomes closer to the RNA polymerase in Figure 3c have been "working" for a longer period of time, so "their" amino acid chains are longer. Each of the tRNAs is linked to a specific amino acid. They recognize their "turn" in protein synthesis with the help of an anticodon, which is complementary to the codon on the mRNA. (Diagrams: Anne Kemmling, Goettingen, Germany.)

of the DNA to be synthesized are arranged in the sequence dTTP dATP dATP dGTP dCTP dCPT dTTP, so the DNA polymerase only needs to travel along the sequence and connect these precursors. As a result, the fragment TAAGCCT is formed, which is complementary to ATTCGGA. What has been described here for seven building blocks has to proceed 4.9 million times during replication of strain 536, mentioned above, then the second chromosome is completed. There is a slight problem with what is called the polarity of DNA strands; this will be discussed in the Study Guide.

Obviously, two cells require more RNA molecules than one. RNA has several special features: it contains ribose instead of deoxyribose, and the base thymine (T) is replaced by uracil (U). Since T and U hardly differ in their tendency to form a base pair with A, AU instead of AT is therefore the base pair at the RNA level. The process of RNA synthesis is called transcription. As in replication, the principle is base pairing. There are regions on the DNA that contain the information for synthesis of ribosomal RNAs and of transfer RNAs. They are transcribed, thus yielding additional ribosomal RNAs and transfer RNAs. Furthermore, messenger RNA is synthesized. These molecules travel to the ribosomes and transmit the information for protein synthesis. The information for protein synthesis is organized on the DNA and, after transcription, on messenger RNA in the form of genes. A gene is a segment of the nucleic acids that contains the information for the synthesis of a particular protein. It is defined by a start and a stop signal, which are recognized by the machinery. Thus, at the ribosome, a gene of an exactly defined size gives rise to a given protein. The dynamics of these processes are depicted in Figure 3. Ribosomes like start signals. As soon as start signals are available on the messenger RNA, the ribosomes bind to the messenger RNA. The synthesis of an amino acid chain can then begin. When the stop signal is reached, the ribosomes detach from the messenger RNA and release the amino acid chain. Upon folding, the amino acid chain becomes a protein with enzyme activity.

The conversion of a base sequence into an amino acid sequence is called translation. This process is not as simple as transcription because the sequence of the four bases UAGC at the messenger RNA level has to be translated into a sequence of 20 amino acids. To achieve this, the genetic code evolved. The information is not provided by a single base but by a base triplet, a sequence of three bases that gives the code word for a particular amino acid. Therefore, when a gene on the messenger RNA has a length of 990 bases, this represents the information for the synthesis of a protein consisting of 330 amino acids. The number of possible base triplets is large enough to provide organisms with a sufficient number of code words. Four bases allow 4^3 or 64 possible combinations for triplets. These combinations are used in all of nature (▶Study Guide).

How does cell division proceed?

In the cell there is a special protein called the FtsZ protein. In the center of the mother cell, this protein forms a ring structure that contracts until the membranes touch each other and fuse. This results in the formation of two compartments, in other words, two cells (see Figure 1, left cell). It's fascinating how proteins and structures, not yet known in detail, manage processes essential for life: the distribution of cell constituents such as chromosomes, RNAs, ribosomes, and proteins. In this process, the proper distribution is a prerequisite for formation of two viable cells.

Eukaryotic cells are certainly more complex than prokaryotic ones

Yes, they are. In addition to the nucleus, there are other compartments in eukaryotic cells, including the mitochondria or the Golgi apparatus (Figure 2a). However, the role of ATP and the processes leading to the synthesis of cell constituents – the 20 amino acids and the substances making up the DNA and RNAs – are quite comparable. Processes in the cell nucleus are much more complex than those occurring in the vicinity of the microbial chromosome. The number of bases making up the human chromosomes is about one thousand times larger than that of the *E. coli* chromosome. In humans, this information is sufficient for the synthesis of approximately 100 000 proteins. However, in both cases the underlying genetic code is identical. There are significant differences when we examine the localization of genes on human chromosomes in comparison to the *E. coli* chromosome. The latter – also true for other microbes – is packed with genes. The genes are strung together and there is very little "wasted space" in terms of sequences. By contrast, the intergenic space between the genes on human chromosomes is extremely large. Often there are millions of bases between two genes, and the function of these bases is not yet clear. Maybe this is yet another secret of life that must be brought to light. In addition, a eukaryotic gene differs in one important aspect from a prokaryotic gene. We recall that 990 bases on a prokaryotic genome correspond to 330 amino acids. In higher organisms, the messenger RNAs are much larger. They contain introns that are inserted into the messenger RNA, giving it a mosaic-like structure. In a process called splicing, these inserts have to be removed before the messenger RNA can actually function as a matrix for protein synthesis. When taken together with the cell differentiation processes, the most remarkable features of higher organisms, it becomes clear that there are tremendous differences between prokaryotes and eukaryotes.

Nevertheless, I stick to my statement: Bacteria are organisms like you and me.

> We need to know
> We will know
>
> Inscription on the tombstone of David Hilbert, mathematician

Chapter 3
My name is LUCA

I beg your pardon, but isn't that the beginning of a song about "the girl who lives upstairs?"

I don't mean that Luca. Here LUCA stands for the Last Universal Common Ancestor, the living organism that was the mother of all organisms on Earth.

Who was LUCA?

Before this question can be discussed, we need to know if the definition of a species as we know it for animals and plants can also be applied to bacteria.

It was Mrs. Fanny Angelina Hesse whose recommendation made it possible to solidify growth media for bacteria. She suggested using the gelatinizing substance agar (also called agar agar), which was introduced in Robert Koch's laboratory in 1884. The use of agar allowed bacteria to grow on the surface of growth medium, like on the surface of pudding or a slice of bread. On such a surface, clusters of bacteria originating from one cell are formed. These are called colonies. When a tiny portion of a colony of *Escherichia coli* is transferred to fresh agar, new colonies of cells of *E. coli* are recovered (Figure 4). It could then be concluded that the definition of a species could also apply to bacteria. Therefore, elephants arise from elephants, oaks from oaks; *E. coli* yields *E. coli*, and *Staphylococcus aureus* yields *Staphylococcus aureus* and not a bacterium with completely different properties.

In the past 120 years, thousands of bacterial species have been isolated and their properties have been described. However, for many years their evolutionary (phylogenetic) relatedness remained unknown. Of course, efforts were made in numerous laboratories to learn something about this relationship. It was necessary to identify bacterial species on the basis of their properties and to develop strategies to control those causing disease. The actual breakthrough, however, occurred just a quarter of a century ago.

Discover the World of Microbes: Bacteria, Archaea, and Viruses, First Edition. Gerhard Gottschalk.
© 2012 Wiley-VCH Verlag GmbH & Co. KGaA. Published 2012 by Wiley-VCH Verlag GmbH & Co. KGaA.

16 | Chapter 3 My name is LUCA

Figure 4 Colonies of a bacterium on agar growth medium in a Petri dish. When a minute amount of cells is streaked out (starting at 9 o'clock on the dish), cells are so separated that round colonies can develop from single cells. Such a cell community then represents a clone because it originates from one cell. Depending on its size, a colony may contain between 50 and 500 million cells. (Photograph: Anne Kemmling, Goettingen, Germany.)

I have a question regarding Mrs. Hesse. What is agar agar and how did she get the idea to use it to solidify growth media for bacteria?

Fanny Eilshemius was born in 1850 in Laurel Hill, New Jersey (USA), the daughter of German immigrants. While traveling through Europe, she met a medical doctor, Walter Hesse, whom she married. Walter Hesse had a strong interest in bacteriology. He grew bacteria on gelatin surfaces and was very disappointed that the gelatin melted so easily. Fanny remembered a recipe she had from Dutch friends who had lived on the island of Java. They used agar agar, a polymer extracted from algae to solidify deserts. Agar is ideal because hot solutions containing 2 percent agar solidify at about 50 °C and they only melt again upon boiling. In addition, most bacteria do not degrade agar, so the introduction of agar provoked a revolution in microbiological laboratories. Bacteria could then conveniently be grown on agar surfaces in Petri dishes and the properties of pure cultures, those containing only one species, could be studied.

What about the breakthrough you mentioned?

In the 1960s a procedure for sequencing DNA fragments was worked out by the British molecular biologist Frederick Sanger. He developed an elegant method to shorten DNA fragments base by base and to determine whether the last base in the fragment is T, C, G, or A (see Chapter 31 for details). This method was so ingenious that Frederick Sanger was awarded the Nobel Prize (together with Walter Gilbert and Paul Berg), his second one – see Chapter 28 for the first one. Progress was enormous, and it is no exaggeration to say that now more than a billion base sequences are determined every day using this and more advanced methods. At the University of Illinois in Urbana, the microbiologist Carl Woese adopted Sanger's method to learn something about the phylogenetic relationship of bacterial species. He chose the 16S-rRNA to be sequenced. This RNA is present in all bacteria because it is essential for the formation of the ribosomal protein synthesis factories. A milestone in this research was reached in 1977: Carl Woese and his coworkers published their research paper on the "molecular approach to prokaryotic systematics." The relatedness between bacterial species was determined on the basis of differences in the sequence of 16S-rRNA. At this point, a sensational experiment appeared on the horizon. Ralph S. Wolfe, another eminent microbiologist, was working next to the laboratory of Carl R. Woese. He did pioneering research on methane-producing microorganisms, especially on the biochemistry of the pathways leading to the production of methane. Of course, Ralph Wolfe and Carl Woese talked about their work. They decided to collaborate, but let's have Ralph Wolfe (Champaign-Urbana, Illinois, USA) tell us about the results of their experiments:

> "Early in his career, Woese had studied the ribosome and was convinced that this organelle was of very ancient origin, that it had the same function in all cells, and that variations in the nucleotide sequence in an RNA of the ribosome could reveal evidence of very ancient events in evolution. He chose the 16S-rRNA and developed a similarity coefficient that could be used to compare the relatedness of two different organisms. By 1976 he had documented the 16S-rRNAs of 60 bacteria.
>
> The research program of Ralph Wolfe concerned the biochemistry of methane formation by methanogenic bacteria, an area poorly studied because of the difficulties of cultivating the organisms. Techniques were developed for culture of cells in a pressurized atmosphere of hydrogen and carbon dioxide. By 1976 the structure of two unusual coenzymes, coenzyme M and Factor 420 (a unique deazaflavin), and their enzymology had been elucidated.
>
> The conjunction occurred with an experiment designed to examine the 16S-rRNA of methanogens. The pressurized atmosphere technique proved ideal for containing the high level of injected radioactive phosphorus to label the rRNA of growing cells. The two-dimensional chromatograms of the labelled 16S-rRNA oligonucleotides from the first experiment were so different from anything previously seen, that Woese could only conclude that somehow the wrong RNA had been isolated. The experiment was carefully repeated, and this time with the same results: Woese declared, "Wolfe, these organisms are not bacteria!" "Of course they are, Carl; they look like bacteria." "They are not related to anything I've seen." This experiment marked the birth of the archaea!"

This is an important contemporary testimonial. It led to the conclusion that a third form of life exists on our planet, the archaea, in addition to the eukaryotes (animals, plants, fungi) and the bacteria. Ralph Wolfe describes why his colleague Carl Woese chose 16S-rRNA, then he describes his own research during which the techniques for growing large amounts of methane-producing microorganisms were developed. These microorganisms were used for biochemical investigations that led to the discovery of novel "methano-vitamins" such as coenzyme M or factor 420.

The studies in Woese's lab involved growing the methanogenic organisms in the presence of radioactive phosphate. The 16S-rRNA then contained radioactivity because of the phosphate bridges present in the molecule. Fragments were isolated, sequenced, and compared with the 16S-rRNAs of *E. coli* and other organisms. When they compared the sequences obtained, differences were encountered that could not be explained. Both Carl Woese and Ralph Wolfe were speechless. The sequences of all the bacterial species studied before were written in essentially the same language and contained a number of deviations, which gave insight into the distance between two bacterial species in the phylogenetic tree. But what the researchers now discovered was that parts of the text were deleted and parts were written in another language, say in Hebrew. This was something unique, and this also proved to be the case when the 16S-rRNA of other methanogenic organisms was sequenced.

Could you please explain a bit more about these differences in the 16S-rRNAs?

Let us imagine a large mosaic, for instance, the triumphal march of Dionysos in one of the Roman houses in Paphos (Cyprus) discovered in 1962. Like 16S-rRNA, mosaics also consist of thousands of building blocks. If we change something in the border surrounding the scene, this will have little effect on the general impression of the mosaic. It is the same with the sequence of the 16S-rRNA of bacteria. Trends as well as a few variations in the sequences can be recognized, making it possible to determine the relatedness of the organisms from which the 16S-rRNAs were isolated. However, if Dionysos were to be replaced in the mosaic by a mythological priest, then the number and color of the mosaic tiles would be quite different. It is impossible to derive one figure from the other, so a common ancestor must be postulated that gave rise to Dionysos, on the one hand, and to the priest, on the other—an ancestor such as LUCA.

Let's now look at the sequences of the 16S-rRNAs of *Escherichia coli* (1542 bases long) and *Bacillus licheniformis* (1548 bases) as representatives of the bacteria and of *Methanosarcina mazei* (1474 bases), *Archaeoglobus fulgidus* (1492 bases), and *Methanosphaera stadtmanae* (1480 bases), representing the archaea. The alignment of these sequences results in a beautiful picture (Figure 5), which was done with the aid of a computer program that searches for maximum correspondence of the sequences. In order to attain this sequence homology, the computer takes sequences apart and introduces gaps. Several archaeal "gaps" are apparent, including some large ones, but there are also a few bacterial gaps. Identical sequences

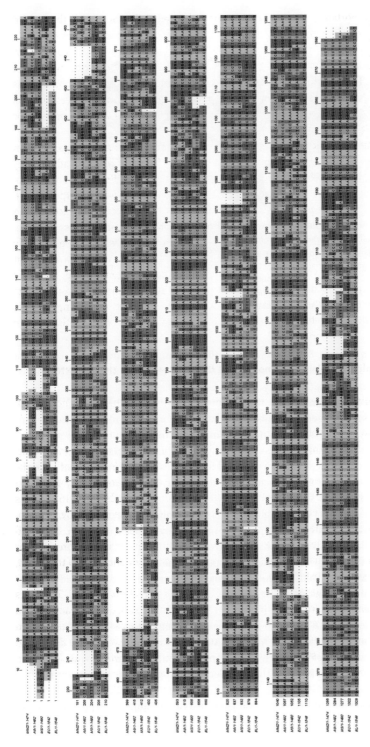

Figure 5 Base sequence of the 16S rRNAs of three archaea (*Methanosarcina mazei*, *Archaeoglobus fulgidus*, and *Methanosphaera stadtmanae*) and two bacteria (*Escherichia coli* and *Bacillus licheniformis*). The depiction is based on a ClustalW-alignment under standard conditions (W.A. Larkin *et al.*, Bioinformatics 23, 2947, 2007). Visualization was done with Jaliew using the standard color code for nucleotides. (C. Clamp *et al.*, Bioinformatics 20, 416, 2004). Gaps were inserted into the sequences to achieve maximal agreement. (Adaptation: Antje Wollherr, Goettingen, Germany.)

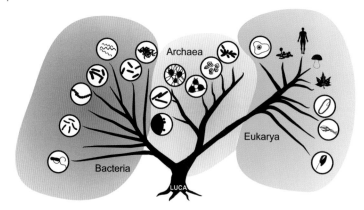

Figure 6 Phylogenetic tree of the three domains of all living organisms, depicted clockwise from left to right: Bacteria: *Aquifex, Bacteroides,* cyanobacteria, proteobacteria, Spirochaeta, Bacilli, green filamentous bacteria. Archaea: *Nanoarchaeum equitans* (small dots) *attached to Ignicoccus hospitalis, Thermoproteus, Pyrodictium, Methanococcus, Methanosarcina,* halophilic archaea. Eukarya: *Entamoeba,* mucilaginous fungi, humans, fungi, plants, ciliates, trichomonads, diplomonads. LUCA (at the bottom) stands for the Last Universal Common Ancestor. (Diagram: Anne Kemmling, Goettingen, Germany.)

can be seen around 810, 1440, and 1540. Looking at Figure 5 as a whole, the impression is that the upper three sequences are related as well as the lower two. These differences made history.

Carl Woese recognized the tremendous importance of his discovery. In the case of the methanogenic organisms, he had sequenced the 16S-rRNA of an organism that phylogenetically does not have much in common with the bacteria. This new domain of organisms was originally called archaebacteria; later the word "bacteria" was omitted, so now we speak of archaea. The terms methanobacteria or methane-forming archaebacteria are no longer used; instead, they are called methanoarchaea.

When these results were published, many microbiologists opposed such views. But there was also a lot of support, for instance by the German microbiologists and molecular biologists Otto Kandler, Wolfram Zillig, and Karl Otto Stetter, who later made important contributions supporting the archaeal concept. In the US, acceptance was slow but became enthusiastic when major textbooks adopted the concept of three domains of life.

Now we will jump from the situation in 1977 to the present and look at an actual phylogenetic tree. It is apparent in Figure 6 that, beginning with LUCA, evolution proceeded to the domains of Archaea and Eukarya, from which the domain of Bacteria branched off early in evolution. Not only methane-forming bacteria belong to the archaea but also the so-called extremophilic microorganisms, which will be discussed in Chapter 6.

Readers not so familiar with this concept of evolution will find it difficult to accept the idea that two of the three domains of life on our planet are devoted to microorganisms. The third one is reserved for plants, fungi, animals, and us. We may not forget that bacteria and archaea, on the evolutionary time scale, were by themselves during most of the biological evolution on Earth. One might even ask the question why microorganisms allowed the evolution of higher organisms – an interesting but difficult question. But still, what were the properties of LUCA? It can be assumed that it was a bacterial- or archaeal-like organism that lived in the absence of oxygen. LUCA was a fermenting organism or an organism that converted sulfur and molecular hydrogen to hydrogen sulfide. Archaea performing this kind of fermentation are still present on our planet. How LUCA evolved and how our current atmosphere developed with oxygen as the indispensable element will be outlined in the next two chapters.

> The question "what is life"
> is precisely the question
> "what is evolution"
>
> Carl R. Woese

Chapter 4
From the Big Bang to LUCA

The Big Bang occurred 13.7 billion years ago. Guenther Hasinger (Garching, Germany) compressed this incredibly long time scale into one calendar year. During the first eight months the universe was like Hell. On January 5 the first stars appeared as well as the chemical elements, including carbon, oxygen, and nitrogen. In March, quasars reached their maximum and planets began to appear. At the beginning of September, approximately 9.1 billion years after the Big Bang, our solar system evolved. One month (1.1 billion years) later, life developed on our planet. This was 3.5 billion years ago. Plants and vertebrates appeared after December 19, 500 million years ago. *Homo sapiens* showed up on Earth on December 31 at 8 p.m., and Jesus Christ was born just 5 seconds ago. By the way, by January 12 of the second Big Bang year, our planet will become so hot that life no longer will be possible. This of course has nothing to do with man-made global warming, but has a different reason: By then, the sun will be in the process of becoming a red giant. January 12, by the way, will be 447 million years from now.

Experts point out that once our planet had been formed, it experienced a bombardment by comets for half a billion years. This bombardment (in the region of one million hits) was an important event for the upcoming development of the Earth. It had a great impact on the composition of the atmosphere, which at the time consisted of carbon dioxide, hydrogen sulfide, and steam, gases that had been liberated from the Earth's mantle. This composition was greatly altered by the gases brought to our planet by comets, especially molecular nitrogen and ammonia but also noble gases, including argon and helium. As a result, molecular nitrogen was predominant four billion years ago but the composition of the atmosphere still differed from that of our present atmosphere. The most important difference was the lack of oxygen; scientists assume it was present in only trace amounts. Another important difference was, of course, the presence of H_2S and ammonia.

How could comets bring all those gases to the Earth?

Comets have ice caps with large amounts of entrapped gases. Jonathan J. Lunine (Tucson, Arizona, USA) writes that these ice caps may amount to a mass of around 10^{12} tons per comet, which corresponds to approximately 1 millionth of the total water mass on our planet.

Discover the World of Microbes: Bacteria, Archaea, and Viruses, First Edition. Gerhard Gottschalk.
© 2012 Wiley-VCH Verlag GmbH & Co. KGaA. Published 2012 by Wiley-VCH Verlag GmbH & Co. KGaA.

Let's imagine how the world looked four billion years ago: hot, active volcanoes, land masses that later became the continents, steaming oceans, practically no oxygen, the terrible smell of hydrogen sulfide (like rotten eggs) and caustic ammonia. Creating life on Earth under these conditions was a difficult and highly improbable task. Several hundred millions of years passed before something happened. What an incredible time scale with the near-zero probability of such processes occurring at all!

What is your concept of how life appeared on Earth? Didn't Charles Darwin also wonder about this?

Charles Darwin published his fundamental work, *The Origin of Species,* in 1859. The main focus of his book, of course, was the origin of *Homo sapiens* and the animal and plant species. Catchwords such as "bacteria" or "microorganisms" are not indexed in Darwin's work. Microbiology was still in its infancy in Darwin's time, so it is understandable that bacteria were not included in his theory. However, the basic principles of natural selection and the struggle for existence can also be applied to the bacteria, as will be outlined in Chapter 20.

Well, how did it all get started?

For a long time it was assumed that the ancient atmosphere primarily consisted of molecular hydrogen, methane, and ammonia. John B. S. Haldane (1892–1964) in England and Alexander Oparin (1894–1980) in Russia postulated that large amounts of organic compounds were formed by electric discharges. These compounds included amino acids and the building blocks of nucleic acids. This hypothesis inspired Stanley L. Miller (1930–2007) and Harold C. Urey (1893–1981) to perform corresponding experiments in Chicago (Illinois, USA) in the 1950s. Indeed, they were able to demonstrate formation of the building blocks of simple creatures when hydrogen, methane, and ammonia are exposed to electrical discharges. Figure 7 shows the experimental setup with which Miller and Urey produced what was then called the "Miller-Urey Soup." It was assumed that the oceans had been converted in this way, over millions of years, into a watery consommé, a nutritious clear soup.

These ideas became somewhat obsolete when new theories on the composition of the ancient atmosphere evolved: very little or no molecular hydrogen and methane but plenty of molecular nitrogen instead. Nevertheless, it is assumed that self-replicating cells appeared in an aquatic habitat such as a lagoon or a crater lake, in which the building blocks of living organisms had accumulated.

A model habitat would be a crater lake because of the presence of a continuous supply of reactive gases, including carbon monoxide and molecular hydrogen as well as carbon dioxide, hydrogen sulfide, and ammonia. In addition, such a habitat is sympathetic to the theories of Guenter Waechtershaeuser (Munich, Germany). He has proposed that the surface of minerals containing iron and nickel sulfides

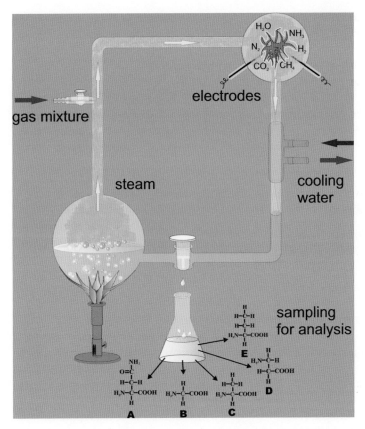

Figure 7 Layout of the apparatus used by Miller and Urey to produce organic compounds from "volcanic gases". Products were identified by gas chromatography. Examples: A, asparagine; B, glycine; C, alpha-alanine; D, beta-alanine; E, alpha-aminobutyric acid. (Diagram: Anne Kemmling, Goettingen, Germany.)

(sulfur-containing ores) in contact with water could have been the birthplace of the first living creatures. Pyrite (FeS_2) is such an ore, which is abundant in nature and often associated with nickel salts.

What are the advantages of such a mineral base?

There are several. Enzyme systems quite often contain iron-sulfur or iron-nickel-sulfur complexes as catalytic centers. In prebiotic times, these ores could have exhibited catalytic activity. Furthermore, such ores have sorptive activities, being able to absorb organic molecules like a sponge. So the products of the Miller-Urey reaction could have been present in much higher concentrations on the surface or in cavities of the ore than in the water of a crater lake as a whole. There is an additional factor: Energy-rich molecules would have to be available in order to get reactions going. It is rather unlikely that ATP (discussed in Chapter 2) was already present at the time to fulfill this task. ATP is a complex compound that probably evolved later. However, thioesters would be a possibility.

$$R\text{-}S\text{-}CO\text{-}CH_3, \text{ a thioester} \quad R = CH_3, \text{ for example}$$

Thioesters are relatively simple compounds that could have been generated in volcanic gases, serving as a basis for synthesis of more complex organic molecules. The pyrite mentioned previously is formed from iron sulfide and hydrogen sulfide:

$$FeS + H_2S \rightarrow FeS_2 + H_2$$

In this context, it is interesting to note that, even today, some archaea are able to grow with molecular hydrogen and elemental sulfur as energy sources:

$$\underset{\text{(hydrogen)}}{H_2} + \underset{\text{(sulfur)}}{S} \rightarrow \underset{\text{(hydrogen sulfide)}}{H_2S} + \underset{\text{(metabolic energy)}}{ATP}$$

One organism of this type is *Thermoproteus tenax*. The utilization of H_2 in the above process involves hydrogenase, an enzyme containing metal-sulfur centers as discussed.

Let's summarize the information so far: we have developed an idea about a hotbed of life, a crater lake containing a dilute broth containing cell constituents continuously generated by reactive gases: Iron-nickel-sulfur surfaces are available, and we have a comfortable temperature, let's say between 50 and 80°C (120–180°F). Proteins may have been formed, then decomposed, and some of them may have been capable of enzyme activity. Millions of years passed but there was something missing, a qualitative leap to a system capable of self-replication and, hence, of establishing a uniform population.

Let's have a look at the three important types of molecules that determine life in the cell: DNA, RNA, and proteins. Which of these is capable of facilitating its own replication? Proteins are not very useful in this respect because they are not easy to copy. They consist of 20 different amino acids, and the amino acid chains are folded to complex structures. It's like trying to copy a crumpled piece of paper. RNA and DNA resemble chains of pearls; they are relatively easy to copy by taking advantage of the principle of base pairing, described in Chapter 2. Of these two types of macromolecules, RNA is more practical than DNA. In addition to the functions already described, RNA has one that was discovered by Thomas R. Cech (Boulder, Colorado, USA) and Sidney Altman (New Haven, Massachusetts, USA) in 1982: it actually exhibits enzyme activity. Some types of RNA are able to cut other RNA molecules at distinct sites. Following their synthesis, some RNAs even form a loop and cut off part of their own molecule further down the chain. You could compare it with a tree that forms a saw at the very top, then bends down and trims its own branches. The importance of this discovery by Thomas Cech and Sidney Altman was so profound that they were awarded the Nobel Prize. In the meantime, many examples of the catalytic capabilities of RNA have been discovered. The RNAs of this type are called ribozymes.

Why is this property so important?

This may indicate that RNA molecules were the first to develop the property of self-replication. These discoveries were so fascinating that Walter Gilbert proposed the existence of an RNA world during a period of biological evolution. How this world worked and how it pushed evolution forward has been described by the Nobel laureate Manfred Eigen (Goettingen, Germany), who together with Peter Schuster (Vienna, Austria) formulated the hypercycle in 1979. Manfred Eigen wrote in a manuscript,

> "What now is the hyper cycle? I have deduced this term from the reaction mechanism which comprises a hierarchic superposition of several cycles, so it is a cycle of cycles. The word "hyper cycle" is not my discovery, as I first thought. It can already be found in a letter of Carl Friedrich Gauss to his Hungarian friend and colleague Wolfgang Bolyai (1832) in connection with his thoughts on noneuclidic geometry. However, there it is used in an abstract mathematical sense in connection with geometric properties. Our hyper cycle, however, not being in contrast to Gauss' interpretation, describes the dynamic principle of a feedback in the process of the rise of genetic information. This principle can be depicted as a hierarchic superposition of reaction cycles. The simplest case of a hyper cycle could have occurred in the RNA world. The replication of both complementary plus and minus strands has to be catalytically promoted in order to achieve a high production rate. In the RNA world, the catalyst is an RNA molecule, a ribozyme. The replication of the ribozyme gene results in the amplification of the catalyst. The turnover rate contains a quadratic term of the RNA concentration because replication requires both RNA strands, one functioning as the template for copying and the other one as ribozyme."
>
> *(cited with his permission)*

Therefore, the hypercycle with its self-reinforcing properties has its basis in RNA as an information-containing molecule (base sequence) and as a molecule with catalytic properties (ribozyme). In this context, the synthesis of RNA molecules capable of catalyzing their own replication is quite significant. Gerald F. Joyce (La Jolla, California, USA) reports on this achievement,

> "There are now many examples of ribozymes, which were obtained by evolution in the laboratory, that catalyze the RNA-templated joining of RNA substrates. Some of these ribozymes catalyze a single joining reaction (ligation), while others catalyze multiple successive joining reactions (polymerization). Ligation involves the joining of fragments of RNA (oligomers), whereas polymerization involves the joining of individual nucleotides (monomers). Both reactions have special relevance to the origin of life because the underlying chemistry is similar to what would be required by an RNA replicase in the RNA world.

For many years, experimental efforts have focused on attempting to convert an RNA ligase ribozyme to a polymerase, and ultimately to a replicase. An alternative approach is to treat RNA oligomers as the building blocks, and attempt to progress from oligomer ligation to replication. This has recently been achieved in our laboratory in a cross-catalytic system involving two ribozymes that catalyze each other's synthesis from a total of four oligomer substrates. The cross-replicating ribozymes undergo self-sustained exponential amplification at constant temperature and in the absence of proteins or other biological materials. Amplification occurs with a doubling time of about one hour, and can be continued indefinitely.

The next step was to construct populations of various cross-replicating ribozymes that can be made to compete for limited resources within a common reaction mixture. The ribozymes reproduce with high fidelity, but occasionally give rise to recombinants that also can replicate. Over the course of many "generations" of selective amplification, novel variants can arise and grow to dominate the population based on their enhanced fitness under the chosen reaction conditions. The system is not yet capable of open-ended Darwinian evolution resulting in the emergence of novel function. However, self-replication can be made dependent on the ability of the RNAs to recognize particular chemicals in their environment, with their rate of replication determined by how well they execute that recognition function. Ultimately, the system should have the capacity to evolve functions as sophisticated as the replicase itself and to support an RNA-based metabolism. That would amount to the reinvention of the RNA world in the laboratory."

There was plenty of time in the ancient laboratory for the replicase mentioned by Gerald Joyce to have evolved. Then it would have been possible for new RNA variants with altered properties to appear continuously. Assuming that the required RNA building blocks were available, an RNA world could have then established itself. However, limits would have soon been reached.

Let us first look at an RNA virus for comparison. Viruses contain either DNA or RNA as an information memory. One of the simplest RNA viruses is the tobacco mosaic virus (TMV), which is responsible for the mosaic-like structures on the leaves of tobacco plants. The RNA consists of approximately 6200 building blocks (bases). For comparison, the DNA ring of the *E. coli* strain mentioned in Chapter 2 consists of 4.9 million building blocks. If this little RNA world enters a plant cell in the form of TMV, it's obvious what still would have to be provided for virus reproduction. The information for reproducing TMV is available, but the synthesis of amino acids and the production of viral proteins and viral RNA is strictly dependent on the enzymic apparatus of the host cell.

Let us now imagine that the crater lake was one giant RNA-containing cell. The RNA would like to multiply, but building blocks are required. These substances would become limiting because the Miller-Urey reactions would not be effective enough to provide them in the amounts required. The RNA must "learn" how to synthesize its building blocks by itself, but it needs more information in the form of base sequences in order to express the necessary catalytic activities. The RNA molecules grow longer and longer; so the RNA world soon reaches its limits because long RNA molecules are unstable. Millions of years passed. Finally, the management and maintenance of genetic information was passed on to the DNA, which is much more stable than RNA (see Chapter 30). In addition, it was necessary to create a reaction chamber in which a high concentration of building blocks for proliferation could be maintained. Hence, the cell surrounded by a membrane came into being.

Eventually, the crater lake was teeming with plenty of bacteria- or archaea-like cells. They must have been anaerobes, that is, fermenting or H_2S-producing organisms. We only know to a small percentage the species of these organisms, still on our planet in incredibly large numbers. Among these first organisms must have been LUCA, the organism with which the evolution of the three domains of life began.

There's one problem, though. At some point all the nice ingredients of the Miller-Urey Soup must have been eaten up.

You're right. In ancient times as well as now we need a primary producer of biomass. This is mainly the role of plants on the continents and of phytoplankton in the oceans. These organisms convert carbon dioxide (CO_2), together with minerals and water, to cell substance by using sunlight as an energy source. The cell substance thus produced is ultimately the source of food for most other organisms.

How was it in the case of LUCA or his descendants? For a long time, microorganisms of the *Thermoproteus* type fed on H_2 and S. They assimilated all the delicacies in the Miller-Urey soup but also gradually learned to synthesize them on their own when, for example, amino acids became limiting. They grew more and more independent and finally were able to grow with H_2, S, CO_2, and, of course, minerals. It was a sparse life because the conversion of H_2+S to H_2S does not provide much energy for growth. An alternative to microorganisms like *Thermoproteus* are fermenting microbes of the *Clostridium ljungdahlii* type. They grow on mixtures of H_2 and CO plus CO_2 and produce acetate, and they also could have been living creatures from the very beginning. But a breakthrough was needed, and it came when organisms began to take advantage of light. The first type of photosynthesis to appear is called anoxygenic (anaerobic) photosynthesis because it proceeds in the absence of oxygen and is not connected to any oxygen production whatsoever. Organisms of this type still exist, including the green and purple sulfur bacteria. Microorganisms such as the green *Chlorobium limicola* have a photosynthetic apparatus that includes various types of pigments (chlorophylls and carotenoids). They are able to perform the following reactions:

$$H_2S \xrightarrow{light} S + 2H + ATP$$
hydrogen sulfide → sulfur + reducing equivalents + metabolic energy

$$S + 4H_2O \xrightarrow{light} H_2SO_4 + 6H + ATP$$
sulfur + water → sulfuric acid + reducing equivalents + metabolic energy

$$2CO_2 + 8H + nATP \xrightarrow{minerals} \text{cell substance} + H_2O$$

Light energy is used by these organisms to gain ATP and, in addition, to generate reducing equivalents from hydrogen sulfide. Both ATP and reducing equivalents allow bacteria to convert CO_2 into cellular substance. In cells of *Chromatium okenii*, which are not green like the chlorobia but purple, the sulfur drops stored inside the cells temporarily during oxidation of sulfide to sulfate are quite impressive when seen in the light microscope (Figure 8a). Still today, mass accumulations of phototrophic bacteria can be encountered in nature (Figure 8b). In the early days of evolution, the sun had already been discovered as an energy source for biological reactions. This gave evolution a strong impulse, also because a process then evolved in the opposite direction, which enabled sulfate to be reduced in the dark to H_2S by bacteria called sulfate reducers, still present today in large numbers. Well-known sulfate reducers are *Desulfovibrio vulgaris* among the bacteria and *Archaeoglobus fulgidus* among the archaea.

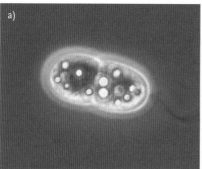

Figure 8 Phototrophic bacteria. (a) Light micrograph of the phototrophic bacterium *Chromatium okenii*. The refractive sulfur globules inside the cells can be seen as well as the tuft of very powerful flagellae (micrograph: Michael Hoppert, Goettingen, Germany). (b) Bloom of the sulfur purple bacterium *Amoebobacter purpureus* at a depth of 7 meters (23 ft) in Lake Mahoney (British Columbia, Canada). (Photograph: Joerg Overmann, Munich, Germany, already published in Brock, Biology of Microorganisms, with permission of author, M. T. Madigan.)

ATP was explained already, but what are reducing equivalents?

The importance of reducing equivalents becomes apparent when we compare the formula of CO_2 with that of glucose, $C_6H_{12}O_6$, as a typical constituent of cells. The CO_2 must be reduced, so reducing equivalents are required to provide the Hs in the glucose molecule. There are willing helpers in the cell, called NAD^+, $NADP^+$, or ferredoxin, that like to be loaded with H; they then transport the Hs to wherever they are needed for reduction processes (▶Study Guide).

Just to sum up, light provides the energy for the synthesis of ATP from ADP and P_i. In addition, it makes the Hs of H_2S available as reducing equivalents for the formation of cellular substance. A light-dependent sulfur cycle with partners such as *Chlorobium* and *Desulfovibrio* species developed, making biomass available to serve as food for the further diversification of the world of bacteria and archaea. Approximately one billion years passed, and the world of anaerobes reached its peak, but then another breakthrough began to change the organismic world completely. Oxygenic (aerobic) photosynthesis evolved from the anoxygenic type, so oxygen was produced instead of sulfur or sulfuric acid.

$$\text{anaerobic} \quad 2H_2S \xrightarrow{light} 2S + 4H$$

$$\text{aerobic} \quad 2H_2O \xrightarrow{light} O_2 + 4H$$

In contrast to H_2S, water was of course available in unlimited amounts. But the difference between these two types of photosynthesis is an enormous one. H_2O is the gold among the hydrogen donors needed for the reduction of CO_2 to biomass. In other words, it is extremely difficult to mobilize the two Hs from H_2O. If this were so easy, the energy problems of the world would have been solved very elegantly by hydrogen technology. H_2S is more like zinc; it provides the two Hs very readily. The cyanobacteria were the pioneers that brought about the transition of anoxygenic to oxygenic photosynthesis. They developed a light-dependent apparatus that could cleave water into oxygen and reducing equivalents. This was the beginning of the oxygen era on our planet. As for the anaerobes, most of them died out, and the remaining ones withdrew into the sediments and the mud where they still proliferate today.

> Oxygen is a nasty stuff
>
> *Bruce Ames*

Chapter 5
O₂

In which context did Bruce Ames say that?

Bruce N. Ames (Berkeley, California, USA) is the inventor of a test named after him, a test that determines whether a certain compound is carcinogenic or not. Ames developed this test in the 1970s and became famous for it. The Ames test is based on the assumption that a compound exhibiting a mutagenic effect (causing damage to the DNA) most probably will also be carcinogenic, meaning it may cause cancer. Bruce Ames had worked on the biochemistry of amino acid synthesis, especially the synthesis of histidine. The microorganism he used for his studies was a strain of the bacterium *Salmonella typhimurium*.

Salmonella typhimurium, doesn't it cause diarrhea?

Correct, but there are strains such as strain LT2 that is only slightly virulent. If the necessary safety regulations are observed, it is possible to work with strain LT2 in a microbiological laboratory without any risks.

A mutant of *Salmonella typhimurium* strain LT2 is used in the Ames test. This mutant carries a gene defect on its DNA and is unable to synthesize the amino acid histidine. Such a strain, called his-minus, is only able to grow when histidine is added to the growth medium. If this strain is treated with a mutagenic agent, for instance, a component of tar, millions of mutations occur in the bacterial culture. Among them are reversions by which a genetic defect is reversed. In the case of the his-minus mutant, a reversion restores the ability of the cell to synthesize histidine, so the cell and its descendants are again able to grow on medium in the absence of histidine. In the Ames test the rates of reversion from the mutant (requiring histidine in the medium) to the wild type (no longer requiring histidine) are compared in the absence (control) or in the presence of compounds to be tested. The Ames test is also of great importance because it resulted in a significant reduction in the number of animal tests.

What does this have to do with oxygen?

Ames realized that it was not sufficient to solely test a compound to evaluate its carcinogenic potential. It had to be taken into account that compounds may be modified in the human body. Therefore, tests were included in which the compounds to be tested were treated with liver extract to simulate such modifications. This liver extract contains hydroxylases, enzymes by which oxygen is introduced into compounds. The oxygen-containing compounds formed exhibit in many cases a higher carcinogenicity than the original substances. The principle of this test is shown in Figure 9. Let's take benzopyrene as an example, which is present in tar but also in cigarette smoke. When benzopyrene is tested as such, then after incubation with the liver enzymes for 30 minutes, its mutational power and thus its carcinogenicity increase dramatically.

Oxygen is a nasty stuff, and it plays an ambivalent role in nature. Human beings, animals, and plants require oxygen for respiration. We can only survive a few minutes without oxygen. The metabolism of respiring organisms is optimally adapted to the oxygen concentration of our atmosphere, 21 percent. Higher oxygen concentrations can be very dangerous. In chemistry class at school, it is a popular

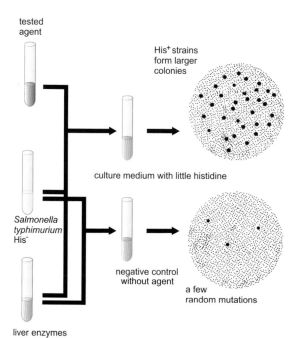

Figure 9 Examination of the carcinogenicity of compounds with the Ames test. Slight growth of the his-minus mutant on the Petri dishes can be seen (background growth). Three revertants grew on the control dish, but many more grew when the tested compound was present, indicating a positive test. (Diagram: Anne Kemmling, Goettingen, Germany.)

experiment to expose a glowing wooden splint to a pure oxygen atmosphere and watch it go up in flames. Many also remember the tragic accident of Apollo 1 on January 27, 1967, in which three astronauts were killed within seconds by a flash fire in an oxygen-enriched atmosphere.

But even the oxygen we inhale is not completely harmless to us. The danger lies in the fact that the oxygen molecule O_2 not only forms water but also to a certain extent radicals and hydrogen peroxide during respiration.

superoxide radical $•O_2^-$

hydroxy radical $•OH$

hydrogen peroxide H_2O_2

The dot in the O_2 forms shown above represents a single electron. Compounds of this kind are called radicals, and they are very reactive. Radicals–in this context we speak of ROS (reactive oxygen species)–attack all sorts of molecules in our cells. They damage DNA and enzyme systems and they contribute to certain diseases as well as aging. For example, such damage to the mitochondria, causes these power plants of our cells to lose efficiency over the years.

Did the evolved oxygen directly accumulate in the atmosphere?

When oxygen production by the cyanobacteria began, oxygen as such was not present. The whole planet was in a reduced state, with large amounts of hydrogen sulfide, ammonia, and ferrous iron (Fe^{2+}) present in the oceans, in the atmosphere, and on the continents. Nearly 1.5 billion years elapsed before a complete change was achieved. Hydrogen sulfide became sulfate, ammonia was converted to nitrate, and Fe^{2+} to Fe^{3+}, which precipitated in large amounts as ferric oxides. Only when practically nothing was left to be oxidized did oxygen slowly accumulate in the atmosphere. As a result, an ozone layer accumulated, so UV light no longer hit the Earth's surface with full intensity. This laid the foundation for the evolution of organisms on the continents. Earth became the blue planet. The transition from an anaerobic to an aerobic world left early traces of life in the form of stromatolites (Figure 10), which the geobiologist Joachim Reitner (Goettingen, Germany) describes as follows:

> "Indeed, stromatolites are the oldest biosignatures on Earth, going back 3.5 billion years. They represent remains of biofilms of microorganisms. The slime matrix of the biofilms was cemented together with sediments and/or authigenic minerals such as calcite or aragonite to form a laminated rock structure. With the beginning of the aerobic world approximately 2.5 billion years ago, the world of biofilms was predominated by cyanobacteria, from which the stromatolites originated. They formed gigantic structures reaching a thickness of 1000 meters and covering 1000 km². The growth of stromatolites was governed by photosynthesis, by the cation concentration (Ca^{2+}, Mg^{2+}), as well as the alkalinity caused by the high bicarbonate or

Figure 10 Early proterozoic stromatolites formed 2.2 billion years ago in the eastern Andes south of Cochabamba (Bolivia). The dark bands consist of cyanobacteria in carbonate-containing sediment layers. (Photograph: Joachim Reitner, Goettingen, Germany.)

carbonate concentrations of the proterozoic oceans. The proterozoic O_2 production was largely controlled by the cyanobacteria-dominated stromatolites. The dominance of these systems lasted for nearly 1.4 billion years and was terminated by a global climate catastrophe, the so-called Snowball Earth events with a total glaciation of the Earth around 300 million years ago. Following this critical interval of the Earth's history, algae and later higher plants played a growing role in O_2 production. Stromatolites develop at present in some extreme habitats, e.g., in soda or salt lakes but rarely in marine environments because current iron concentrations of seawater do not favor the calcification of biofilms."

Proterozoic oceans, by the way, only contained microorganisms because higher organisms had not yet evolved.

Now I understand what stromatolites are and that bacteria and archaea solely dominated the Earth for about 3 billion years. The cyanobacteria are responsible for the primary production of oxygen. After the Snowball Earth mentioned by Joachim Reitner, algae and plants contributed more and more to oxygen production.

But now I would like to learn something about organisms that use oxygen like I do.

As already mentioned in Chapter 4, the transition to the aerobic world was achieved by the cyanobacteria, which caused a sort of revolution within the organismic world. Many anaerobes died because they were unable to cope with the toxic byproducts of oxygen and to acquire the machinery needed to take advantage of oxygen in respiration. Still today, organisms like the methanoarchaea or many clostridia are killed when exposed to oxygen. Other organisms "invented" a respiratory apparatus localized in the cytoplasmic membrane, which enabled them to couple the reduction of oxygen to water with the generation of ATP. This respiratory apparatus is quite similar to the one present in our own mitochondria, in which oxygen is also consumed and reduced, providing us with ATP. Obviously, the aerobic organisms had to deal with the toxic byproducts of oxygen reduction mentioned above. This led to the evolution of enzyme systems that were capable of rapidly detoxifying the radicals and H_2O_2. Enzymes of this type include superoxide dismutase and catalase; the former detoxifies the superoxide radical and the latter converts H_2O_2 into water and molecular oxygen. As a matter of fact, we humans take advantage of similar systems. This is something we also have learned from the bacteria.

Independent of the evolution of higher organisms, microbes still play key roles on our planet. We shall start with their roles in some extreme habitats.

Some like it hot

I.A.L. Diamond and Billy Wilder

Chapter 6
Life in boiling water

The press most likely did not highlight the report of Thomas Brock (Bloomington, Indiana; later Madison, Wisconsin, USA) in 1969 on the isolation of a bacterium from a pond in Yellowstone National Park. This bacterium, *Thermus aquaticus*, grows at 70 °C (nearly 160 °F). Every molecular biologist or microbiologist is familiar with the enzyme Taq polymerase, which was isolated from *T. aquaticus* and has proved essential for the PCR technique used for analysis of trace amounts of DNA (see Chapter 25).

Growth at 70 °C sounds pretty exciting, but it's far from the 100 °C (212 °F) at which water boils!

Just wait a few minutes. Brock then isolated an organism that grows at 80 °C, *Sulfolobus acidocaldarius*. Among microbiologists, these reports aroused an interest in microorganisms growing at high temperatures, so sites in the immediate vicinity of volcanoes or geysers, where it spits and sputters, became the hunting grounds of microbiologists. Two German microbiologists from Munich, Wolfgang Zillig (1925–2005) and Karl Otto Stetter, were especially eager to isolate and characterize microorganisms growing in boiling water. First, they studied the sulfur-oxidizing *Sulfolobus* species from the Solfatara volcanic crater near Naples, Italy, then they concentrated on hot springs in Iceland. It is no easy task to take samples at such hostile sites and to then demonstrate in the samples the presence of living organisms that can be grown in cultures and are able to yield a sufficient cell mass for detailed studies. Zillig and Stetter obtained spectacular results, especially the realization that solely organisms of the *Archaea* domain are able to exist in boiling water. As for the members of the domain *Bacteria*, 80 °C is already quite hot. Nevertheless, there are species such as *Aquifex pyrophilus* and *A. aeolicus* that make it up to 95 °C, but boiling water is wholly and solely the habitat of some species of archaea.

The first "catches" of Zillig and Stetter were, for instance, *Thermoproteus tenax* and *Methanothermus fervidus*. These organisms and related ones grow at up to 97 °C. Ultimately it was Karl Otto Stetter, who really likes it hot when it comes to sampling sites, who dived to the hot seabed near the island of Vulcano, one of the Lipari Islands south of Naples. The fascinating result was the isolation of *Pyrodictium occultum*, which grows at the incredible temperature of 110 °C. But

Discover the World of Microbes: Bacteria, Archaea, and Viruses, First Edition. Gerhard Gottschalk.
© 2012 Wiley-VCH Verlag GmbH & Co. KGaA. Published 2012 by Wiley-VCH Verlag GmbH & Co. KGaA.

let's have Karl Stetter (Regensburg, Germany) tell us how he took the sample and how he finally isolated *P. occultum*:

> "The question of life beyond the conventional sterilization of water by boiling had really fascinated me for some time. Of course, any habitat of organisms growing above 100 °C had to be under an elevated pressure in order to remain a liquid. To search for organisms of this sort, I took advantage of a family vacation in 1981 on the island of Vulcano. I took samples from the sea floor close to Porto di Levante at a depth of 5 to10 meters where temperatures were up to 110 °C. Back at my laboratory at the University of Regensburg, we performed growth experiments using various energy sources and temperatures. In experiment PL-19, I used artificial sea water, a volcanic gas mixture (molecular hydrogen, carbon dioxide, and hydrogen sulfide) as well as sulfur inoculated with material from the sampling site and incubated under pressure at 105 °C. After two days, I already saw with the naked eye some unusual, hazy material covering the sulfur granules. Under the microscope I saw novel disc-shaped cells that were connected by very thin threads. It took us a year to show that this isolate was able to grow at 110 °C. It was named the "hidden fire network" (*Pyrodictium occultum*) and it is shown in Figure 11. This was the first report that life was possible above 100 °C, which was completely unexpected, especially in the presence of high salt concentrations in the ocean."

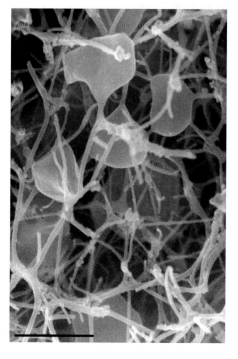

Figure 11 *Pyrodictium occultum*, scanning electron micrograph. The flat, irregularly shaped cells are connected by a network of protein filaments. (Reinhard Rachel and Karl Stetter, Regensburg, Germany.)

So, a specially equipped laboratory was established in the microbiology department at Regensburg to allow isolation and investigation of heat-loving archaea, now called hyperthermophilic archaea. It is fascinating to watch a growing culture of archaea in a vessel surrounded by boiling water. These microorganisms are so adapted to high temperatures that they usually don't even grow at temperatures below 80 °C. The world record in terms of growth temperature is held by *Pyrolobus fumarii*, which grows at 113 °C but is unable to grow below 90 °C; where it's simply too "cold."

It certainly wasn't easy to obtain these breathtaking results. Where else have microbe hunters been successful?

It really became a "hot" issue. Microbiologists got in touch with geologist colleagues because they were interested in hot sampling sites other than Yellowstone and the geysers in Iceland. Other extreme sites, those that were very acidic, alkaline, or salty, also became very attractive. One example is the Russian peninsula Kamchatka with its twenty-eight active volcanoes and hot springs. Several expeditions of microbiologists to Kamchatka came up with a rich variety of "new" organisms. The hot springs of Japan and the Azores were also visited by microbiologists, as well as the Wadi El Natrun in Egypt, an alkaline habitat with salt lakes and lagoons. The list of spectacular isolates of archaea is long, only a few will be mentioned here. *Picrophilus torridus* was isolated from hot acidic soil near Kawayu, on the Japanese island of Hokkaido (Figure 12). This archaeon is one of the most unusual living creatures on Earth. It grows at 60 °C and at a pH value of 0.7. In other words, in hot dilute sulfuric acid, caustic enough to burn our hands. The circular chromosome of *P. torridus* was sequenced in our laboratory. It consists of 1.5 million base pairs and is only one-third the size of the *Escherichia coli* genome.

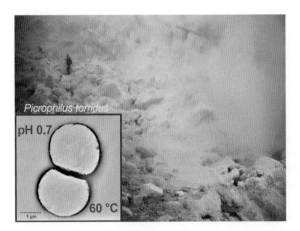

Figure 12 *Picrophilus torridus* and its habitat. The organism was isolated from a sample taken close to the volcano of Kawayu (Hokkaido, Japan). Christa Schleper (University of Vienna) is in the left background. (Photograph: Gabriela Puehler, Munich, Germany.)

Genes on the *P. torridus* chromosome can be identified with the conventional methods of molecular biology, but it proved very difficult to use this information for making enzymes. Obviously, *P. torridus* is an eccentric result of evolution.

A fascinating archaeal isolate was fished out of the Wadi El Natrun by microbiologists. It was named *Natronomonas pharaonis*; its habitat is a warm 45 °C, soda-containing brine with a pH of 9.5–10. Finally, an isolate from the Great Salt Lake in Utah (USA) should be mentioned, *Haloraptus utahensis*, which grows at 55 °C in saturated sodium chloride solution. Lagoons with a high salt content attract attention from the air because of their bright red color, for instance, the southern part of San Francisco Bay. These lagoons contain dense populations of haloarchaea, which have high concentrations of carotenoid pigments to protect themselves from the sun. Several of them also contain a fascinating purple compound, bacteriorhodopsin (see Chapter 8).

It is hard to believe: on the one hand, there are hard-boiled eggs with completely denatured protein, and, on the other hand, there are living creatures that grow vigorously under such conditions. Does anyone know why there are such stark differences? Why are archaea but no bacteria found in boiling water?

Those are good questions. For one thing, the composition of the cytoplasmic membranes differs between bacteria and archaea. The importance of this membrane for all living organisms was pointed out in Chapter 2. In bacteria and in all higher living organisms, the membrane is composed of so-called phospholipids, which are fat-like compounds. In archaea, however, they consist of special ether lipids that are more stable and better adapted to higher temperatures than phospholipids. At high temperatures, the DNA needs to be stabilized as well, which is achieved by binding proteins. In addition, the enzymes must be so engineered that they are active at temperatures at which we normally boil eggs, broil fish, or kill bacteria in the pasteurization process. Their existence can be considered a crowning achievement of evolution. The question, of course, is why these enzymes are so heat-resistant. Let's see what an expert, Gregory Zeikus (East Lansing, Michigan, USA), has to say about this:

> "Enzymes from hyperthermophilic microbes have very subtle structural differences from those of mesophiles or psychrophiles. These subtle structural differences allow hyperthermophilic enzymes (also called thermozymes) to be stable and active at very high temperatures (i.e. at 80 °C to 110 °C), but are generally less active at low temperatures (i.e. at 5 °C to 40 °C). A thermozyme shares the same general architecture and fold as that of catalytically similar mesozymes or psychrozymes. Thermozymes differ because they are more rigid, compact, and have fewer water accessible amino acids like asparagine and glutamine, which deaminate at high temperatures. Different metabolic types of thermozymes achieve a more rigid and compact structure by different means. The addition of a single salt bridge/ion pair at the enzyme's terminus can greatly enhance thermal

stability and prevent unfolding. Some thermozymes contain more hydrogen bonds than catalytically similar mesozymes. Enzymes that are stabilized by metal binding, like calcium-containing alpha amylases, come with an additional tightly-bound zinc in *Pyrococcus furiosus*. Dimeric enzymes from hyperthermophiles have evolved unique ways to enhance active dimer stability at high temperature. For example, xylose isomerase from *Thermotoga neopolitana* contains an additional proline in the dimer interface. Here, proline puts a bend in the protein, which rigidifies it and also exposes a surface aromatic amino acid for an additional hydrophobic bond to further enhance dimer stability."

This area still requires a great deal of research. Another question is why don't we find archaea that compete with bacteria in milk, in juices, or during the degradation of starch in soil. It cannot be ruled out that they once existed in such habitats but were superseded by bacteria. Only the rims of the domain of Archaea remain, much like the crater rim forming a circle of islands in Santorini (Greece). In their habitats, however, they are the masters.

> Where Sodom and Gomorrah reared their domes and towers
> that solemn sea now floods the plain
> in whose bitter waters no living thing exists . . .
>
> Mark Twain,
> *The Innocents Abroad, or The New Pilgrims' Progress*

Chapter 7
Life in the Dead Sea

The title above is that of a publication appearing in "Nature" in 1936. The article was written by B. Wilkansky, who later called himself Ben Volcani. In this publication, the author reports that he had seen bacteria when he inspected Dead Sea water under the microscope.

Is there really life in the Dead Sea?

To the naked eye, life is absent there. It is amusing how on the so-called Madaba Mosaic Map (Figure 13), a possibly frightened fish, upon reaching the mouth of the river Jordan, immediately turns back before entering the Dead Sea. The inhospitality of the Dead Sea lies in its extraordinarily high salt content, around 340 g/l. Sodium chloride is the predominant salt in ocean water, whereas in the Dead Sea it is magnesium chloride. One liter of Dead Sea water contains 188 g magnesium chloride, 91 g sodium chloride, and various potassium salts and bromides. The water level of the Dead Sea is the lowest point on Earth, at present 423 meters below sea level. This value corresponds to that reported in 1800. Then it began to rise, reaching 391 meters below sea level around 1900. Since then it has been falling continuously. If the Dead Sea water were to mix completely, as was the case in 1979, a liter would weigh 1.238 kg. Its density is much higher than that of the human body, which easily floats on its surface, so taking a dip in the Dead Sea is an exciting experience.

A totally mixed Dead Sea is indeed dead because there are no organisms capable of photosynthesis under these rather hostile conditions. In other words, biomass cannot be produced from CO_2 by using the energy from sunlight. In the oceans, such a biomass production sets a food chain in motion that reaches from the tiny phytoplankton all the way to the whales. Life in the Dead Sea is triggered when strong rainwater inflow causes salinity to decrease in the upper water layer. This is a relatively rare event that happened recently in 1980 and 1992. As a result, the drop in salinity caused a decrease in water density from 1.23 to 1.17 in the upper five meters, allowing the algae *Dunaliella* to develop an intensive bloom. Between blooms, these algae are present in a dormant form. Whenever the density of the water drops below 1.21, the algae are revitalized; they multiply to a cell density of up to twenty thousand cells per ml. Because of evaporation, the density of the

Discover the World of Microbes: Bacteria, Archaea, and Viruses, First Edition. Gerhard Gottschalk.
© 2012 Wiley-VCH Verlag GmbH & Co. KGaA. Published 2012 by Wiley-VCH Verlag GmbH & Co. KGaA.

Chapter 7 Life in the Dead Sea

Figure 13 The Madaba mosaic from the sixth century, part of the floor of the early Byzantine Church of St. George in Madaba, Jordan. This is the earliest cartographic depiction of Jordan and Israel. The walled city of Jerusalem can be seen below the Dead Sea. (Photograph: Daniel Graepler and Marianne Bergmann, Archeological Institute of the University of Goettingen, Goettingen, Germany.)

water slowly increases until a point is reached at which the organisms lose their viability, undergo lysis, and release considerable amounts of glycerol into the water. Living *Dunaliella* cells accumulate glycerol in order to counteract the osmotic pressure caused by the salinity of the Dead Sea. Without this high intracellular glycerol concentration, water would leave the cells, and they would dry out and shrink.

This liberated glycerol is the basis for a microbial food chain in the Dead Sea. Microorganisms able to live in the Dead Sea use glycerol as a nutrient, allowing them to grow and multiply. Microorganisms isolated from the Dead Sea include the archaea *Haloferax volcanii* and *Haloarcula marismortui* and the bacterium *Sporohalobacter lortetii*. But let's hear what the microbiologist Aharon Oren (Jerusalem, Israel) has to say. He is an expert on Dead Sea microorganisms:

> "When I started my studies of the Dead Sea in 1980, we knew very little about the types of halophilic microorganisms living in its salt waters and the dynamics of their populations. Surprisingly, very little work has been done on the microbiology of the lake since the pioneering studies of Ben

Volcani in the 1930s. The year 1980 was a special year for the Dead Sea, as dense blooms of green algae and red archaea developed in the upper water layers, following inflow of large amounts of rainwater. Water and sediment samples collected during the period yielded a number of novel organisms, including the red archaeon *Halorubrum sodomense*. I have been monitoring microbial processes in the Dead Sea and studied the properties of its biota ever since. The even denser bloom in 1992 provided us with new opportunities to deepen our understanding of the microbiology of this unique lake that presents the most demanding challenges to organisms adapted to life under the most hostile conditions. However, microbial blooms in the Dead Sea are rare events and most of the time the lake can almost be considered a sterile environment."

Strong rainfalls in 1992 are of interest because they may have been connected with the eruption of the volcano Mount Pinatubo in the Philippines the year before. This was the largest eruption of a volcano since 1912, ejecting nearly 20 billion tons of sulfur dioxide into the stratosphere. Because of oxidation reactions, the resulting sulfate aerosol strongly affected the climate (warmer winters and cooler summers) as well as photosynthesis in 1992 and 1993. The aerosol belt surrounding the Earth caused the proportion of diffuse light to increase; so photosynthetic activity and CO_2 uptake increased. This had a dramatic effect on coral reefs: because of improved growth conditions, algae and cyanobacteria overgrew the corals, killing them off to a considerable extent. This effect and the conditions in the Dead Sea in 1992 show how sensitive our Earth is to disturbances. This should alarm us when we consider the continuous increase of the CO_2 concentration in the atmosphere. This topic will be covered in Chapter 14.

All this is very interesting. Are there other sites on Earth where you wouldn't expect to find life?

An extreme habitat of a completely different kind is found at the deep ocean floor far away from continents. Generally speaking, the floor of oceans is more like a desert. It is cold and pitch dark, and – depending on its actual water depth – the hydrostatic pressure can be more than 100 atmospheres (1500 pounds per square inch). Deep-sea life feeds on the detritus formed by dead organisms sinking into the abyss. But there are oases of life in the deep sea, for instance, along the oceanic ridges where water with a temperature of up to 400 °C (750 °F) and a high hydrogen sulfide content emerges from the seafloor. These hydrothermal vents were first discovered in 1977 in the Pacific Ocean at a depth of 2600 meters (8530 feet), near the Galapagos Islands. At some distance from these vents, where the temperature drops below 50 °C (120 °F), a fantastic underwater world consisting of gigantic worms more than one meter in length was discovered. One species is called *Riftia pachyptila* and consists of tubes with bright red plumes of gills (Figure 14). The lifestyle of these organisms was described in 1981 by research teams of the microbiologist Colleen Cavanaugh (Cambridge, Massachusetts, USA) and her

Chapter 7 Life in the Dead Sea

Figure 14 Dense population of the giant tube worms *Riftia pachyptila*. They are up to 2 meters (7 ft) long and 6 cm (2 in) in diameter. They harbor hydrogen sulfide-oxidizing bacteria. Galapagos, seafloor spreading zone, 0° 48′N; depth, 2550 meters (8400 ft). (Holger Jannasch [1927–1998], with permission of Nordrhein-Westfaelische Akademie der Wissenschaften, Duesseldorf, Germany.)

colleagues, Holger Jannasch (1927–1998) and John Waterbury (Woods Hole, Massachusetts, USA) with Meredith L. Jones (1926–1996) and Stephen L. Gardiner (Smithsonian Institution, Washington, D.C., USA).

What do these worms feed on?

It is pitch dark down there, so the organismic world must be independent of photosynthesis. Let's ask Colleen Cavanaugh for an answer:

> "The giant tubeworms – the "poster child" of deep-sea vents, with their long white tubes and brilliant red plumes – were truly a nutritional puzzle as the adults are completely mouthless and gutless. The vent communities were thought to be sustained on organic matter produced by free-living chemosynthetic bacteria. Upon hearing a description of the anatomy of the tube worms noting the presence of sulfur crystals in their tissues, I realized that these amazing invertebrates were being fed from the inside by symbiotic bacteria! These intracellular symbionts oxidize inorganic sulfur compounds to drive autotrophic carbon fixation – effectively making these bacteria like "chloroplasts" using sulfur instead of sunlight for energy. The host, in turn, provides its symbionts access to the necessary nutrients and energy substrates for chemosynthesis – notably simultaneous delivery of sulfide and oxygen via the tubeworm vascular system and its remarkable hemoglobin. Indeed, at deep-sea vents, chemosynthetic symbioses in these worms and all of the major macrofauna form the base of the food chain. Holger Jannasch, a leader in vent microbiology, and John Waterbury, a microbiologist

and electron microscopist, provided early encouragement and continued rigor in my development of the experimental proof that bacteria existed within the worm. Given our insights from studying such hydrothermal vent symbioses, we now recognize that similar chemosynthetic symbioses have evolved widely in diverse sulfidic and methanogenic environments (e.g., coastal sediments, seeps, vents) and these have now been described worldwide for numerous invertebrate and protist groups. Studies of vent symbioses have thus redefined perceptions of eukaryote-bacteria nutrition, ecosystem function, and evolutionary innovation."

This is a great discovery: these bacteria have what is called a chemolithotrophic way of life, which is based on the chemical conversion of inorganic compounds (lithos = stone), in this case, the oxidation of hydrogen sulfide with oxygen to elemental sulfur and further to sulfate.

Is this the only chemolithotrophic way of life?

In Chapter 12 we will read about the Russian microbiologist Sergej Winogradsky (1856–1953). He investigated the process of oxidation of ammonia (NH_3) to nitrate (NO_3^-) by bacteria. He then coined the term chemolithotrophy and designated these organisms as chemolithotrophic bacteria. They live from the energy produced by the oxidation of inorganic compounds, in this case, the oxidation of ammonia with oxygen. At the hydrothermal vents, it is not ammonia but hydrogen sulfide that is oxidized by bacteria. In Winogradsky's term, "troph" means "to feed on." Of course, these bacteria need a source of cell carbon in addition to metabolic energy; here, it's carbon dioxide. The life observed at the hydrothermal vents and at many other sites on Earth is a fascinating way of life that is a domain of bacteria and archaea, since no higher organisms are able to live on H_2S, O_2, and CO_2. Aren't they an interesting couple, the worms and the H_2S-oxidizing bacteria? The bacteria produce some sugar-like compounds from CO_2, which serve as food for the worms, in addition to the bacterial biomass. The worms protect "their" bacteria and provide them with oxygen and H_2S. Giant clams and mussels also live together with H_2S-oxidizing bacteria in the neighborhood of hydrothermal vents, which really are sensational. In the meantime, hydrothermal vents also have been discovered in the Atlantic and in the Indian Ocean. These are magnificent biotopes in the pitch darkness of the oceans, but their glory can only be seen in the floodlight of a research submarine.

What else is exciting at the seafloor?

You probably have heard of the ice-like gas hydrates that predominantly consist of methane and are found on the ocean floor. These are the largest methane sources on Earth. Gas hydrates, however, are relatively unstable, and their occurrence depends on low temperatures and high pressures. If the temperature increases and the water depth is less than 600 meters (1970 feet), a spontaneous

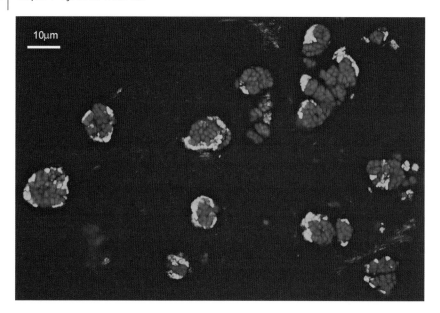

Figure 15 Anaerobic oxidation of methane by a consortium consisting of methanoarchaea (colored red) and sulfate-reducing bacteria (colored green). (Photograph: Antje Boetius, Bremen, Germany.)

liberation of methane is possible. Fortunately, not much of the greenhouse gas methane escapes from the seafloor to the surface. Microorganisms utilize it as an energy source and have established various sorts of oases in oceans, called "cold seeps". Until recently it was assumed that the oxidation of methane can only be accomplished by oxygen-consuming bacteria. But how can we explain the methane oxidation observed by geologists in the sea floor in which oxygen was strictly absent? Now we know that there is also an anaerobic oxidation of methane, a process in which methane is respired with sulfate as a sort of substitute for oxygen. What we know is summarized by Antje Boetius (Bremen, Germany):

> "The so-called cold vents or methane seeps were discovered in the deep sea shortly after the discovery of the hydrothermal vents. They represent fascinating ecosystems in which the chemical energy of methane is the basis for life for a great variety of microorganisms – but also for animals. Surprisingly, the underwater worlds around the cold vents sometimes resemble those of hydrothermal vents, including the presence of tubeworms and mussels. But where does the sulfide supporting the chemosynthetic life come from? We were able to demonstrate that the hydrogen sulfide found in large amounts in methane-rich areas of the seafloor is a result of the activity of billions of archaea, which together with bacteria oxidize methane with sulfate. Consequently, the sulfate is reduced to hydrogen sulfide. Both the archaea and the bacteria form cell aggregates as shown in Figure 15.

Biogeochemically and ecologically, this is an extremely important process. The methane-oxidizing archaea control the emission of methane from the seafloor and are one of the key players in the regulation of the greenhouse gas methane on Earth."

The diversity of processes at the seafloor is fascinating. At the hydrothermal vents, hydrogen sulfide of volcanic origin is the basis of a food chain all the way to one-meter-long tube worms. In methane seeps, where methane and sulfate are the basis and hydrogen sulfide is the product, a variety of organisms also flourishes under these conditions.

at hydrothermal vents: $S^{2-} + 2\,O_2 \rightarrow SO_4^{2-}$

at methane seeps: $CH_4 + SO_4^{2-} \rightarrow CO_2 + S^{2-} + 2\,H_2O$

S^{2-}, sulfide; SO_4^{2-}, sulfate, CH_4, methane.

> The immediate principles of living bodies
> would be, to a degree, indestructible if,
> of all the organisms created by God,
> the smallest and apparently most useless
> were to be suppressed.
> And because the return to the atmosphere
> and to the mineral kingdom
> of everything which had ceased to live
> would be suddenly suspended,
> life would become impossible
>
> *Louis Pasteur*

Chapter 8
Bacteria and archaea are everywhere

If microbial life is present in habitats such as hot mudpots or fetid, poisonous sulfur springs, then it is reasonable to assume that there is practically no place on Earth without bacteria and/or archaea. They are everywhere, having been carried by air and water to the most remote sites on our planet. The only places free of bacteria have undergone heating or ultrafiltration to kill or remove the microorganisms present. Bacteria are just simply omnipresent and they may multiply when nutrients and proper conditions for growth are available. As a matter of fact, our Earth is primarily occupied by bacteria and archaea. According to calculations done by the microbiologist William Whitman (Athens, Georgia, USA) and his colleagues, our planet hosts more than 10^{30} microbial cells. They contain around 500 billion tons (500 Gt) of carbon, which exceeds the total carbon of all animals by a factor of 100. These figures defy our imagination: 100 tons of bacteria per ton of elephants and other animals. It's just incredible. Let's have William Whitman tell us how the total number of microbial cells on Earth was even calculated in the first place:

> "When I joined the faculty of the Department of Microbiology at the University of Georgia in 1982, I set about isolating new strains of the methanoarchaeon *Methanococcus* for developing genetic systems. Bill Wiebe, a microbial ecologist and professor in my department, agreed to help me obtain samples from the salt marshes at the university's marine station at Sapelo Island. At the same time, Bill and I also performed a number of experiments on the role of methanogenesis in these marshes. About that time, I read that the number of *E. coli* on Earth was about 10^{20} cells, and it was natural to wonder how many methanococci were present as well. However, considering how little we knew about the microbiology of salt marshes, Bill and I concluded it would be easier to estimate the number of all prokaryotes on Earth than the number of any particular free-living species. We didn't do much with this idea until a decade later when Dave

> Coleman, a process-oriented soil ecologist, renewed our interest in the problem.
>
> Because prokaryotes are virtually everywhere in the biosphere, we initially thought it might be impossible to estimate their total numbers. After reviewing the literature in detail, it became obvious that the total number would be dominated by the subsurface, seawater, and soil. In the end, we estimated the total number of prokaryotes at 5×10^{30} cells, containing 5×10^{17} grams of carbon or 10^{17} grams of nitrogen. We also estimated that the total number of new cells made each year was about 2×10^{30}. These numbers told us two things. First, cells live by catalyzing chemical reactions, and these reactions form the basis of the chemical nature of the biosphere. The more cells, the bigger the reactions they catalyze. Since the amount of prokaryotic cell material was comparable to that of plants, the prokaryotes must be making a similar contribution to the chemical composition of the biosphere. Secondly, the large numbers of prokaryotes make it easy to explain their enormous diversity. For instance, mutations that are rare in the relatively small populations of eukaryotes would be extremely common in the prokaryotes."

Aren't these numbers overwhelming? We will come back to some of William Whitman's conclusions in Chapter 14.

Could you go back and outline how bacteria were found in the first place?

Bacteria were discovered by Antonie van Leeuwenhoek (1632–1723) in 1676. He had designed a simple microscope that allowed him to look into the microcosm. He already had described erythrocytes and sperm. Not much changed after his discovery because researchers of the time and the public even more so found it difficult to accept the idea that the changes occurring in meat or sugar solutions were because of bacterial activity. These processes, especially alcoholic fermentation, were at the time explained by mechanical interactions. Georg Ernst Stahl (1659–1734) wrote;

> "A body already subject to decomposition will very easily promote decomposition of another body that is yet unaffected. A body already in internal motion will also induce this internal motion in another one that still is inactive but susceptible. (Cited by Haehn, Hugo: Biochemie der Gaerungen, Walter de Gruyter und Co., Berlin 1952, p. 21)"

So contact causes interior motion to be passed on, which then results in fermentation or decomposition. This theory had to be disproved. To do so, Lazzaro Spallanzani (1729–1798) and Theodor Schwann (1810–1882) carried out careful experiments. They showed that infusions of seeds or other materials were not subject to putrefaction when they had been boiled for at least 45 minutes and when these infusions had only been in contact with air that had been heated for quite some time. Very slowly the idea was accepted that these infusions contained germs that were responsible for putrefaction. It was not until Louis Pasteur (1822–1895) and Robert Koch (1843–1910) performed their pioneering studies that bacteria and other microorganisms such as yeast were recognized as being responsible for fermentation and putrefaction.

Nearly 10 000 bacterial and archaeal species have been isolated since then and their properties described. However, it has been estimated that more than a million species are to be found on our planet. Most of them cannot be isolated and properly described because they have withstood the enormous efforts of microbiologists to grow them in the laboratory.

How do microbiologists know that there are so many microbial species?

You may remember the discoveries of Carl Woese described in Chapter 3. He employed the sequence of the 16S rRNA to classify microbial isolates so that they could be incorporated into the phylogenetic tree. One can say that the 16S-rRNA sequence is the picture ID of a bacterial or archaeal species. Just imagine that you see a few beautiful bacterial cells under the microscope, but you are unable to find the right conditions to culture them in order to determine their properties. It is possible to determine the picture ID by employing special methods, which will be described in detail later (Chapter 25). These methods are so effective that 16S rRNA from a few lysed cells can be multiplied and finally sequenced. Such a sequence can then be added to the phylogenetic tree, even though the actual physiological properties of the microorganisms from which the RNA originated remain a secret. Branches and branches have thus been added to the phylogenetic tree, but the microorganisms they represent remain in a coma. All we have is their picture IDs. A very impressive database of 16S-rRNA sequences has been set up. This will be introduced by the microbiologists Karl-Heinz Schleifer and Wolfgang Ludwig (both from Munich, Germany):

> "When Carl Woese recognized that the 16S rRNA was suitable as an informative phylogenetic marker, sequence determinations were still difficult and time-consuming. Due to the rapid development of sequencing techniques and largely automated equipment, the situation now is entirely different. During the last five years, the number of available 16S/18S-rRNA sequences has increased tenfold. The sequences are deposited in public databases and they are available as reference sequences for phylogenetic analysis. In addition to the sequence databases (EMBL, www.ebi.ac.uk; GenBank, http://www.ncbi.nlm.nih.gov/Genbank), special ribosomal RNA databases have been set up. Here, sequence data are available that have been edited for phylogenetic analyses and taxonomic assignment (ARB-Silva, www.arb-silva.de; RDP, rdp.cme.msu.edu; greengenes, greengenes.lbl.gov). The actual version of the ARB-Silva database contains more than 1.3 million entries."

Those are astronomical figures: more than a million different 16S/18S rRNAs. Just incredible! Why, by the way, is it 16S/18S?

S is a unit of size of a macromolecule. The 16S rRNA of bacteria and archaea corresponds to the larger 18S rRNA of the eukarya, the plants and animals. Why don't you click one of the cited databases in order to get an impression of the treasures that have been accumulated over the years?

Now you must get around to the importance of the various microbial species.

The global role played by bacteria and archaea makes it possible to complete the cycles of matters on Earth. These cycles are driven by the generation of biomass in the photosynthesis process (Figure 16). Actually, two more or less separate cycles of matter operate on our planet, one on the continents and the other in the oceans. In the latter, cyanobacteria and algae proliferate in the sunlight and form the phytoplankton that serves as the basis of ocean life, all the way to the turtles, sharks, and whales. On the continents, the plants are eaten by insects, which in turn will serve as food for small birds, which then become the prey of eagles, buzzards, and hawks. Cattle grazing on vegetation provide humans with milk or meat. In food chains, part of the biomass is oxidized to CO_2 by respiration but the large part remains as dead plant or animal material, fortunately not forever. If we look around, we may see a

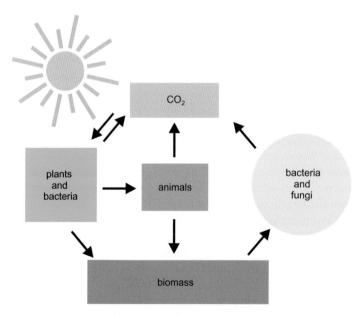

Figure 16 The carbon cycle. Driven by photosynthesis, biomass is formed from CO_2 and minerals. As a result of various food chains, CO_2 and minerals are regenerated in respiratory processes and fermentations. A more detailed cycle is depicted in Figure 29. (Diagram: Petra Ehrenreich, Goettingen, Germany.)

transient accumulation of leaves, straw, and windfall fruits, and perhaps dead fish in a nearby lake. Eventually, all of this material will disappear. This process is faster with apples, cherries, and sugar beets than it is with straw and leaves. Dead animals and their stench also disappear after a short period of time, as do the floral colors and fragrances. So the winters are gray and brown, and a new green landscape with colorful blossoms will appear next spring. It is the same around the world, especially in the zones that are productive with respect to biomass formation. Particularly in soil it is the job of bacteria and archaea, of fungi as well, to decompose biomass and convert it to CO_2 and minerals, which will be channeled into a new round of the cycle of matter. Fungi are especially important in the mineralization of wood because they have enzyme systems capable of degrading lignin, an essential component of wood. But bacteria and archaea are present in soil, ditches, creeks, lakes, and oceans; at high and low temperatures, and at various pH values. Not only are they present everywhere but they are at work everywhere as well.

A great variety of organic compounds is found in nature. Many of these compounds occur in large amounts: cellulose, starch, and proteins, just to mention a few. Others such as dyes, alkaloids, or flavors are present in relatively small amounts. It is the principle of biological degradation that all these compounds serve as food for the microbial community, but it is not so that one particular bacterial species is able to degrade all of these compounds. There is a pronounced specialization. There are organisms that have specialized on proteins, carbohydrates, or organic acids, whereas a few are highly specialized and grow solely on uric acid, nicotine, or naphthalene and related compounds. The availability of degradable biomass, of course, varies in different habitats, as does the density of the microbial population. It is scant in deserts but extremely high in a field of sugar beets. When wading through the Wadden Sea tidelands, most people are unaware that much of the material oozing between their toes consists of bacteria. We could go on with more examples, but one thing is for sure: Without bacteria and archaea, the biomass on our planet would accumulate incredibly, making it impossible for higher organisms to live. If bacteria and archaea were to go on strike, the Earth would soon look like a megacity without a functioning garbage-collection service.

In bacteria and archaea there is an impressive abundance of enzyme systems with which various organic compounds are degraded under diverse conditions in nature. A great number of pathways for ATP generation have evolved, making microbial metabolism a wide field, so understanding it requires a good knowledge of biochemistry and physiology. Some of the various pathways employed by microbes will be outlined in further chapters of this book. Three of the processes that have evolved in archaea and bacteria and are essential for life in general will be highlighted here: charging the membranes by proton translocation, ATP synthesis at the membranes, and CO_2 fixation. These essential mechanisms and processes were adopted by plants and animals from archaea and bacteria.

As pointed out in Chapter 2, the charged cytoplasmic membrane is a miracle of evolution: charging the membrane–negative inside and positive outside–and maintaining this state involves transporting protons or, in some cases, sodium ions from the cytoplasm to the cell's exterior. This is achieved by respiration or by

Figure 17 Salt works on the island of Lanzarote, in the foreground a salt evaporation pond with reddish haloarchaea. (Photograph: Dieter Oesterhelt, Munich, Germany.)

photosynthesis. One of the best-studied light-driven proton pumps is bacteriorhodopsin. It is present in certain halophilic archaea and contributes to the wonderful purple color of salt lakes and salterns (Figure 17). It charges the membrane of these archaea by proton translocation. Bacteriorhodopsin was discovered by Dieter Oesterhelt in the laboratory of Walther Stoeckenius in San Francisco. I asked Dieter Oesterhelt (Martinsried, Germany) to summarize the discovery and the function of bacteriorhodopsin:

> "Having just received tenure at the University of Munich in 1969, I had the chance to do research in the United States for one year. Biological membranes, electron microscopy, and the city of San Francisco, taking these together, the choice of the laboratory was easy, that of Walther Stoeckenius.
>
> A goal of my research was to learn something about the violet color of halobacterial membranes, which are still called the purple membrane. Some useful information on the physical properties was obtained, but problems arose when the lipids were extracted from the purple membrane. The color changed from purple to yellow. Daily discussions with Allan Blaurock and a chemical analysis led to the conclusion that the pigment was without any doubt the vitamin A aldehyde, retinal; it was bound to an unknown protein. Soon after this, the pigment was named "bacteriorhodopsin". Had the archaea already been known by then, the pigment certainly would have been called "archaerhodopsin".
>
> Walther's first comment on the exciting news was: "Only a stubborn chemist could claim the existence of retinal in a prokaryote." But then he eagerly collaborated on this project. What could this molecule be good for? Its role was unraveled in 30 years of worldwide research. Bacteriorhodopsin, also found recently in bacteria as proteorhodopsin, is a light-driven proton pump and it represents a second type of photosynthesis."

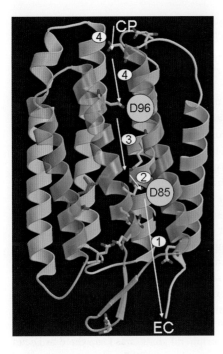

Figure 18 The path of protons through the bacteriorhodopsin (BR) pump, from the inside (CP) to the exterior (EC). The seven helices of BR span the membrane. The critical amino acids of the light-mediated vectorial catalysis are aspartic acid 85 (D 85) as proton acceptor and aspartic acid 96 (D 96) as proton donor. The key steps of proton movement are: (1) proton release, (2) proton transfer following rearrangement of retinal, caused by a light impulse, (3) backfolding of retinal and proton uptake, (4) reprotonization, that is, proton uptake by aspartic acid 96 after it has provided the proton to the retinal in step (3). (Model: Dieter Oesterhelt, Munich, Germany.)

The first type of photosynthesis is the one carried out by green plants, cyanobacteria, and phototrophic bacteria, to be described in Chapter 9. However, the second type, as it occurs in haloarchaea, is mechanistically quite well understood, especially with respect to the way protons travel from the inside to the outside. The molecule of bacteriorhodopsin is depicted in Figure 18. The protein skeleton consists of seven loops that span the membrane. The cytoplasmic side (inside) of the membrane is at the top (CP); the exterior side (EC) is at the bottom. Retinal is located near the center of the membrane (close to the number 2). Upon illumination, retinal folds so that H$^+$ is released via D85 to the outside. Then, retinal reassumes its initial position and takes up one H$^+$ from the inside via D96. So retinal works like a switch, in which a light-dependent reaction transports protons from the cytoplasmic side to the exterior space, thus generating what is called the proton motive force. The amino acids D85 and D96 participating in the proton translocation are aspartic acid residues of the protein skeleton.

How does the cell take advantage of this proton motive force generated by proton pumps?

First, the proton motive force passes nutrients via specific transport systems into the cell, nutrients such as sugars and amino acids but also phosphate or magnesium ions. Secondly, probably the main function, it drives the synthesis of ATP from ADP and inorganic phosphate. The very complex enzyme system capable of coupling the reverse flow of protons into the cell with ATP synthesis is called ATP synthase or F_1F_0-ATPase. It is anchored in the membrane. This system is unsurpassed with respect to ATP synthesis and is ubiquitous in nature. Here it is described by Volker Mueller, (Frankfurt/Main, Germany):

> "ATP synthase is an unusual enzyme. It is the smallest electric motor in nature. This enzyme consists of a membrane domain, which transports ions along their potential like water running downhill, and a cytoplasmic domain, where ATP is synthesized from ADP and inorganic phosphate (Figure 19). Both domains are connected via a central shaft. In addition, there is a second shaft located on the periphery. The rotor is inserted in the membrane. Depending on the organism, it consists of 10–15 rotor blades. The ion flow through the membrane sets the rotor in motion. Ions jump onto the rotor blades and promote the rotation as does water in a watermill. The rotation of the rotor is transmitted to the central shaft, which has a small "nose" at its upper end. The cytoplasmic domain consists of six subunits arranged like slices of an orange. When the nose hits the space between two slices, ATP is released from the enzyme and, at the same time, another molecule of ATP is synthesized from ADP and inorganic phosphate in the adjacent space between two slices. The third space between two slices remains unoccupied. During rotation, these three spaces go through three stages: ATP synthesis from ADP and inorganic phosphate, ATP release, and unoccupied. The job of the peripheral shaft is to keep the orange "in position". Therefore, the ATP synthase is a manifold converter: an electrical potential is converted to mechanical work (rotation) and then to chemical work, the synthesis of the fuel of life, ATP. It is an ingenious principle, a unique nanomachine that is already present in the simplest organisms."

And this nanomachine is breathtakingly beautiful. The structure depicted in Figure 19 is not the product of fantasy but is actually based on structural analysis and kinetic data, for which John E. Walker (Cambridge, UK), Paul D. Boyer (Los

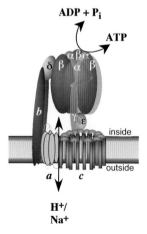

Figure 19 Composition of the ATP synthase and arrangement in the cytoplasmic membrane. Subunits c are the rotor blades; subunits γ and ε form the central shaft. The "orange slices" mentioned in the text are the three α (alpha) and the three β (beta) subunits. The peripheral shaft consists of subunits b and δ. (Model: Volker Mueller, Frankfurt/Main, Germany.)

Angeles, USA), together with Jens Skou (Aarhus, Denmark) received the Nobel Prize in 1997. Speaking of Nobel Prizes, the discovery of the proton pumps requires an addendum. It was Peter Mitchell (1920–1992) (Bodmin, UK) who conceived the chemiosmotic mechanism, as we now call the process in which the membrane is charged by proton translocation. He received the Nobel Prize in 1978 for this intellectual achievement.

So we have taken a look at two of the basic processes essential for life, membranes charged by proton translocation and ATP synthesis at the membranes. What was the third process?

The elucidation of the third process – CO_2 fixation, the fundamental step in biomass formation – takes us to the Californian University at Berkeley in the 1940s and 1950s. It is there that the decisive experimental breakthrough was achieved, not only because of the ingenious researchers working there but also because of the excellent research environment. Just try to imagine what the problem was: the green algae *Chlorella* grow in the light with CO_2, but how to identify the molecule to which CO_2 is attached and the products of this reaction? The first Cyclotron had been operating in Berkeley since 1929, making it possible to generate radioactive isotopes and carry out so-called tracer experiments. It is as if dollar bills were to be marked with a dye that is only visible under UV light, making it possible to identify and trace the path of these banknotes. So if CO_2 were to be labeled with a radioactive isotope, then its way through the *Chlorella* cell could be traced. This is difficult enough because hundreds of radioactive compounds have to be separated, purified, and identified. But what could be the source of the radioactive carbon? The opportunity to find the answers opened up in Berkeley.

In nature, there are two stable (nonradioactive) carbon isotopes, C_6^{12} and C_6^{13}. The first one, with an atomic weight of 12 makes up 99 percent of all carbon compounds and the second one (atomic weight 13), nearly 1 percent. Two radioactive carbon isotopes with relatively short half-lives were generated in the Cyclotron. One, C_6^{11}, has a half-life of 21 minutes and was actually employed for tracer experiments, but quite tedious because the radioactivity decreased by 50 percent every 21 minutes. It was a real breakthrough when Samuel Ruben (1913–1943) and Martin D. Kamen (1913–2002) were able to produce considerable amounts of C_6^{14} at Berkeley. They wrote in a 1941 issue of *Physical Review*,

> "Two five-gallon carboys filled with saturated solutions of ammonium nitrate were placed in the region opposite the deflector of the Berkeley medical cyclotron. The solutions were irradiated with neutrons during a period of ~six months. Approximately 40 000 µ amp. hr. [microampere-hours] of deuteron bombardment at 16 MeV [megaelectron volts] of various targets (principally beryllium and phosphorus) took place in this time. A chemical analysis was performed in which carbon monoxide, carbon dioxide, methane, carbon, methanol, cyanide, and formaldehyde were added as carriers and then separated and analyzed for radioactivity. After

> vigorous and prolonged shaking, the gaseous compounds were pumped off. The gases were oxidized to carbon dioxide and absorbed in Ca(OH)$_2$. The CaCO$_3$ was very [radio]active, containing 10^5 counts/min. By the chemical methods described above the activity was shown to be isotopic with carbon."

So, a saturated solution of ammonium nitrate was bombarded with neutrons for six months. The following reaction took place:

$$N_0^1(\text{neutron}) + N_7^{14}(\text{nitrogen}) \rightarrow C_6^{14}(\text{carbon}) + H_1^1(\text{proton}) + Q_1(\text{energy})$$

In this process, the chemical element carbon is formed from the chemical element nitrogen. Radioactive carbon 14 is considered a "soft" emitter that allows relatively safe handling. The half-life is 5700 years, so there is no time pressure while doing experiments with carbon 14.

An experiment in two 20-liter carboys (glass bottles) made its mark on the whole era. With the aid of ^{14}C-labeled compounds, the world of metabolism finally could be explored. We are now familiar with hundreds of metabolic pathways and thousands of intermediate products thanks to the application of ^{14}C-labeled compounds. The first biological experiment with ^{14}CO$_2$ was performed by Horace A. Barker (1907–2001), Sam Ruben, and Martin D. Kamen in Berkeley. They were able to show that methanogenic organisms (see Chapter 13) form radioactive methane from radioactive carbon dioxide.

In the late 1940s, a hard-working *Chlorella* research team was formed under the leadership of Melvin Calvin (1911–1997), Andrew A. Benson, and James A. Bassham in Berkeley. This team unraveled the secret of the conversion of CO$_2$ into biomass. Green algae (*Chlorella*) were confined to a glass vessel, called lollipop because of its shape. The algae were illuminated, and radioactive CO$_2$ in the form of bicarbonate was added. After a few seconds, samples were withdrawn and analyzed, applying state-of-the-art techniques of the 1950s. The first breathtaking result was the initial appearance of radioactivity from the CO$_2$ in 3-phosphoglycerate, a compound with three carbon atoms, H$_2$O$_3$–P–O–CH$_2$–CH(OH)–COOH. The radioactivity was located in the carboxyl group, –COOH, at one end of the molecule. A few years of hard work elapsed until the substance combining with the radioactive CO$_2$ could be identified. The over 93-year-old Andrew A. Benson refers us to his 1951 publication in the Journal of the American Chemical Society (volume 73, p. 2971), in which he wrote,

> "The intermediates involved in carbon dioxide fixation by plants are largely phosphorylated hydroxy acids and sugars. A compound observed during the first few seconds of ^{14}CO$_2$ photosynthesis in all the plants investigated in this laboratory has now been identified as ribulose (adonose) diphosphate."

This was the final, long-sought result. Ribulose 1,5-bisphosphate, as it is called today, was the decisive compound to be identified in this research work. The equation is as follows:

$$C_5 + CO_2 = 2\,C_3$$

ribulose   two molecules of
1,5-bisphosphate   3-phosphoglycerate

Which enzyme carries out this reaction?

Andrew Benson (Santa Barbara, California, USA) asked himself that same question many years ago. He comments on his work,

> "... What I consider my greatest discovery. With great enthusiasm and excitement, I and Jacques Mayaudon of the Catholic University of Louvain discovered that the carboxylating enzyme of photosynthesis is identical with the major protein of leaves isolated by Sam Wildman of Caltech and UCLA, 'Fraction 1 Protein.'"

This "carboxylating enzyme" is now called ribulose 1,5-bisphosphate carboxylase, easy to remember as "rubisco," for short.

Is it really important to remember rubisco?

You have to take into account that almost the whole biomass on our planet, with few exceptions, is based on the rubisco reaction, not only the fossil energy carriers (e.g., coal and oil) but also the current microbial biomass mentioned before (500 Gt), the forests with around 600 Gt, as well as humans, animals, flies, and so on. There would be no gasoline or petrol without rubisco! This is more or less only a snapshot because living biomass is continuously being produced and decomposed. Rubisco, whose name sounds more like a Latin American dance, is the most common enzyme on Earth.

These were thus the three highlights. Charged membranes, a prerequisite for organismal life, are simply everywhere. A number of different ion pumps have evolved to charge the membranes, one wonderful example being bacteriorhodopsin. As for CO_2 fixation, the overwhelming role of rubisco is not diminished by the fact that the biomass of methanoarchaea and a number of phototrophic and anaerobic bacteria is not the product of rubisco activity. In these cases, other reactions are involved in the conversion of CO_2 to biomass (▶Study Guide).

The activity of ATP synthase is just as spectacular. Consider for a moment the transition from a world solely inhabited by prokaryotic organisms to a prokaryotic/eukaryotic world in which larger cells, the ancestors of plants and animals, developed. These cells had an enormous energy requirement. The best way to meet this requirement was to engulf cyanobacteria or respiring bacteria and take advantage of their capacity for ATP production, either light or oxygen dependent. The origin of chloroplasts and mitochondria, the sites of ATP production, has been explained by what is known as the endosymbiontic theory. This is such an exciting theory that it will be discussed in detail in a statement by Jack Meeks (Chapter 9) and in Chapter 27, as well.

For the photosynthetic CO_2 reduction only those hydrogen compounds can serve as H-donors in which the hydrogens can be sufficiently activated by the organism. Then it is quite conceivable that organisms exist which cannot use H_2O as a H-donor because they cannot activate the H in this compound sufficiently. These organisms might, however, be typically photosynthetic in case some other H compound is present containing the hydrogen in a form in which these organisms can bring about a sufficient activation

Cornelis B. van Niel

Chapter 9
The power of photosynthesis, even in almost complete darkness

This title probably means that some bacteria are able to "see" tiny amounts of light.

Correct, but first we need to see how microorganisms take advantage of bright sunlight.

Photosynthesis had a decisive impact on evolution. When the world was still anaerobic, the ability to convert light energy into metabolically useful energy provoked a first revolution. It was not only the generation of ATP from light but also the possibility to oxidize compounds such as hydrogen sulfide or ferrous ions (Fe^{2+}) as sources of the reducing power required for CO_2 reduction. So sulfate and ferric ions (Fe^{3+}) became available for the first time, and these compounds allowed new microbiological processes to evolve, such as those carried out by the sulfate-reducing bacteria. The second revolution occurred with the emergence of oxygenic photosynthesis and, with it, the development of the aerobic world (see Chapter 5). These processes should be examined more closely.

You described the bacteriorhodopsin of some archaea as a light-driven proton pump, the second type of photosynthesis, in Chapter 8. Are you now talking about the first type of photosynthesis?

Yes, and this type is much more complex. It's found all over the world and involves the phototrophic anaerobic bacteria; the oxygen-evolving cyanobacteria; the algae in oceans, lakes, and streams; and all the green plants on the continents. All of these organisms have a machinery called the photosynthetic apparatus. When we look at the flora around us, at the flowers, the meadows and the forests, we are of course aware that all this is the result of photosynthesis, but normally we don't think about the machinery required to capture light and take advantage of its energy. The conversion of light energy into other forms of energy is no easy task.

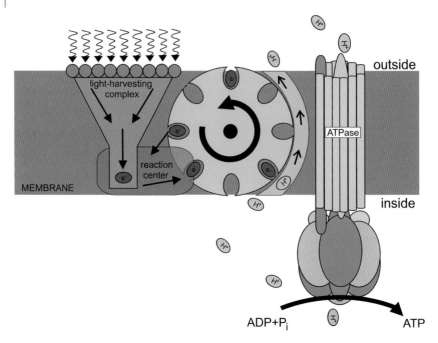

Figure 20 The photosystem as it appears in purple bacteria carrying out anoxygenic photosynthesis. Light is harvested and the excitation energy is channeled to the reaction center in which electrons (red) are released. They are pushed through a cycle of carriers and, driven by this cycle, protons (yellow) are translocated from inside to outside. The electrons return to the reaction center (cyclic electron flow). The resulting proton gradient is then used for ATP synthesis. The driving force is the high redox energy of the electrons released, which is lost stepwise in the cycle.

Think of the sophisticated technologies that have been developed to "harvest" light in photovoltaic plants.

The photosynthetic apparatus consists of several components (Figure 20). Light is absorbed by carotenoids that are part of the light-harvesting centers. The light energy is then converted into a flow of electrons that hit the reaction center pigments. They consist of special chlorophyll molecules. The reaction center then initiates a cyclic electron flow, which functions like the wheel of a water mill. It pumps protons from one side of the cytoplasmic membrane to the other, and the resulting proton gradient can be coupled to the synthesis of ATP as discussed in Chapter 8.

This is the principle of anoxygenic photosynthesis as carried out, for example, by the brightly colored purple bacteria (Figure 8). Not all microorganisms carrying out anoxygenic photosynthesis contain beautiful, purple-colored carotenoids. Other groups contain yellowish carotenoids capable of absorbing shorter-wavelength light that reaches greater depths in bodies of water. This, however, is not the only adaptation of the green sulfur bacteria for growth in deeper layers of water or where it's turbid. Furthermore, these microorganisms contain super antennae, chlorosomes, which enable them to absorb the tiniest amounts of light.

They don't look through binoculars, as the purple bacteria do, but through a telescope. They are even so efficient that they are able to thrive in the Black Sea. Let's ask Joerg Overmann (Braunschweig, Germany) under which conditions green sulfur bacteria can grow in this habitat:

> "Green sulfur bacteria are much less flexible than any other phototrophic bacterium with regard to their metabolism: they obligately depend on light to obtain energy, on carbon dioxide as carbon source, and on sulfide to feed electrons in their photosynthetic system and to ultimately reduce that carbon dioxide and form progeny. Because of their narrow ecological niche, natural populations of green sulfur bacteria are only found in certain lakes or sandy beaches where microbial processes lead to sulfide production in deeper layers still reached by underwater light. Green sulfur bacteria thus face a dilemma – they are confined to lower light levels than any other phototrophic organism. This is probably the reason, why green sulfur bacteria have developed the largest photosynthetic antenna of all photosynthetic organisms to absorb light: it has been determined that a single cell contains about fifty million bacteriochlorophyll molecules.
>
> In the Black Sea, the inflowing river water cannot mix with the heavier, deeper water that originates from the Mediterranean. As a result, only the upper 100 m contain oxygen, algae, crustacea, jellyfish and fish, etc., whereas the entire 2100 m below that are completely free of oxygen and higher life forms. Consequently, the Black Sea currently represents the largest anoxic water body on Earth. Our recent measurements of underwater light intensities employing especially sensitive quantum meters revealed unprecedented low values. The light intensities at 100 m depth equals the light of a small candle seen at a distance of 60 m in a pitch dark night. Yet, one type of phototrophic green sulfur bacterium has been found to be capable of growing even under these extreme environmental conditions.
>
> After earlier anecdotal reports on the occurrence of different anoxygenic phototrophic bacteria in the 1950s and 1970s, a U.S.–Turkish expedition in May 1988 detected traces of bacteriochlorophyll e, a unique pigment of green sulfur bacteria, at about 100 m depth. This represents the deepest occurrence of these bacteria so far. In the following years, we could show that only a single type of green sulfur bacteria occurs in this environment and were able to isolate and study it. The Black Sea strain is able to increase its photosynthetic antenna to twice the size of its relatives, thereby increasing the light energy harvested even further. Most notably, however, the cells are capable of surviving very long periods in the dark because they require much less maintenance energy than any other bacterium investigated so far. Recently, we found that the Black Sea bacterium is actually distributed across the major part of the Black Sea. Thus, this bacterium holds the world record for low-light adaptation, low maintenance energy requirement and probably represents the largest contiguous population of phototrophic bacteria known to date."

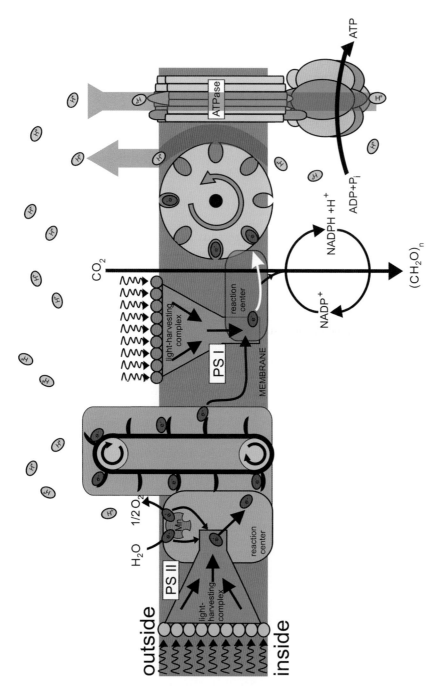

Figure 21 Oxygenic photosynthesis requires the interplay of two photosystems. PS I generates electrons (red) that travel through the carrier cycle but are also used to provide reducing power for the reduction of CO_2. This results in a lack of electrons that no longer reach the reaction center of PS I. The pool of electrons is replenished by the action of PS II. Ultimately, these electrons come from water, which is cleaved by a manganoprotein complex. As mentioned in Figure 20, the redox energy is important. At the reaction center of PS II, it is so high that water can be oxidized to oxygen and the electrons are pulled all the way through the carrier system to reduce $NADP^+$.

What an impressive report: photosynthesis at a depth of 100 meters, but performed by a highly specialized species of the green sulfur bacteria.

As already mentioned, the transition of H_2S to H_2O as source of reducing power was a dramatic step in evolution that was taken with the emergence of the cyanobacteria on our planet. They represent the first important group of organisms capable of carrying out an oxygenic photosynthesis.

What is the mechanistic difference between anoxygenic and oxygenic photosynthesis?

There's a big difference. The anoxygenic photosynthesizers have one photosystem. This system is not potent enough to remove the two hydrogen atoms from the water molecule. As already mentioned, water is the gold among the suppliers of reducing power. Two photosystems are required, and their combined action makes the whole system extremely powerful. It is like going from a propeller engine to a jet engine. Instead of air, photosystem II sucks in electrons and generates a manganese-containing protein so deficient in electrons that it can separate water into oxygen, protons, and electrons (Figure 21). Recent findings indicate that the photosystem of the green sulfur bacteria is the ancestor of photosystem I and that photosystem II originated from the purple bacteria. Isn't that a miracle, two types of bacteria performing anoxygenic photosynthesis coexisted for hundreds of million years. Combination of their photosystems in one cell and further evolution took place. As a result, cyanobacteria appeared performing an oxygenic photosynthesis. They are the pathfinders of eukaryotic phototrophs, of algae and plants, and their importance cannot be overexaggerated. We have asked an expert on cyanobacteria, Jack C. Meeks, (Davis, California, USA), to summarize the importance of the cyanobacteria during evolution and today:

> "Cyanobacteria are defined by their oxygenic photoautotrophic mode of energy and carbon metabolism. They are the most nutritionally independent organisms in the biosphere, requiring only light, water, CO_2 and a few macro- (N, P and S) and microinorganic nutrients for growth. To appreciate the ecological role of cyanobacteria, it is important to recognize that, although ATP is the primary cellular energy currency, reduced carbon compounds (biomass) are the major ecological energy currency in supporting growth of fermenting and respiring organisms. Fueled by an energy source from outside of the biosphere (sunlight), anoxygenic photosynthesis, utilizing H_2S and H_2 as electron donors, was (and is) an efficient mechanism for the production of reduced carbon, and the emergence of the two different types of reaction centers of anoxygenic photosynthesis undoubtedly led to the diversification of fermentative and respiratory metabolism in other organisms. However, the abundance of water as an electron donor, and its essential requirement as the solvent of life, gave oxygenic photosynthesis by cyanobacteria an enormous competitive advantage. Moreover, oxygen, the waste product of the photolysis of water, is a

biological toxin because of its interaction with reduced biomolecules and conversion to reactive oxygen species (ROS, see Chapter 5), which damage DNA and proteins. To this day, the vast majority of anoxygenic phototrophic bacteria cannot grow as phototrophs in the presence of oxygen. Thus, cyanobacteria improved their competitive advantage by converting illuminated habitats from an anoxic to oxic state and essentially eliminating their photosynthetic bacterial competitors in the oxic environments. Many cyanobacteria also reduce (fix) atmospheric N_2 to NH_3 (see Chapter 11), thereby enhancing both their nutritional independence and the value of their metabolites. In this regard, they also show their nutritional needs to be simpler than their progeny eukaryotic algae. From an ecological perspective, nutritional independence also implies that cyanobacteria are less susceptible to biological selective pressures than are nutritionally dependent organoheterotrophs, or even chemolithoautotrophs.

The photosynthetic production of oxygen by cyanobacteria was, arguably, the most dramatic event in the continued evolution of life on Earth. Its effect on ancestral microorganisms was discussed in Chapter 5. One can speculate on its impact in the evolution and diversification of eukaryotes. Eukaryotic cells have many unique characteristics that distinguish them from the bacteria and archaea. With respect to one character, it now appears that all existing eukaryotes have, or had, mitochondria. In some anaerobic eukaryotic microorganisms, the mitochondria have been reduced in structural and functional complexity, and exist as membrane-enclosed hydrogenosomes or mitosomes. Phylogenetic analyses indicate that mitochondria arose following an endosymbiotic association between an unknown anaerobic phagocytic partner cell and a respiring bacterium. A question to ask is: what was the selective pressure that stabilized such an endosymbiosis? Much attention has been given to the advantage of respiratory electron transport coupled to ATP synthesis; i.e. energy metabolism. Clearly, mitochondria have evolved elaborate translocation mechanisms for small (ATP, ADP and other metabolites) and large (proteins) molecule communication with the partner cytoplasm and nucleus. An alternative hypothesis suggests an oxygen protective role in mitochondrial endosymbiosis. Since the mechanisms for detoxification of ROS appear to be universal in bacteria and eukaryotes, the oxygen protective role has credence, with the genetic information transferred from the tolerant endosymbiont to the intolerant eukaryote nucleus. In this scenario, a eukaryotic cell with a stabilized mitochondrial endosymbiont could tolerate the oxygen produced by cyanobacterial primary producers.

A second, sequential, endosymbiotic event with a cyanobacterium then led to the chloroplasts of algae and plants. This event greatly reduced the competition for organic substrates by the partner cell, which could now live on CO_2 as carbon source. This competitive advantage would have stabilized the endosymbiosis.

These two endosymbiotic events in response to the selective pressures of oxygen tolerance and reduced-carbon/energy acquisition could have resulted in quite rapid diversification of the two early eukaryotic lines. The microbial photoautotrophic line led to terrestrial plants, with their highly successful light-harvesting morphology. The selective pressure on the microbial organoheterotrophic line was for external nutrient (biomass) acquisition, ultimately leading to the evolution of highly effective means for harvesting prey."

We learned a lot from Jack Meeks' contribution. It is not only ATP synthesis that stabilized the endosymbiosis ultimately leading to animals and to plants. It is also the oxygen-protective role and, in the case of plants, the ability to grow on CO_2 as carbon source. Further aspects of endosymbiosis will be discussed in Chapter 27.

> For creatures your size I offer
> a free choice of habitat,
> so settle yourselves in the zone
> that suits you best, in the pools
> of my pores or the tropical
> forests of arm-pit and crotch,
> in the deserts of my fore-arms,
> or the cool woods of my scalp.
>
> *Wystan Hugh Auden*

Chapter 10
Man and his microbes

Without the bacteria in our tummies, we would be one kilogram (2.2 pounds) lighter.

Are you serious? Do you really mean that the bacteria living in our bodies weigh that much?

Correct. This weight corresponds to a bacterial dry weight of approximately 100 grams. Bacteria not only reside in your tummy but also in your nose, mouth, and throat, and on your skin. It is hard to believe, but a healthy human body consists of approximately 10^{13} cells as well as nearly 10^{14} bacterial and archaeal cells. Archaea don't play a big role, but two species of methanoarchaea are usually found in our intestines, where they effectively produce biogas. An overview of our microflora is presented in Figure 22.

Let's begin with our skin, which is densely populated by bacteria. There are nearly 10^5 bacterial cells per cm², so a penny on the back of your hand would cover about 100 000 bacterial cells. They are especially dense between adjacent skin cells. Bacterial proliferation on our skin is limited because skin is relatively dry. Nevertheless, more than 20 different bacterial species have been detected on our skin; two of these are *Staphylococcus epidermidis* and *Propionibacterium acnes*.

Doesn't the second one cause acne?

Yes and no. We are covered with cells of this bacterium, but acne does not occur on all parts of our body nor occur at every age. Several factors have to coincide, such as the hormonal changes during puberty and the clogging of hair follicles with sebum, a wax-like substance. This results in the formation of a favorable habitat for *P. acnes*. For a more detailed description of acne, let us turn to Holger Brueggemann (Goettingen/Berlin, Germany), who was involved in the total sequencing of the genome of *P. acnes*:

Discover the World of Microbes: Bacteria, Archaea, and Viruses, First Edition. Gerhard Gottschalk.
© 2012 Wiley-VCH Verlag GmbH & Co. KGaA. Published 2012 by Wiley-VCH Verlag GmbH & Co. KGaA.

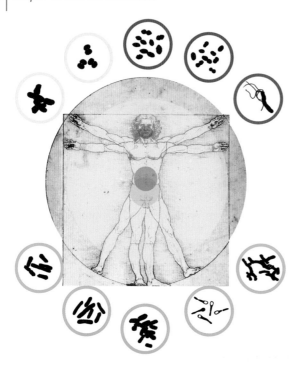

Figure 22 Microbes and men. Locations of microflora are: skin (yellow): *Propionibacterium acnes*, *Staphylococcus epidermidis*; mouth (purple): *Streptococcus salivarius*, *Streptococcus mutans*; stomach (red): *Helicobacter pylori*; intestine (green): *Bifidobacterium*, *Clostridium difficile*, *Escherichia coli*, *Eubacterium rectale*, *Bacteroides fragilis*. (Diagram: Anne Kemmling, Goettingen, Germany.)

"The importance of P. acnes for the formation of inflammatory skin acne cannot be recognized by simply applying the Henle-Koch postulate because the presence of this bacterium does not inevitably lead to acne. The pronounced response of the immune system to rapidly proliferating P. acnes suggests an opportunistic-pathogenic character of this organism. Especially the deciphered sequence of the genome of P. acnes provided new insights into the (pathogenic) lifestyle and the survival strategies of P. acnes on the human skin. The genome sequence provided information on the potential of this bacterium for the enzymic breakdown of constituents of dermal tissue, the development of the virulence traits as well as the interaction with the immune system, which contributes to the inflammatory efflorescence (skin proliferation). Currently, new therapy forms for acne are discussed on the basis of these findings. These are vaccine-based strategies, e.g., against dominant markers on the bacterial surface as well as the use of specific growth inhibitors that could replace the application of broadband antibiotics. Incidentally, it is now investigated if P. acnes is the causative agent of another disease. This bacterium is frequently detected in diseased prostate tissues. Like *Helicobacter pylori* in the stomach, P. acnes could contribute to the development of prostate cancer."

A few explanatory remarks may prove helpful. The Henle-Koch postulate will be discussed in Chapter 22. Opportunistic-pathogenic refers to a microorganism that

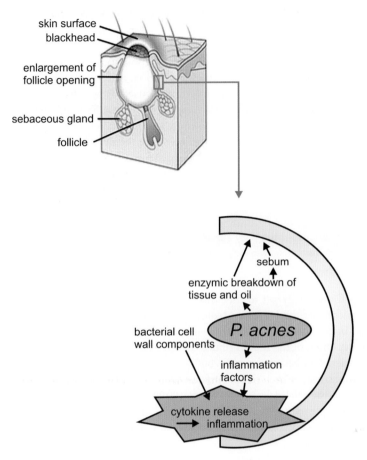

Figure 23 Diagram of the possible role of *Propionibacterium acnes* in the emergence of acne vulgaris. Top: an enlarged sebaceous gland. Bottom: possible interaction of *P. acnes* with surrounding tissue cells. Arrows pointing upward indicate that secreted enzymes degrade tissue, proteins, and fats, thereby providing nutrients to the bacteria. Downward arrows indicate inflammatory reactions caused by factors secreted by *P. acnes*. (Diagram: Holger Brueggemann, Goettingen, Germany.)

normally is not a pathogen but may cause disease when there are favorable conditions for its proliferation. What *P. acnes* is able to do is depicted in Figure 23. On one hand, it secretes enzymes that degrade tissue components and provide nutrients for the bacteria; on the other hand, it causes an inflammatory response that supplies it with water, minerals, and additional nutrients.

Tears serve to protect the eyes from microbial invasion. This fluid contains a number of components that kill bacteria or inhibit their proliferation. The most important one is lysozyme, which destroys bacterial cell walls and therefore kills bacteria. Eye inflammations develop whenever there is a disturbance of the tear film composition. Conjunctivitis is regularly caused by staphylococci but also by the very dangerous *Clamydia trachomatis*.

The nasal fluid is of the same composition as the tears and protects the respiratory tract from infection. But as we all know, it is not always possible to maintain the relatively low colonization of nose, mouth, bronchia, and lungs. Viral infections can be followed by severe bacterial infections caused by streptococci, *Haemophilus influenzae*, or even *Pseudomonas aeruginosa*. A proliferation of these organisms and of *Neisseria meningitides* in the nose and mouth gets out of control, penetrating barriers and leading to inflammation of the meninges (meningitis).

The urogenital tract may also be populated by bacteria. In addition to lactic acid bacteria, yeasts such as *Candida albicans* can be found in this habitat. Inflammations are often caused by special strains of *Escherichia coli*, which will be discussed in Chapter 29.

So far, you've only discussed the body surfaces that normally are not very densely populated in healthy humans. They don't account for the one thousand grams of bacteria mentioned.

Yes, but we will eventually get to the "inside story," as it was called by Laurie Comstock, Professor at the Harvard Medical School, after a short remark about the microflora in our mouth. We find three major habitats for bacteria in the oral cavity, the moist lining of the mouth (mucosa), the teeth, and the saliva. Saliva is a source of nutrients for bacterial growth, but saliva also contains components that inhibit the colonization of bacteria on surfaces such as the teeth. Saliva is essentially a very dense bacterial culture with about 100 000 000 (100 million) cells per ml. A human being produces approximately 750 ml of saliva per day, so at the same time we swallow approximately 75 000 000 000 (75 billion) bacteria daily.

What kind of bacteria do we swallow?

The microflora of the mouth is dominated by streptococci. Most of these species are benign. In addition to a dozen *Streptococcus* species, more than thirty additional bacterial species can be found in our mouth.

How are bacteria involved in the formation of dental plaque?

Bacteria have developed mechanisms to prevent being washed off by saliva. They form so-called biofilms, more or less sticky layers consisting of sugar molecules, with the bacteria inside. Dental plaque is a problem because the streptococci residing in the biofilm produce lactic acid that directly attacks the tooth enamel.

By the way, the sugar (saccharose from sugar cane or sugar beets) in our food favors the development of plaque because bacteria such as *Streptococcus salivarius* cleave the sugar into glucose and fructose. These bacteria grow on the glucose and produce lactic acid. At the same time, the fructose is polymerized to the macromolecule levan, which contributes to the biofilm on the surface of the teeth. Plaque development is less favored when syrups from starch, enzymically converted first to glucose and then to a mixture of glucose and fructose, are used in our food.

What is the difference?

The difference is as follows: saccharose is a disaccharide in which glucose and fructose are linked chemically. When this disaccharide is cleaved by microorganisms, the glucose produced serves as growth substrate and fructose is polymerized. In glucose/fructose mixtures, both sugars are utilized for growth.

Now we come to the inside story, that of the great diversity and the various activities of microorganisms primarily in the large intestine, which has moved to the center of interest of microbiologists, geneticists, and medical doctors. This is where most of the one thousand grams of bacteria within us is located. There is practically no other habitat in nature where a bacterial population reaches such a high density. As we all know, our well-being and health is very much affected by the composition of the intestinal flora. We often hear about the intestinal bacterium *Escherichia coli*, because it is the best-studied organism. However, in the large intestine, *E. coli* represents less than one thousandth of all bacterial cells present therein. Since *E. coli* is most easily isolated and identified, it is used as an indicator of fecal contamination. The main "bugs" in the intestine are *Bacteroides* and *Eubacterium* species, for example, *Bacteroides fragilis* and *Eubacterium rectale*. Anaerobic bacteria often escaped culture-dependent detection because they used to be cultured in the presence of air, and the oxygen in the air is toxic to many anaerobic microorganisms.

So most people do not know which bacteria contribute to the microbial community in our intestine?

Correct, *Escherichia* coli is present, but it represents a very small proportion thereof, but it's still a little more complicated. We speak of the human microbiom when considering the intestinal bacteria as a whole. It is represented by hundreds of bacterial species, and the total number of different genes in them exceeds the genes of the human genome by a factor of 100. Admittedly, these are mainly genes that code for enzymes required for the degradation of the various compounds arriving in the intestine. These compounds are the large remainder of food ingredients that escape digestions by human enzymes and absorption in the small intestine. The composition of the intestinal microbiota adjusts to the substrate supply, so it differs in vegetarians as compared to people who regularly eat meat. There also exists a correlation between obesity and the kind of "intestinal microbes" present. Pharmaceuticals also have a great effect: for example, treatment with antibiotics may lead to a depletion of the beneficial bacteria and a preponderance of the pathogen *Clostridium difficile* in the intestinal microbiota (see Chapter 29).

How do we humans benefit from the intestinal microflora?

Above all, our microbiota suppresses the settlement of pathogenic bacteria in our intestine. We just mentioned *C. difficile* as an example of a pathogen that takes over if the beneficial microbiota is depleted. Another role of the microflora is the detoxification of poisonous compounds arriving in the intestinal tract. Last but not least, we benefit from the nutrients resulting from microbial activities in the intestine, also vitamin K. Michael Blaut (Potsdam-Rehbruecke, Germany) will give us a more detailed description:

> "One of the major tasks of the intestinal microbiota is to convert indigestable carbohydrates (dietary fiber) to acetic, propionic, and butyric acids. Butyric acid provides 70 percent of the energy required by the epithelial cells of the large intestine whereas acetic acid serves as energy source in the peripheral tissues. Propionic acid is an important building block for gluconeogenesis in the liver, i.e. the synthesis of sugars. Many substances found in plant-derived foods are bioactive, i.e. they have a health-promoting effect. However, some of them first need to be activated by the intestinal microbiota to become biologically active. For example, linseed and rye contain polyphenolic compounds, such as secoisolariciresinol and matairesinol (very complicated names), which are converted to enterodiole and enterolactone. The latter presumably are preventively active against breast and prostate cancer.
>
> The intestinal microbiota also affects the metabolic fate of pharmaceuticals. Hydrophobic (fat-soluble rather than water-soluble) compounds are oxidized in the liver and then linked by so-called phase-2 enzymes to glucuronic acid molecules or sulfate, to make them water soluble. The majority of these conjugated compounds is discarded in the urine via the kidneys, but a considerable amount appears in the intestine together with bile fluid. There, the conjugated compounds are hydrolyzed by bacteria. The products, again hydrophobic, are reabsorbed and transported back to the liver where they are retransformed into conjugated compounds that reappear in the intestine. This process, the enterohepatic circulation, results in a longer retention time of any such compounds.
>
> The bile acids are also subject to enterohepatic circulation, but they are primarily conjugated with glycine and taurine. In addition to hydrolyzing the bile acid conjugates, intestinal bacteria modify the sterol skeleton of the bile acids, leading to secondary bile acids. There is good experimental evidence that secondary bile acids are tumor promoters in colon cancer.
>
> The intestinal microbiota affects not only the metabolism of the host but also the development and maturation of the immune system. It plays an important role in the maturation and proper functioning of the innate and the adaptive immune system. The intestinal microbiota helps the immune

system to distinguish between pathogenic and harmless bacteria. The immune system learns to fight pathogens and to tolerate nonpathogenic bacteria and food antigens. Failure of this kind of tolerance leads to inflammation. Oral tolerance is perturbed in humans suffering from inflammatory bowel diseases such as ulcerative colitis and Crohn's disease. In addition, the intestinal microbiota supports the barrier function of the intestinal epithelium that prevents the growth of pathogenic bacteria. It is noteworthy that the intestine contains approximately 70 percent of all immune cells of the body."

This is a very competent account on the importance of our intestinal microbiota. The information is challenging, but it is important to know that the intestinal bacteria not only grow at the expense of dietary fiber and other materials but also fulfill important functions.

After all this information, I have quite a different view of my insides. But how does the composition of the intestinal microbiota differ in babies? This difference is apparent to parents who have changed their baby's diapers, or nappies.

It is interesting that the intestinal microbiota of babies is dominated by bacteria that are considerably less important in adults: *Bifidobacterium* species and related bacteria. These are lactic acid bacteria that produce large amounts of acetic acid in addition to lactic acid. In newborns, the microbiota originates from the mother and the environment. Within the first days of life a microbial community dominated by bifidobacteria develops, especially in breast-fed babies. Bifidobacteria in infants are favored by certain oligosaccharides present in human milk and by the low pH. In addition, the presence of certain amino acid containing compounds promote the growth of these microorganisms. After weaning, the intestinal microbiota of the infant gradually shifts toward that of an adult. This process may take two to three years.

So in an ideal world you don't think about gastric ulcers and diarrhea?

Of course, but first the good side of bacteria, their role in various habitats outside our body, has to be made clear. Their bad side will be discussed toward the end of this book, in Chapter 29.

> It's a very odd thing, as odd as can be
> that whatever Miss T. eats turns into Miss T.
>
> Walter de la Mare

Chapter 11
Without bacteria there is no protein

That can't be true – I get my protein from steaks, chicken, and milk.

Yes, but where does this protein come from? Proteins come from plants, so we think of cattle grazing in lush meadows. In order to be able to synthesize proteins, plants require what is called bound nitrogen. This is nitrogen in the form of ammonia (ammonium ions), NH_3 (NH_4^+), or in form of nitrate (NO_3^-). The largest inexhaustible source of nitrogen in nature is nitrogen gas, which makes up 78 percent of our atmosphere. This nitrogen, N_2:

$$1N \equiv N1$$

is inert, not readily reacting with other elements. Neither plants nor animals are able to convert N_2 into bound nitrogen. The triple bond between the nitrogen atoms is not easy to crack. The conversion of molecular nitrogen from our atmosphere into bound nitrogen is only possible in three steps, first of all, the Haber-Bosch process based on a catalyst developed by Fritz Haber (1868–1934). It catalyzes at high temperatures (450 °C) and high pressures (300 bar) the conversion of N_2 into ammonia:

$$N_2 + 3\,H_2 \rightarrow 2\,NH_3$$

The enormous technical problems involved were solved by a team at the German chemical company BASF under the direction of Robert Bosch. The industrial Haber-Bosch process has been in use since 1913. It is the basis for the production of fertilizers and is considered one of the greatest scientific-technical achievements at the beginning of the twentieth century. In recognition of their achievements, Fritz Haber and Robert Bosch (1861–1942) were awarded the Nobel Prize.

What was going on long before that? Small amounts of N_2 can be converted into bound nitrogen by lightning in the atmosphere. The nitrogen oxides formed are further converted to yield ammonia. In prebiotic times on our planet, as mentioned before, electrical discharges in the atmosphere were quite important because they led to an accumulation of bound nitrogen. However, this source of bound nitrogen became limiting when microbial life started to flourish.

Discover the World of Microbes: Bacteria, Archaea, and Viruses, First Edition. Gerhard Gottschalk.
© 2012 Wiley-VCH Verlag GmbH & Co. KGaA. Published 2012 by Wiley-VCH Verlag GmbH & Co. KGaA.

Chapter 11 Without bacteria there is no protein

What lies between lightning and the Haber-Bosch process?

It is an exclusive achievement of the bacteria, and to a lesser extent of the archaea, to have "learned" to convert atmospheric nitrogen into bound nitrogen. In other words, plants and animals are not capable of using free nitrogen, N_2. The layman may think of fertilization as a process involving animal manure or plants (green manuring); in such cases, the bound nitrogen is just transferred from one organism to another. Solely the bacteria and archaea, but not all species, are able to synthesize a system of enzymes that catalyzes the Haber-Bosch reaction, but at ambient temperatures and without H_2 under high pressure. This system is called nitrogenase. It has an ingenious structure, so because of its importance it will be described somewhat in detail. The structural biologist Oliver Einsle (Freiburg, Germany) has written,

> "The reaction center of the nitrogenase is able to cleave the extraordinarily stable N_2 molecule without requiring extreme conditions as in the Haber-Bosch process. This is achieved by means of one of the most complex biological metal centers we know, by the iron-molybdenum cofactor. In the heart of this enzyme, the nitrogen molecule is bound in a highly symmetric skeleton consisting of one molybdenum atom and seven iron atoms that are interconnected by sulfur atoms. As a result, a miniaturized bioreactor is formed in which geometries, bond lengths, and angles are much more precisely modulated than would be possible in an industrial process with our current technologies. Moreover, in the case of nitrogenase, nature has implemented a nanomachine, the precision and properties of which we have not completely understood until now. The cleavage of the stable N_2 molecule is accomplished supposedly because one cleavage product, a single nitrogen atom, is bound in the center of the cofactor even more firmly than N_2 by itself. As a result, the other nitrogen atom is easily released in the form of NH_3 but, in compensation, a vast amount of energy has to be invested in order to get rid of the second nitrogen atom in the form of NH_3."

Figure 24 gives you an impression of the active center of the nitrogenase. The metals molybdenum and iron are shown along with the firmly bound nitrogen in the center as well as the iron-sulfur clusters and the carbon compounds surrounding the center. Everything is embedded in a protein structure. Another protein docks onto the nitrogenase and provides reducing power for the Haber-Bosch reaction. The nitrogenase becomes "fluffed up" with reducing power, brimming over with potential activity. However, in this state, the nitrogenase is so full of energy that it would even be able to react spontaneously with oxygen, resulting in its own inactivation. So it is absolutely essential that microorganisms shield and thus protect their nitrogenase from oxygen. Only then can the Haber-Bosch reaction be carried out. Oliver Einsle has mentioned that the N_2 molecule is held firmly within the iron-molybdenum cofactor. He goes on to say that the two N atoms are

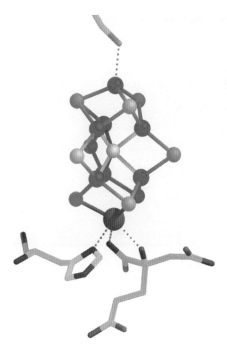

Figure 24 Active center of nitrogenase. The iron-molybdenum cofactor consists of molybdenum (orange), iron (gray), and sulfur (yellow) with the central nitrogen atom (blue). The cofactor is connected to the protein via two amino acids that affix the metal center only at its ends. In addition, an organic molecule, homocitrate, is linked to the center (bottom right). (Model: Oliver Einsle, Goettingen, Germany.)

reduced one after the other, the first one relatively easily and the second one with a high input of energy. It is helpful to imagine that one of the two N atoms is held firmly by the cofactor, like in a vise, whereas the second one flounders around and is therefore amenable to reduction. What about the *tour de force* by which the enzyme manages to extract the second N atom from the cofactor, then reduce it to NH_3 as well? This mechanism has not yet been explained.

That nitrogen fixation requires the trace element molybdenum was discovered by Hermann Bortels (1902–1979). He analyzed the composition of the catalyst used in the Haber-Bosch process. Upon noticing the presence of molybdenum, he looked whether molybdenum salts would stimulate N_2 fixation by bacteria and found this to be the case. Incidentally, there are two additional types of nitrogenase in microbes: one contains vanadium instead of molybdenum and the second one is an all-iron nitrogenase devoid of both molybdenum and vanadium. How these nitrogenases work is still unknown to scientists.

So, bound nitrogen becomes available because some microbial species are able to utilize N_2, reduce it to NH_4^+, and incorporate the latter into proteins. The biomass of these organisms serves as a nitrogen source for all other organisms on Earth. The rumen, the "first stomach" of ruminants, is a habitat in which N_2 to a large extent is bound by microorganisms. It is a nearly oxygen-free (anaerobic) environment full of reducing power. Microorganisms grow there and produce the microbial proteins, which together with plant proteins serve as a source of bound nitrogen for cattle, sheep, and goats.

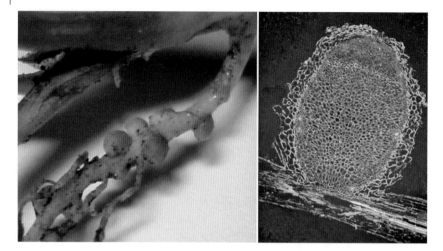

Figure 25 Symbiotic nitrogen fixation. Left: root tubercles (Christine Hallmann, Goettingen). Right: cross-section of a root tubercle filled with N_2-fixing bacteroids. (Photograph: Anne Kemmling, Goettingen, Germany.)

What about green manuring?

Not every plant species is suitable for green manuring. Those preferred are legumes such as lupines, beans, clover, and alfalfa. The rhizosphere (root area) of these plants is the site of symbiotic nitrogen fixation. This is a fascinating process. The plant, let's say alfalfa, provides at its root hairs chemical signals that are recognized by a distinct bacterial species, in this case by *Sinorhizobium meliloti*. These bacteria recognize specific plants as partners. They attach to the surface of the root hairs, which they are allowed to penetrate. Inside the root, the bacteria begin to grow and proliferate; they also induce cell division in the plant that leads to root tubercle formation (Figure 25). Something amazing then occurs in these tubercles. Under self-abandonment, the "Sinos" convert themselves into irregularly shaped cells called bacteroids. In these bacteroids, nitrogenase is formed, N_2 is reduced, and bound nitrogen in the form of amino acids is provided to the plants via their supply routes. The plants also play an important role: they embed the bacteroids in a pink substance called leghemoglobin. This substance provides just enough oxygen to the bacteroids to allow respiration but at the same time to avoid damage to the nitrogenase. Symbiotic nitrogen fixation by the legumes is an excellent example for a way in which plants put bacteria to work for production of bound nitrogen. Alfred Puehler from Bielefeld (Germany) explains the importance of symbiotic nitrogen fixation in our world by using questions and the appropriate answers:

> "Q: What is the role of symbiotic nitrogen fixation in agriculture?
>
> A: Approximately 300 kg of bound nitrogen per year and hectare (10 000 m² or around 2.5 acres) are introduced into an alfalfa field by symbiotic nitro-

gen fixation. Compared with the nearly 150 kg nitrogen fertilizer required for an optimal harvest of wheat per year and hectare, this figure underlines the importance of symbiotic nitrogen fixation in agriculture.

Q: Why didn't all plants acquire symbiotic nitrogen fixation?

A: The existing symbiotic systems apparently are sufficient to bring enough bound nitrogen into our ecosystem. There was no evolutionary pressure to furnish more plant species capable of symbiotic nitrogen fixation. Legumes are the pioneer plants in nitrogen-poor soils and they are followed by plants incapable of hosting nitrogen-fixing bacteria.

Q: Is it considered symbiosis when the bacteria obviously are enslaved and eventually converted into a nitrogen-fixing factory?

A: This symbiosis is profitable for the plants. They are provided with substantial amounts of bound nitrogen. As a reward, the microbial partner receives nutrients that allow it to grow in the nodules. The bacteroids, however, are no longer able to multiply, but a strong increase in the number of *Sinorhizobia* in the soil of an alfalfa field has been observed, so there is some benefit for the bacterial partner as well."

I see. Are there other plants besides legumes that serve as partners in microbial nitrogen fixation?

Of course, and I am certain that further discoveries will be made in this field. A number of woody shrubs or trees and herbs host filamentous bacteria of the genus *Frankia* that are able to fix N_2. It is interesting that some frugal pioneer plants, such as sallow thorn or oleaster, belong to this group. The seaweed *Azolla*, inhabited by cyanobacteria (*Anabaena azollae*), is also able to fix N_2. Finally, N_2-fixing bacteria live in the rhizosphere of a number of plants, for example, members of the genus *Azospirillum* in tropical grasses and species of *Azoarcus* in Kallar grass, a pioneer plant growing in salt marshes of Central Asia. Furthermore, it should be stressed that a number of microbes fix N_2 for their own purposes; in other words, the nitrogen in their proteins originates from the N_2 in the atmosphere. When they die, their bound nitrogen serves as a valuable source for other organisms. This is especially the case with cyanobacteria. They are real pioneers, for instance, on islands of volcanic origin. They are unique because they carry out photosynthesis and build up their cell material from CO_2, N_2, and minerals. N_2-fixing rice plants, however, are still a utopian dream.

It's fantastic that bacteria and archaea provide bound nitrogen, a further important prerequisite for plant and animal life.

This is still true despite the fact that our world is now overfertilized. Nearly two-thirds of the bound nitrogen still originates in the nitrogenase factories of bacteria and archaea, and the remaining third is produced in large-scale reactors of Haber-Bosch facilities.

> Victory belongs to the most persevering
>
> *Napoleon Bonaparte*

Chapter 12
Napoleon's victory gardens

There was quite an increase in the demand for saltpeter after the invention of gunpowder. The term saltpeter refers to certain salts of nitric acid, including ammonium nitrate or potassium nitrate. Since the 1820s, these salts were imported from South America or produced in Europe. The white blooms appearing on the brick walls of dung- or cesspits were observed to consist of saltpeter. Corresponding beds were prepared with soil, limestone, and nitrogenous materials, including urine, blood, and animal waste. These beds were aerated and, after some time, the upper layer was removed and extracted with water. Evaporation of this aqueous solution yielded the long sought-after saltpeter.

Which microbial processes are involved?

Ammonia is produced by microbial decomposition of organic materials. This ammonia is then oxidized to nitrate by two groups of microorganisms, called nitrifiers. The first group is able to form nitrite ($NaNO_2$), which members of the second group convert to nitrate, $NaNO_3$. These microorganisms are extremely unpretentious. They require ammonia as starting material, oxygen for the oxidation reactions to nitrite and further to nitrate, and CO_2 for synthesis of cellular material. The Russian microbiologist Sergei Nikolaevich Winogradsky (1856–1953) was the first to study them and describe their amazing chemolithoautotrophic way of life—on a diet consisting solely of NH_3, O_2, CO_2, and minerals. What a sparse life!

What does this have to do with Napoleon?

During the Napoleonic wars, saltpeter was very much in demand for gunpowder production. The farmers were even forced to cultivate numerous "victory gardens" as described above for the production of saltpeter, and all of Napoleon's soldiers were expected to urinate at specifically designed piles in the army camps.

Are these bacteria still important today?

Of course, and wherever ammonia is present, it will be oxidized by these microorganisms to nitrate when oxygen is available. This nitrification process promotes

Discover the World of Microbes: Bacteria, Archaea, and Viruses, First Edition. Gerhard Gottschalk.
© 2012 Wiley-VCH Verlag GmbH & Co. KGaA. Published 2012 by Wiley-VCH Verlag GmbH & Co. KGaA.

88 | Chapter 12 Napoleon's victory gardens

the corrosion of buildings, bridges, and sculptures because nitric acid is a strong, highly corrosive acid. Moreover, there is an excess of ammonia in agricultural areas due to overfertilization with organic (e.g., manure) or chemical fertilizers. This supply of ammonia is highly appreciated by the nitrifiers. They subsequently produce large amounts of nitrate that is washed out and then enters the groundwater. This is extremely problematic because a certain amount of nitrite is always present in addition to the nitrate. Nitrites are toxic salts of the mutagenic nitrous acid (HNO_2). Furthermore, when nitrate enters the surface waters it promotes algal blooms, which in turn may lead to fish death.

My head is buzzing with all these conversions. And what happens to the nitrate?

In principle, it is quite simple because all the reactions mentioned are combined to a cycle known as the nitrogen cycle (Figure 26). We start with atmospheric nitrogen, N_2, which gives rise to the formation of ammonia in a biological process or in the chemical Haber-Bosch process known for nearly 100 years (see Chapter 11). Large amounts of ammonia are required for growth of all organisms on Earth but there is an overproduction of ammonia, which is oxidized to nitrate by nitrifying bacteria. Another very important process has developed during evolution by

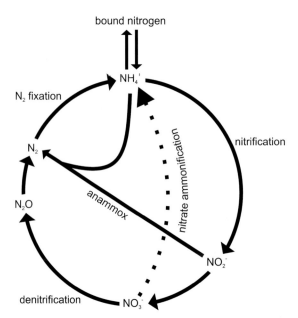

Figure 26 The nitrogen cycle. N_2 fixation yields ammonia (NH_4^+ = ammonium); bound nitrogen is the nitrogen in amino acids or in nucleic acid bases. Nitrification leads to nitrate via nitrite whereas N_2 again can be produced by denitrification; depending on the conditions, this process is associated with the liberation of varying amounts of N_2O (laughing gas). Nitrite and ammonium can be converted to N_2 via the anammox process. The nitrate ammonification is also shown. (Diagram: Anne Kemmling, Goettingen, Germany.)

which nitrate is reconverted back to molecular nitrogen. Microbiologists call this process denitrification, and this is the process that completes the nitrogen cycle. Denitrification occurs in nature in habitats in which anaerobic conditions prevail, in other words, where very little oxygen is present. In such habitats, there are always bacterial species that utilize nitrate as a substitute for oxygen. Not water, as in the case of oxygen, but nitrogen is ultimately produced as the reduction product of nitrate. By the way, this process yields the byproduct N_2O, an anesthetic called laughing gas. Large amounts of N_2O escape into the atmosphere and contribute to the greenhouse effect (see Chapter 14).

I have heard of a process called "anammox" in this connection. It sounds like the name of a drug, but apparently it has something to do with the consumption of ammonia.

Anammox is a very important process in the nitrogen cycle. It was only recently discovered in 1986. Let's ask the discoverer of the anammox bacteria, Gijs Kuenen from Delft (The Netherlands), to introduce us to this process:

> "Anammox stands for <u>An</u>oxic <u>Am</u>monium <u>Ox</u>idation with nitrite (or nitrate) as the oxidant and nitrogen gas as the product (Figure 26). It is clear that the anammox reaction represents a short-cut in the nitrogen cycle. We discovered that this combined nitrification/denitrification reaction is carried out by peculiar and very slowly growing anammox bacteria with doubling times between 12 and 14 days. They have a specialized compartment in the cell, the anammoxosome, where ammonia oxidation takes place and the highly toxic hydrazine (N_2H_4, known as a rocket fuel) is formed as an intermediate. This is why the anammoxosome is surrounded by a very dense membrane. Like the nitrifiers, the anammox bacteria are chemolithotrophs that derive their cell material from CO_2. In order to grow, the anammox bacteria require the simultaneous presence of ammonia and nitrite and hence they are commonly found in nature at the interface of aerobic (oxic) and anaerobic (anoxic) conditions, just below the aerobic nitrifiers. It is now clear that anammox plays a very important role on a global scale, both in fresh and marine waters and sediments. In the marine environment, anammox may contribute up to 50 percent to the global recycling of nitrogen. Since 2005, the anammox bacteria are applied in large-scale reactors designed for cost-effective and energy-saving nitrogen removal from wastewater."

Anammox is a very attractive process because it helps eliminate bound nitrogen in our overfertilized world. Both nitrite (NO_2^-) and ammonium ions (NH_4^+) are required. As mentioned by Gijs Kuenen, both compounds are available at the interface of aerobic and anaerobic bodies of water and sediments.

The nitrogen cycle is of global importance. In addition to natural input, the cycle is "fed" with industrial ammonia fertilizers used in agriculture. The ammonia eventually will end up in the environment, thus boosting part of the cycle. Traditionally, this is, at least to some extent, compensated in sewage plants with large basins in which conditions favor nitrification. The nitrate-containing fluid then flows into chambers in which anaerobic conditions prevail, allowing denitrification and production of nitrogen gas. That's how it is in the modern world: substantial resources are invested in the conversion of atmospheric nitrogen to ammonia, but also in the reverse process as well. Fortunately, the latest developments in the application of anammox may circumvent the costly removal of nitrogen in the form of nitrate and ammonia from waste water.

> Labor to keep alive in your breast
> that little spark of celestial fire
> (called conscience)
>
> George Washington

Chapter 13
Alessandro Volta's and George Washington's combustible air

Actually it was Padre Carlo Campi who was curious about the gas that bubbles up from lake sediments. He wanted to know whether the gas was inflammable; to his surprise, it was. Campi reported this to his friend Alessandro Volta (1745–1827), since they wanted to carry out some experiments together. Unfortunately, Padre Campi became ill. So Volta took a boat alone along the shore of the Lago Maggiore. He pushed a pole into the lake sediment to release some gas, which he collected in a glass container. Back on land, Volta too found that the gas was combustible. He reported his observations in a letter to Carlo Campi in 1776 and attached a sketch of his experiment (Figure 27a).

Apparently, George Washington (1732–1799) together with Thomas Paine (1736–1809) did similar experiments in the United States, as Douglas Eveleigh, New Brunswick, USA, wrote to me,

> "Early citations to Combustible Gas (Methane) in North America:
>
> In North America, first Benjamin Franklin and later George Washington and Thomas Paine discussed combustible gas (methane). Franklin heard of the flammability of marsh gas when he was journeying across Jersey in 1764, but seems not to have mentioned it until Joseph Priestly, the great English chemist jogged his memory ten years later. Several instances are noted, including that a Dr. Finley performed a combustion experiment in November, 1765, which was read at the Royal Society, UK (but did not get published in the Royal Society Transactions). A second more detailed experiment was by Washington and Paine on November 5, 1783, in which they proved while working from a boat in the Millstone River, Rocky Hill, New Jersey, that the flammable marsh substance was a gas and not a floating bituminous substance. While awaiting the news of the end of the revolutionary war, which was defined by the signing of the Treaty of Paris in September 1783, the Congressional Congress met in Princeton, New Jersey, with George Washington staying close by at the Berrien Farm on

Chapter 13 Alessandro Volta's and George Washington's combustible air

Figure 27 (a) The "burning air" sketch included in a letter by Alessandro Volta to Carlo Guiseppe Campi. In this letter, Volta describes an experiment he had carried out on November 3, 1776, at the Lago Maggiore. The letter is dated November 14, 1776. From: Carlo Paolini: *Storia de metano*, Milano, 1976. (b) biogas, liberated in the Eel Creek, Woods Hole, USA. (Photograph: Gerhard Gottschalk, Goettingen, Germany.)

the Millstone River at Rocky Hill. Washington and Paine performed the methane experiment, though not documented until much later as detailed by Paine in his classic observations on "The Cause of Yellow Fever." Without Paine's continuing interest to find the cause of yellow fever, the Washington-Paine methane experiment would have gone unrecorded."

What a story – due to slow boat travel between Europe and America, George Washington was waiting for the signed treaty, so he had time to perform the biogas experiment. So this experiment is connected with the days surrounding the birth of the United States of America! It is still a nice experience to repeat the experiment of Volta and Washington. A course on microbial diversity takes place in Woods Hole Research Center every year, and part of the course (introduced by Ralph Wolfe, see Chapter 3) is to wade into the nearby Eel Creek at night in order to release the combustible air, which is collected in a stoppered funnel. When the stopper is removed and the gas ignited, the ensuing jet of flame is quite impressive (Figure 27b).

The combustible air is biogas, so can I assume it is produced by bacteria?

Indeed, it is biogas that bubbles up from the sediment, a mixture of methane and carbon dioxide produced by the methanoarchaea, mentioned in Chapter 3. The annual methane production by these microorganisms is in the region of one billion tons. Six hundred million tons of this are subsequently released into the atmosphere. It is, however, important to point out that archaea, but not bacteria have the ability to produce methane.

Is this the only source of methane in nature?

Methane is also formed in chemical processes. Gases of volcanic origin contain methane. In addition, living plants have been reported to release methane into the atmosphere. No doubt, when it comes to massive methane production, the methanoarchaea are the major players. The one billion tons mentioned is difficult to imagine. Even if the methanoarchaea were able to produce one thousand times their own weight in methane annually, then at least one million tons of methanoarchaea would have to be at work worldwide.

What is the source of all this methane?

The most important sources are the sediments of rivers and lakes, as well as swamps, marshlands, and deeper layers of soil. These habitats are more or less free of oxygen, which is a prerequisite for methane production. Much less methane is produced in ocean sediments, which can easily be explained. Only in freshwater environments do the methanoarchaea carry out the terminal steps of the so-called anaerobic food chain. Let's imagine a dead fish in a lake. The organic material of the fish is decomposed by various bacteria. The products are amino acids, fatty acids, and carbohydrates, which serve to feed the bacteria. The fish goes into solution, so to speak. Fermentation processes such as lactic acid fermentation then take over. Gases such as H_2 and CO_2 are produced, but also compounds containing methyl groups, including acetic acid (CH_3COOH), methanol (CH_3OH), and methylamine (CH_3–NH_2). From a chemical standpoint, these methyl groups are not too distant from methane, CH_4. Methanoarchaea have specialized in producing methane from these CH_3-containing compounds. The overall equations for methane production look very simple, like that for acetate:

$$CH_3 - COOH \rightarrow CH_4 + CO_2$$

However, the processes involved are highly complex. A unique methane world arose during evolution. A special set of enzymes and coenzymes enables the methanoarchaea to convert such substrates into methane. These microorganisms also "learned" to reduce CO_2 to methane with H_2 as the reducing agent:

$$CO_2 + 4H_2 \rightarrow CH_4 + 2H_2O$$

The conversion of CH_3 groups to CH_4 is especially complicated. The involvement of nickel in this process was discovered by the microbiologist and biochemist Rolf Thauer (Marburg, Germany). Here he reports how this was discovered:

> "One of the initial problems in studying the biochemistry of microbial methane formation was that the methanogens grew very slowly with the nutrients supplied. However, in order to study microorganisms, biochemists require grams or, even better, kilograms of microbial cells. This was difficult to achieve with methanogens. In growth experiments in 1979 with methane producers we had isolated from the sewage plant in Marburg (Germany), we noticed that the organisms grew much better when the injection needles supplying H_2 and CO_2 were made of steel instead of glass. Steel contains, in addition to iron, the metals chromium and nickel. Since iron was already present in the growth medium, the effects of chromium and nickel were examined. To our surprise, the addition of a nickel salt to the medium was so beneficial that the methanogens grew nearly as well as the intestinal bacterium *Escherichia coli*. It was surprising because it had never been reported that microbial growth is dependent on the presence of nickel. So this metal had never been included in the cocktail of salts and nutrients used to grow microorganisms. It was known that plants require nickel for synthesis of the enzyme urease, which cleaves urea into CO_2 and ammonia.
>
> We used a radioactive nickel isotope, ^{63}Ni, with a long half-life to trace nickel in cells of methanogenic archaea. More than 50 percent of the nickel taken up by the cells was incorporated into a low molecular weight yellow compound. In cooperation with the research group of Albert Eschenmoser at the ETH Zurich [Swiss Federal Institute of Technology] (Switzerland), the compound was identified as a nickel tetrapyrrole (Figure 28). We now know that this compound, called coenzyme F_{430}, is directly involved in the catalysis of methane formation. Further investigations revealed that nickel is also required for the synthesis of enzymes that activate H_2, the so-called hydrogenases, as well as of enzymes involved in the reduction of CO_2 to CO (CO dehydrogenase) and in the reversible synthesis of acetyl-CoA from CO, a methyl group, and CoA (acetyl-CoA synthase). Nickel is not only important for methanoarchaea but is also required by many other microorganisms. For example, *Escherichia coli* has three hydrogenases that contain nickel."

This report by Rolf Thauer has taken us from the "metal needle effect" to the discovery of coenzyme F_{430} and thereby right into the heart of methanogenesis.

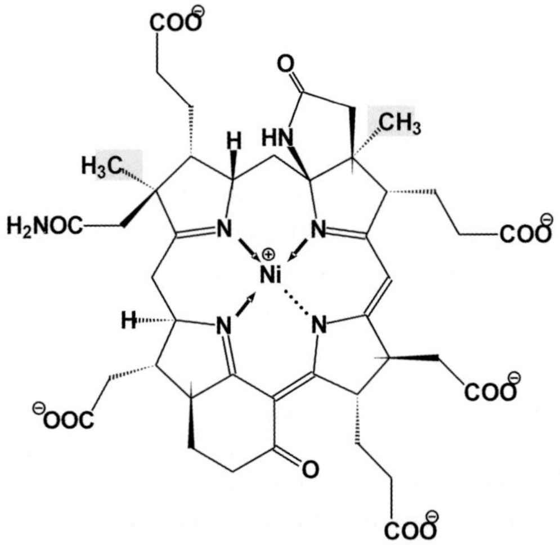

Figure 28 Structure of the nickel-tetrapyrrole coenzyme F_{430}. This compound is part of the machinery that catalyzes the methane-forming step in methanoarchaea. It is structurally related to heme (iron-tetrapyrrole), chlorophyll (magnesium-tetrapyrrole) and vitamin B_{12} (cobalt-tetrapyrrole). (Drawing: Rolf Thauer, Marburg, Germany.)

What a beautiful molecule! Its nickel center together with other ligands and components catalyzes the liberation of methane from so-called methyl-coenzyme M, not in minute amounts but in up to one billion tons per year.

Coenzyme M–S–CH$_3$ + HS-coenzyme B → CH$_4$ + coenzyme M–S–S-coenzyme B

All methyl groups, whether arising from acetic acid, methanol, or the reduction of CO_2 with H_2, are first directed by enzyme systems to form methyl-coenzyme M. In addition to coenzyme M, another very complex compound is required, coenzyme B. These two important compounds, the coenzymes M and B, were discovered by Ralph Wolfe and coworkers at the University of Illinois, Urbana, USA (see also Chapter 3). The release of methane takes place under the directorship of the enzyme methyl-coenzyme M reductase with its coenzyme F_{430}. In this reaction, the two coenzymes are converted into a so-called heterodisulfide, which is then cleaved with H_2 or other agents to return the coenzymes M and B to their original states (▶Study Guide). This complex set of reactions leading to methane is one of the central enzyme systems on our planet. Since methane is a gas, it escapes at its site of formation. This is important because the cell carbon of organisms essentially is released as two different gases, carbon dioxide and methane, at water and soil surfaces and ultimately into the atmosphere. However, not all of the methane escapes: part of it is oxidized to CO_2, as shown in the next chapter.

Mud is not the only environment in which methanoarchaea are happy. They also reside in the intestinal tract of higher organisms: in termites, horses, camels, humans and, most noteworthy, in the rumen of cattle. The rumen of an adult cow is a 120-liter fermentation reactor working with an efficiency yet to be achieved on a technical scale. By and large, the conversions are the job of bacteria and protozoa (small eukaryotic organisms) living anaerobically in the rumen. You have to imagine how a cow feeds on the grasses of a pasture, which we would consider to consist of dietary fiber. Primarily in the rumen, the compounds containing cellulose, hemicellulose, and pectin present in grass and leaves are decomposed by bacteria and fermented. Of course, the cow also wants to benefit from these processes, so it prevents further decomposition of the fermentation products to methane. Especially organic acids such as acetic, propionic, or butyric acids are taken up so efficiently by the cow that bacteria and methanoarchaea are unable to compete for these substrates. Starting with these microbial fermentation products, the cow grows and makes the milk, steaks, and leather that we like so much. Nevertheless, every day an adult cow produces around 160 liters of methane, which is formed from H_2 and CO_2 and released directly into the atmosphere by belching. The biogas arising from cows or from lakes has a slightly different history. Biogas from cows comes from CO_2 and H_2, whereas biogas from lakes arises primarily from the methyl groups of acetate, methanol, and trimethylamine.

Why is less methane produced in oceans?

Well, ocean water contains in addition to table salt, which is sodium chloride, large amounts of sodium sulfate. This sulfate supports a massive development of sulfate-reducing bacteria in the sediments. These organisms consume fermentation products such as H_2, acetate, and their precursors, for example, ethanol or lactate. They produce large amounts of hydrogen sulfide (H_2S) from sulfate. As a result, practically nothing is left for the methanoarchaea in ocean sediments, so methane production is not a dominating process. H_2S, by the way, is very toxic, but fortunately only very little enters the atmosphere. As it ascends from the practically oxygen-free sediments, it reaches oxygen-containing waters where it is oxidized by bacteria to sulfur and then to sulfate.

The activities of man have significantly enlarged the habitats suitable for methanoarchaea. The anaerobic digesters of sewage plants should be mentioned in this context. There, the biomass consisting primarily of bacteria from the various water-purification steps is fermented by bacteria and, in the end, biogas is produced by the methanoarchaea. In big cities, several hundred tons of bacterial biomass are channeled into anaerobic digesters every day. The biogas produced there is utilized for the cogeneration of heat and electricity. Another important site of biogas production is the area around the roots of rice plants, the so-called rhizosphere. Rice is grown on vast areas, mostly in the 130 million hectares of flooded rice fields on our planet. Approximately 600 million tons of rice are harvested annually worldwide. Irrigation provides excellent microbial growth conditions in the rhizosphere. As a result, oxygen is consumed and a gigantic playground for methanoarchaea is established. The methane produced does not ascend through the water in the form of gas bubbles but diffuses through the stems of the rice plants from which it directly enters the atmosphere. Nearly 17 percent of the methane produced annually is released from wet rice fields, so these fields are essentially an ocean of methane-emitting chimneys.

Methane is everywhere. One ton of methane corresponds to 1.5 million liters of methane, which multiplied by the annual one billion tons gives an incredibly large volume. Methane-oxygen mixtures are explosive; they are often responsible for the tragic accidents in mines. Methane is also called swamp gas, the *ignis fatuus* or flickering bluish light in swamps that in the past has attracted hikers at night and led them onto treacherous terrain. Methane does not burn by itself. In swamps, ignition is due to phosphorous-hydrogen compounds produced along with the methane and these serve as the ignitor upon contact with oxygen. Garbage dumps are another site of methane production where methane increasingly is being collected and put to use.

Until recently, methane as a microbial product was only known to experts. Fortunately, it has received much attention in the past few years, so nearly everyone now knows what biogas is.

> Aerem corrumpere non licet
> Polluting air is not allowed
>
> *(Corpus Iuris Civilis, body of laws of the East Roman Emperor Justinian I)*

Chapter 14
Microbes as climate makers

Mankind has been unscrupulous in polluting our environment for centuries, and this has resulted in severe damage. Our environment consists of three of the four elements, according to the Greek philosopher Empedocles: water, soil, and air (and fire). By the second half of the twentieth century, many rivers and lakes were in critical condition. They were about to die, meaning they would become completely anaerobic. Because the load of organic material was enormous, bacteria grew on a massive scale and consumed all the dissolved oxygen. Since oxygen diffuses very slowly from air into water, anaerobic zones developed in bodies of water nearly all the way up to the surface, with disastrous results for fish. In addition, annoying odors developed, so sewage plants were built to relieve these problems. There, masses of bacteria are at work. Aeration and agitation of the water to be purified leads to the bacterial conversion of most of the organic materials present into CO_2. In addition to the water that has been treated or clarified, the so-called activated sludge remains. It consists primarily of dead microorganisms, and this sludge is transferred to anaerobic digesters where it is converted to biogas as already discussed in Chapter 13. Thales of Miletus (624–546 BC) wrote, "Water is the principle of all things. Everything is created from water and everything returns to water." This is correct. However, in our modern society too much is being returned along with the water, so we have to take remedial action. That is why we need sewage plants all over our planet.

In addition, soils to a large extent have been contaminated with all kinds of oils and solvents. Here, the concerted input of bacteria and nutrients can provide some relief. However, when it comes to heavy metals, even the bacteria don't have mechanisms to remove them.

Not only bacteria but humans, despite all their technological capabilities, are more or less helpless when it comes to coping with changes in the composition of our atmosphere. The only effective solution in this case is prevention. Prohibiting the use of hydrochlorofluorocarbons (HCFCs) as expanding agents, for example, in aerosol sprays, is a good example. These compounds are very inert, so they are not easily degraded. While they were still being used in aerosol sprays, the HCFCs had a tendency to concentrate in the extremely cold layer of air above Antarctica. During Antarctic summers, radical chain reactions were induced by exposure of the HCFCs to ultraviolet radiation from the sun, and this process

Discover the World of Microbes: Bacteria, Archaea, and Viruses, First Edition. Gerhard Gottschalk.
© 2012 Wiley-VCH Verlag GmbH & Co. KGaA. Published 2012 by Wiley-VCH Verlag GmbH & Co. KGaA.

accelerated to a more or less "oxidation orgy." The ensuing consumption of ozone has resulted in what is known as the "ozone hole." A ban on the use of HCFCs has helped eliminate this cause of ozone consumption in the stratosphere, so that the ozone layer that protects us from ultraviolet radiation can gradually recover. This will be a lengthy process that may not be completed before 2070.

The alarming changes in our atmosphere are related to its content of carbon dioxide (CO_2), methane (CH_4), and nitrous oxide (N_2O; laughing gas). In some places, the increase in the CO_2 concentration of the atmosphere is not considered a serious problem. However, you only have to look at how this concentration has changed in the past 600 000 years. The CO_2 concentration in ancient times can be determined today in air bubbles entrapped in ice core samples, as shown in the book by the Nobel laureate Al Gore, *An Inconvenient Truth*. Up to the modern age, the CO_2 concentration fluctuated between 180 and 260 ppm (parts per million). Then a seemingly unstoppable increase began. In 1978, the CO_2 concentration was approximately 335 ppm; now it has reached 380 ppm (Figure 29a). According to some projections, a value of 600 ppm may be reached by 2050. The CO_2 content of 380 ppm amounts to 780 gigatons of carbon (1 Gt = 1 billion tons), the equivalent of 2850 Gt CO_2. Approximately one-third of this amount (900 Gt CO_2) has been attributed to the burning of fossil fuels (coal, oil, gas) and wood.

Let us discuss these processes in somewhat more detail. There is an eternal cycle of CO_2 fixation and CO_2 production on our planet: the conversion of CO_2 into biomass and the reconversion of this biomass into CO_2 by way of the various food chains on Earth. This cycle is not completely closed. A small percentage of biomass remains in the sediments of waters and in deeper layers of soil as compounds that are biologically hardly degradable under the given circumstances. Over millions of years this has led to an accumulation of fossil carbon compounds, coal, oil, and gas. On the continents, this process is fostered by long-lived biomass in the form of wood. It consists of cellulose, hemicellulose, and lignin, and is subject to slow biological degradation. In a forest left to itself, some of these materials will always remain in the deeper layers of soil in the form of humus, which is permanently removed from the carbon cycle. In addition, climate catastrophes and geological changes during the Earth's history have resulted in carbonization of whole forests.

According to the figures of the ecologist Christian Koerner (Basel, Switzerland), 600 Gt of carbon are present on the continents in the form of biomass, and 80 to 90 percent of this is wood. This is a real treasure, so it must be our primary goal to maintain this highly valuable, long-lived biomass on our planet. Large-scale slash-and-burn deforestation to reclaim land for agricultural purposes cannot be completely prevented, but a balance must be maintained between the use of wood for various purposes and its production in forests. What is required is a sustainable worldwide biomass management. For this reason it is important to take microbial activities into account. We may recall the calculations of William Whitman, Chapter 8. The microbial world, 5×10^{30} cells containing 500 Gt of carbon, represents almost the same amount of biomass as wood. This biomass is highly active: 40 percent of it (200 Gt of carbon) is generated and decomposed every year.

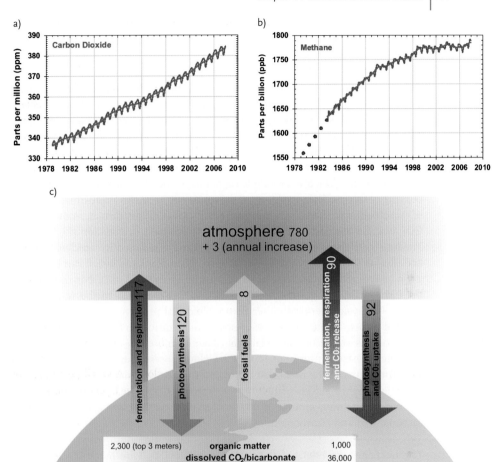

Figure 29 Increase in the CO$_2$ and methane concentrations in the atmosphere and the amounts of carbon cycled annually between atmosphere and oceans as well as continents. (a) Increase of CO$_2$ in ppm (parts per million) between 1979 and 2008. (b) Increase of methane in ppb (parts per billion) between 1984 and 2008. The "zigzag" fluctuations of the curves reflect seasonal variations. (c) The carbon cycle above the continents and above the oceans. Data in gigatons (Gt) of carbon. Total carbon in the atmosphere is given; all other data show annual fluxes. (Sources: NOAA, National Oceanic and Atmospheric Administration; SIO, Scripps Institution of Oceanography; G.A. Olah, A. Goeppert and G.K.S. Prakash: *Beyond Oil and Gas: The Methanol Economy*, Wiley-VCH Verlag GmbH [2006]; R.A. Houghton: "Balancing the global carbon budget," Annu. Rev. Earth. Planet Sci. 35, 313–347 [2007].)

That is a gigantic amount, but how does it contribute to the global CO$_2$ cycle?

The microbial biomass is omnipresent in the Earth's ecosystem. It behaves like a superorganism that is always hungry. If we feed it more plant biomass, more CO$_2$ or CH$_4$ is produced, depending on the conditions. As the permafrost soils thaw, it will be a feast for the superorganism. The ensuing release of CO_2 and CH_4 may well go beyond our worst expectations. It has to be taken into account that the entrapped carbon in these soils exceeds all the carbon present in the atmosphere in the form of CO_2.

The conclusion is that we had better keep CO_2 production under control because global warming may lead to even more global warming, with disastrous effects.

Let's now look at the carbon cycle with the help of the figures shown in Figure 29c. As already mentioned, the 2850 Gt of CO_2 in the atmosphere at present corresponds to 780 Gt of carbon. This amount fluctuates annually because of the vegetation periods (Figure 29a). However, it is also sensitive to the changes just mentioned: the decrease in long-lived biomass and the input of CO_2 from the burning of forests. The vegetation on the continents removes approximately 120 Gt of carbon annually from the atmosphere and 117 Gt of this carbon, in the form of CO_2, is again released into the atmosphere by biomass degradation (Figure 29c). The small difference of 3 Gt represents the carbon retained in wooden biomass and humus. The growth and proliferation of phytoplankton, which can be looked upon as the pastures and forests of oceans, removes 92 Gt of carbon annually in the form of CO_2. A given amount of this CO_2 is dissolved in the water of the oceans – a physicochemical rather than a biological process – so altogether nearly 90 Gt of carbon is returned to the atmosphere by diffusion from the oceans.

The oceans are on our side since they take up more CO_2 from the atmosphere than they release. Unfortunately, this is a slow process, but an additional 2 Gt of carbon remains in the oceans in the form of CO_2 or bicarbonate, or as biomass, which undergoes sedimentation. Aside from this, the ocean is a gigantic CO_2 reservoir containing around 36 000 Gt of carbon as CO_2 and its bound forms, carbonic acid and bicarbonate.

You mentioned the 120 Gt of carbon, in the form of CO_2. How much CO_2 is that?

The atomic weight of carbon is 12 and the molecular weight of CO_2 is 44, so 120 Gt of carbon correspond to 440 Gt of CO_2. It is customary, and also practical, to relate all of these changes in the environment to carbon content rather than to CO_2 content.

One figure has yet been mentioned. It overcompensates the carbon loss by humification processes in soil or by sedimentation of biomass in oceans. It is the burning of fossil carbon, by which 8 Gt of carbon are added to the atmosphere in the form of CO_2 annually. In fact, nearly 20 percent of this or around 1.6 Gt of carbon comes from deforestation.

Are you generally against biofuel to prevent land being used for "energy crops?"

Not at all. We are against approaches in which forests are converted into cultivation areas for biofuel without considering the long-term effects. No compromises should be made regarding our ability to feed the world and preserve environmental quality.

The increase in the CO_2 concentration of the atmosphere is one problem. The other one is methane. Its concentration in the atmosphere has tripled within the last three centuries, reaching a value of almost 1.8 ppm (parts per million). For a long time, methane increased parallel to the world population. However since 1998, the rate of increase has slowed; this has been explained by a decrease of methane production in wetlands (Figure 29b). The methane concentration in the atmosphere is significantly lower than that of CO_2 (0.47 percent thereof). However, the greenhouse effect triggered by methane is ten times greater. This means that the contribution of one methane molecule to the greenhouse effect is equivalent to that caused by ten CO_2 molecules. Methane is unstable under the conditions in the atmosphere. Solar radiation triggers its oxidation to CO_2, so methane has a half-life of approximately 10 years in the atmosphere. After 10 years, any given methane content in the atmosphere will be cut in half as long as no "fresh" methane is added. But more and more methane is being released because there is a correlation between world population growth and increase in atmospheric methane. This correlation exists because growth of the human population also leads to an increase in the number of ruminants, such as cattle, sheep, goats, and camels. The rumen of a cow is a very efficient fermentation chamber where the concentration of microorganisms is in the region of 10^{10} cells per ml. It is inherent to the fermentation processes in the rumen that large amounts of methane are produced. This was already discussed in Chapter 13.

Supposedly the greenhouse effect of methane is 10, but sometimes a value of 23 is given.

The value on a molar basis is 10; as mentioned above, the effect of 1 mole of methane is compared with that of 1 mole of CO_2. Sometimes a GWP value is given, the global warming potential, but this value has a weight basis: 1 kg of methane is compared with 1 kg of CO_2, so the ratio is 23.

How much methane is released into the atmosphere?

We speak of 600 million tons of methane per year worldwide, based on calculations of the Nobel laureate Paul Crutzen. He received the Nobel Prize together with Mario J. Molina and Frank Sherwood Rowland for their discovery of the ozone hole. Around 30 percent of this methane comes from animals, primarily from ruminants, and 10 percent from rice fields. This makes 240 million tons per year, and this amount will increase along with the world population.

Isn't it possible to do something about the release of methane?

There were attempts to repress methane production in the rumen by adding inhibitors, but that didn't work. Release of methane is an essential process in the rumen, and cattle cannot thrive without it.

The traditional wet cultivation of rice could be replaced if varieties that grow in relatively dry soil became available. This would eliminate the playground for methanoarchaea. Another problem lies in the wetlands. If plenty of nutrients are available, microbial conversions will automatically lead to methane production. However, their opponents, the methane-oxidizing bacteria, are also present. They settle in layers in which the ascending methane encounters molecular oxygen. This group of bacteria can be recognized because their names always start with "methylo", such as *Methylomonas* or *Methylobacterium*. They oxidize methane before it can escape into the atmosphere. How effective this is depends on the presence of populations of methanoarchaea and methylobacteria and their respective activities.

Finally, we come to N_2O. Like methane, N_2O is a typical product of microbial activity. It is released primarily from soil when nitrogenous compounds (fertilizers) are present in excess. In 1978, the N_2O concentration in the atmosphere was 0.30 ppm, and now it is in the region of 0.32 ppm. This doesn't seem to be a lot, but its residence time in the atmosphere is approximately 100 years, and the global warming potential (GWP) of N_2O is 296, compared to the value for CO_2:1. Therefore, around 5% of the greenhouse effect is caused by N_2O. An expansion of the areas for cultivation of biofuel crops, including rapeseed and corn, will provoke a further increase in the release of N_2O caused by bacteria.

So is it justified to call microbes climate makers?

We think so because CH_4 and N_2O are primarily microbial products. The microbial breakdown of biomass releases much more CO_2 than do the respiratory processes of eukaryotic organisms. You should keep in mind the gigantic amount of microbial biomass present on our planet, and this biomass, the superorganism, is extremely active.

> My proposal, which I made officially to Mr. William Clayton, Under-Secretary of State for Economics, was to ferment maize – of which millions of bushels were available in the United States and Canada – and convert it into butyl alcohol and acetone by my process.
>
> *Chaim Weizmann, Trial and Error*

Chapter 15
How a state was founded with the aid of *Clostridium acetobutylicum*

The name of the bacterium *Clostridium acetobutylicum* is closely connected with the name of the first president of the state of Israel, Chaim Weizmann (1874–1952). He worked as a chemist at the University of Manchester and, around 1913, he developed a strong interest in biological chemistry and bacteriology. He traveled to Paris a number of times to deepen his knowledge of bacteria. Following Pasteur's tracks, he developed an interest in fermentations, especially in processes for the production of synthetic rubber. In his memoirs, he wrote:

> "The obvious approach to the problem was to find a method for the synthetic production of isoprene and for its polymerization to a rubber. The easiest raw material I could think of was isoamyl alcohol, which is a byproduct of alcoholic fermentation, but as such was not available in sufficiently large quantities. I hoped to find a bacterium which would produce by fermentation of sugar more of this precious isoamyl alcohol than does yeast – one was not yet aware of the fact that isoamyl alcohol is not a fermentation product (of sugar), but is formed by degradation of the small amounts of protein invariably present in a fermenting mash. In the course of this investigation I found a bacterium which produced considerable amounts of a liquid smelling very much like isoamyl alcohol. But when I distilled it, it turned out to be a mixture of acetone and butyl alcohol in very pure form. Professor Perkin advised me to pour the stuff down the sink."

Fortunately, Weizmann did not follow this advice. During August of 1914, when the British Ministry of Defense invited all scientists in the country to report on discoveries of military interest, Weizmann proposed the acetone-butanol fermentation. After some time Weizmann received a reply from the British Admiralty, which was concerned about the serious shortage of acetone in the country. Acetone was required for the production of cordite, a smokeless gunpowder. Weizmann saw the authorities at the British Admiralty and negotiated, among others, with the First Lord of the Admiralty, who at that time was Winston Churchill. Weizmann wrote:

Discover the World of Microbes: Bacteria, Archaea, and Viruses, First Edition. Gerhard Gottschalk.
© 2012 Wiley-VCH Verlag GmbH & Co. KGaA. Published 2012 by Wiley-VCH Verlag GmbH & Co. KGaA.

"Mr. Churchill was brisk, fascinating, charming and energetic. Almost his first words were: 'Well, Dr. Weizmann, we need thirty thousand tons of acetone. Can you make it?' I was so terrified by this lordly request that I almost turned tail. I answered: 'So far I have succeeded in making a few hundred cubic centimeters of acetone at a time by the fermentation process. I do my work in a laboratory. I am not a technician, I am only a research chemist. But, if I were somehow able to produce a ton of acetone, I would be able to multiply that by any factor you chose. Once the bacteriology of the process is established, it is only a question of brewing. I must get hold of a brewing engineer from one of the big distilleries, and we will set about the preliminary task. I shall naturally need the support of the Government to obtain the people, the equipment, the emplacements and the rest of it. I myself can't even determine what will be required.'

I was given a *carte blanche* [unlimited authority] by Mr. Churchill and the department, and I took upon myself a task which was to tax all my energies for the next two years, and which was to have consequences which I did not then foresee."

Weizmann must have had green thumbs and fingers when he isolated *Clostridium acetobutylicum*. Still today, this bacterium is ranked among the champions when it comes to the fermentation of carbohydrates to acetone and butanol. The acetone-butanol production process was first introduced in the United Kingdom, then in the United States and India. The product used was primarily acetone because there was no application for butanol, which was stored in big tanks. That changed with the onset of the automobile industry in the 1920s. Then butanol was the product favored because it formed the basis for the lacquers needed in large amounts.

Weizmann filed his basic patent in 1915; it was granted in 1919 and was the basis for enormous financial profits. But there was also another important issue in this connection. Weizmann advocated the establishment of a Jewish state in Palestine to give a homeland to the Jews spread over many countries. Along with others, Weizmann brought this request to the attention of the Foreign Secretary, Earl Balfour. This led to his historical proclamation in 1917 and to the foundation of the state of Israel 31 years later, with Chaim Weizmann as its first president.

It's simply incredible that events of historical dimension are actually connected with this fermentation! By the way, what are the raw materials for mass production of acetone and butanol?

Initially, starch from corn (maize) and other grains as well as chestnut flour served as renewable resources for acetone-butanol fermentation. When the process was introduced in the United States, corn starch became the primary starting material because it is available there in large amounts. Especially in the Midwest of the United States, a gigantic acetone-butanol industry developed, which provided butanol, acetone, and ethanol as basic chemicals for industrial use until around 1950. After that, oil became available in sufficient quantities, so the acetone-

Chapter 15 How a state was founded with the aid of Clostridium acetobutylicum

butanol fermentation was nearly phased out, primarily for economical reasons. However, plants in South Africa and Russia were in operation until recently; in China, they are still running and more are coming on stream every year because of the push to generate biofuels.

Clostridium acetobutylicum belongs to the genus of anaerobic microorganisms able to produce dormant forms called spores. These are quite interesting microorganisms that are unable to respire with oxygen, which can damage or even kill them. Many species belonging to the genus *Clostridium* are soil organisms, but some are pathogenic, such as *C. tetani*, *C. botulinum*, or *C. difficile*. However, *C. acetobutylicum* is completely harmless. In the 1920s it was grown in large tanks, for example, by Commercial Solvents in Peoria (US) (Figure 30). The tanks shown had a capacity of 50 000 gallons (190 000 liters). During fermentation by *C. acetobutylicum*, carbohydrates are first converted to butyric acid, acetic acid, carbon dioxide, and molecular hydrogen. After some time, the fermentation broth becomes too acidic for *C. acetobutylicum*, so the bacteria start to convert butyric acid to butanol; they are also able to produce acetone from certain intermediary compounds. Butanol and acetone are neutral organic molecules, so the pH value of the fermentation broth increases, which is advantageous for the microorganisms. We thus owe production of the solvents butanol and acetone to the presence of a regulatory switch in these organisms. Following accumulation of acid in the microbial environment, the switch, so to speak, is flipped and the acids and sugars are converted to the neutral compounds acetone and butanol.

Figure 30 Acetone-butanol production plant in Peoria, USA, around 1925. Drawing based on a photograph published in: S.C. Beesch, Ind. Eng. Chem. 44, 1677 (1953). (Drawing: Anne Kemmling, Goettingen, Germany.)

The gigantic plants constructed for production of acetone and butanol were already mentioned. After the Second World War, they largely became fermentation ruins. Fermentation products were unable to compete with products synthesized from oil. Now, at the beginning of the twenty-first century, the situation has changed again. Due to the astronomical price of oil, the acetone and butanol produced by *C. acetobutylicum* in a fermentation process are once again economically advantageous. *C. acetobutylicum* and its relatives, still unmatched by any other organisms when it comes to the production of acetone and butanol, will soon be in large fermentation tanks once again. We have asked the microbiologists Peter Duerre, Ulm (DE) and Hubert Bahl, Rostock (DE) to tell us how they see the chances for such a process:

> "The future of industrial acetone-butanol fermentation with *Clostridium acetobutylicum* has already begun. In 2006, a plant for acetone-butanol production was put into operation in Brazil. Furthermore, plants are in operation in China. Especially butanol is a very important product for the chemical industry. In addition to varnish production, it is needed for the production of antifreeze fluids, deicers, nail polish, perfumes, security glass, and various surface coatings, just to mention a few applications. A much more important application is on the horizon: butanol is a much better biofuel than ethanol. Its properties are comparable to those of gasoline, and it can be mixed with gasoline in any desired ratio. Since butanol is produced from renewables, its use is also a contribution to climate protection."

Really, this is a very impressive process. However, it has to be made clear that the operation of such fermentations on a large scale may not be carried out at the expense of the availability of sufficient food for the world population.

You mentioned that clostridia form spores – what does that mean?

Bacterial spores are really fascinating. They are not formed by all bacterial species, but primarily by the bacteria we call *Clostridium* or *Bacillus*. The bacilli will be discussed in a later chapter; they comprise many harmless soil bacteria but also *Bacillus anthracis*, the organism causing anthrax. Spores are heat resistant, even outplaying the hypothermophilic archaea discussed in Chapter 6. Spores are not always present in cells of clostridia and bacilli because they are only formed under certain conditions. We asked Michael Young from Aberystwyth (UK) to give us some information on spores, also in connection with the acetone-butanol fermentation:

> "Bacterial spores are probably the most resilient life forms on Earth. They are minute time capsules and can be recovered from archaeological materials thousands of years old. Spores are in a state of deep dormancy and are formed when the environment deteriorates to a point where it is incapable of sustaining bacterial multiplication. However, spores can very quickly germinate so that bacterial growth can resume when the environment

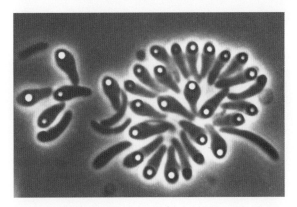

Figure 31 Cells of a *Sporomusa* species carrying spores. (Photograph: Hans Hippe, Goettingen, Germany.)

improves. Spores are difficult to kill and eradicate and the organisms that make them can cause problems in our hospitals (*Clostridium difficile*) and in foodstuffs (*C. perfringens*, *C. botulinum*). Curiously, spore formation by *C. acetobutylicum* is somehow connected with the production of acetone and butanol. We know this because variants that cannot make these chemicals can no longer make spores. Scientists are still trying to understand why this is so. They are also engineering the bacterium to tailor it for a variety of industrial applications, one of which is the bulk conversion of cellulose in waste plant materials (biomass) to butanol (biofuel)."

Spores are fantastic: they are like little fortresses, and it is amazing that they contain all the essentials (DNA, RNA, proteins, and so on) required to establish a bacterial population again after the fortresses have opened their gates, a process we call germination. Cells of the bacterial genus *Sporomusa* are depicted in Figure 31. The spores appear bright and refractile under the light microscope. They contain the essentials mentioned above in high concentrations and, in addition, the calcium salt of dipicolinic acid, which only occurs in spores. This acid is apparently required for the enormous heat tolerance of spores and is also responsible for their brightness.

> Wine is light,
> held together by water
>
> *Galileo Galilei*

Chapter 16
Pulque, wine, and biofuel

Nearly 170 years ago, a short article appeared in *Liebigs Annalen der Pharmacie*, in which chemists made fun of the notion that sugar is converted into alcohol by living organisms. The chemists described little animals that would swallow the sugar, digest it in their stomachs, and secrete alcohol through their intestines and CO_2 through their bladders. They claimed that the bladders of these animals looked like champagne bottles. Chemists at that time believed that the formation of alcohol from sugar was a solely chemical process. The equation for this process had already been set up by Joseph Gay-Lussac (1778–1850) in 1815:

$$\text{glucose} \rightarrow 2 \text{ ethanol} + 2 \text{ carbon dioxide}$$

The fact that ethanol fermentation is a biological process was first generally accepted towards the end of the nineteenth century, largely due to the work of Theodor Schwann (1810–1882) and Louis Pasteur (1822–1895), among others.

Which organisms are of primary interest as ethanol producers?

First of all, we need to mention yeast, especially the various breeds of *Saccharomyces cerevisiae*, also known as baker's yeast. The yeasts are aerobes as long as oxygen is present, so they respire as we do. However, they grow rapidly in solutions containing around 10 percent sugar, which leads to rapid consumption of the dissolved oxygen. As soon as oxygen is depleted in the fermentation broth, a metabolic switch is activated and glucose is degraded to ethanol and CO_2. The pathway by which ethanol is produced is rather simple. Glucose with its six carbon atoms is cleaved into two C_3-compounds. Eventually, two molecules of pyruvate (salt of pyruvic acid) are formed; they each lose CO_2, and the two compounds formed are reduced to two molecules of ethanol (Figure 32a). The reader should take note of the names of two enzymes involved: pyruvate decarboxylase, which catalyzes the formation of the CO_2 molecules, and alcohol dehydrogenase, which finally produces ethanol.

Now we come to the ethanol-producing bacteria. No doubt, yeasts are the champions when it comes to ethanol production from sugars. We will discuss their applications later. The discovery of bacteria able to produce ethanol by fermentation probably began with Paul Lindner (1861–1945), who was director of the

Discover the World of Microbes: Bacteria, Archaea, and Viruses, First Edition. Gerhard Gottschalk.
© 2012 Wiley-VCH Verlag GmbH & Co. KGaA. Published 2012 by Wiley-VCH Verlag GmbH & Co. KGaA.

a)

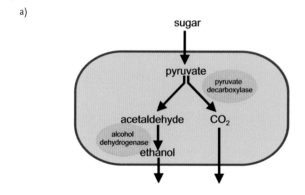

b)

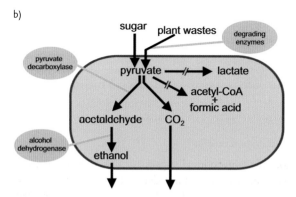

Figure 32 Biofuel. (a) Fermentation of sugar to ethanol and CO_2 as carried out by yeasts and by Zymomonas mobilis. (b) Fermentation of sugar and plant wastes to ethanol and CO_2 by a genetically modified strain of E. coli. It is indicated that the genes encoding pyruvate decarboxylase and alcohol dehydrogenase were genetically introduced into the production strain, likewise the genes for degradation of carbohydrates. These genes are expressed in the cells, and the enzymes thus produced are exported as exoenzymes. Outside the cells, these enzymes degrade plant waste to low-molecular-weight compounds, for example, simple sugars, which can then be taken up by the cells. Furthermore, it is indicated that some metabolic routes are genetically blocked so that primarily ethanol and CO_2 are produced. (Diagram: Petra Ehrenreich, Goettingen, Germany.)

Institute for Fermentation Industries in Berlin (Germany). From 1924 to 1925, he lived in Mexico as a *consultor tecnico*. There he became familiar with the alcoholic beverage "pulque" brewed by Mexican farmers. They collected agave juice, which undergoes a rapid fermentation and produces within 24 hours a stimulating and pleasant beverage. Paul Lindner discovered under the microscope a "happily motile fermentation bacterium," which he called *Thermobacterium mobile*. Later it was renamed *Pseudomonas lindneri*, but now it's called *Zymomonas mobilis*.

This microorganism has specialized in the formation of ethanol from sugars. It is unable to respire with glucose and oxygen as yeast does because it is more or less restricted to carrying out fermentations in the absence of oxygen. The enzymes mentioned above, pyruvate decarboxylase and alcohol dehydrogenase, are both present in Z. mobilis. As a production organism for ethanol on a large industrial scale, Z. mobilis cannot yet compete with yeast, which is very robust when it comes to tolerating acidic conditions. Ethanol production can be carried out under conditions in which lactic acid bacteria are no longer active. Sterilization of the media is not necessary. On the other hand, very costly equipment would be required with Z. mobilis, and the energy costs for sterilization of the fermentation broth and the production plant would be considerable. It's a real shame because the rate of ethanol production by Z. mobilis is higher than by yeast. On top of that, a final alcohol concentration of 20 percent (volume/volume) can be reached – simply not possible with yeast.

Okay, but what are the researchers working on?

They are designing a fermentation process and the necessary equipment to make ethanol production with Z. mobilis economically feasible. Geneticists have implanted genes into the chromosome of Z. mobilis so that it can grow on other substrates, for example, sugars such as pentoses. They are also studying fermentation by immobilized Z. mobilis cells. The organisms are attached to small beads in vessels that can be flushed with sugar solutions, speeding up the conversion of sugar into ethanol so that undesired processes such as lactic acid fermentation don't have a chance to take place. Compared to yeast, bacteria have the advantage of being able to ferment not only conventional sugars such as glucose, fructose, or saccharose (from sugar cane or sugar beets) but also cellulose and many other constituents of wood. These microorganisms thus have the potential to convert plant wastes, for example, into ethanol. This is a challenge that requires the use of genetic engineering.

Besides Z. mobilis, researchers are concentrating on our commensal *Escherichia coli*. This microorganism is not only able to grow with oxygen and nutrients but also in the absence of oxygen, when it carries out fermentation processes leading primarily to the formation of organic acids and very little ethanol. It was a challenging project to convert E. coli into a good ethanol producer. Lonny O. Ingram (Gainesville, Florida, USA) was one of the first to achieve this. The genes for pyruvate decarboxylase and alcohol dehydrogenase from Z. mobilis were transferred to E. coli, and the pathways leading to unwanted acids were blocked. This resulted in an efficient ethanol producer (Figure 32b). In addition, the spectrum of nutrients taken up and degraded by such a production strain was widened by genetic manipulation. This has opened up a field that may eventually lead to ethanol production strains superior to yeast. We asked Douglas Clark (Berkeley, California, USA) to give us his opinion on the perspectives in this area.

> "Lignocellulose is a remarkable material designed by Nature to stand tall and weather the storm in the face of many environmental challenges. It is composed primarily of three materials: cellulose, the most abundant polymer on Earth, hemicellulose, and lignin. These three components

intertwine and reinforce each other to form a very durable composite that requires harsh treatment and/or specialized enzymes for deconstruction. Indeed, deconstructing lignocellulose into monomeric sugars suitable for fermentation into ethanol is a critical bottleneck in the large-scale economic conversion of biomass into biofuel. Fortunately, extremophiles and their enzymes may be of significant help in this regard.

Release of cellulose and hemicellulose from lignocellulose requires pretreatment of the biomass to make it more accessible to enzymatic attack. Once the cellulose is available in a relatively unhindered and less crystalline form, enzymatic hydrolysis by cellulases (or hemicellulases in the case of hemicellulose) proceeds much more quickly. One strategy to lower the cost and improve the efficiency of enzymatic degradation is to hydrolyze cellulose at a relatively high temperature, which would reduce the risk of microbial contamination, facilitate high solids loadings, and possibly provide higher overall reaction rates and greater compatibility of the degradation process with pretreatment conditions. To this end, the isolation of new thermophilic glycosyl hydrolases (e.g., cellulases and xylanases) from thermophiles and even hyperthermophiles is of keen interest. Indeed, cellulases have recently been discovered in bacteria and archaea that inhabit environments where cellulose is apparently in short supply (e.g., deep-sea vents and volcanic hot springs). These enzymes could potentially be used in their native forms or be modified for improved performance by modern tools of protein engineering.

The prevalence of cellulases and related enzymes in thermophiles and other extremophiles remains to be fully determined, but examples of cellulases from extreme environments are beginning to emerge. This is good news for the quest to find enzymes that are better suited to degrade lignocellulose into fermentable sugars under conditions that meet the requirements of a commercial process. Likewise, the microbes that produce such enzymes may prove useful for the fermentation itself, although developing an efficient thermophilic production strain for biofuels remains a formidable challenge. Nonetheless, the discovery of cellulolytic bacteria and archaea in unexpected places, along with the enzymes that support their unusual lifestyle, expands the realm of possibilities for producing biofuels from recalcitrant feedstocks."

A process leading from lignocelluloses to ethanol would be an important step forward.

By the way, how much bioethanol is currently being produced?

As already mentioned, yeast is still the workhorse of bioethanol production. The volume produced in 2008 was 65 billion liters (17.3 billion gallons), 52 percent of this in the USA and 37 percent in Brazil. The principal substrate for ethanol production in the United States is corn starch, which is enzymatically hydrolyzed to glucose and then fermented; in Brazil, it is sugar extracted from sugar cane. Those 65 billion liters really are a lot, but this is only 4.6 percent of the 1.4 trillion (1.4×10^{12}) liters of liquid fuel consumed annually.

It is interesting to compare the volume of bioethanol production with the ethanol present in the beer and wine produced globally. Around 131 billion liters (34 billion gallons) of beer are produced (and consumed!) annually, and 25.7 billion liters (6.8 billion gallons) of wine. For a rough estimation, let's assume that beer contains an average of 5 percent ethanol and wine, 10 percent. These beverages then contain approximately 9 billion liters of ethanol, only 14 percent of the bioethanol produced industrially. When compared, these figures also document the gigantic volume of gasoline produced from oil.

Everybody knows that beer and wine taste very different: the sources of fermentable sugars differ as well as the ingredients responsible for taste. Beer brewing goes back to the Sumerians, the original inhabitants of Babylonia, the land between the Tigris and Euphrates Rivers. Their recipe for making beer is documented on the famous clay tablet dating from 7000 BC on display at the Louvre in Paris. The Sumerians were masters at growing cereal grains, including barley, and they learned early that germinated barley grains resulted in a broth suitable for fermentation. Now we know the reason: germination activates an amylase, which then hydrolyzes the starch present. This is the prerequisite for brewing because yeast is able to degrade the sugars glucose or maltose, but not starch. Maltose, which consist of two glucose moieties, and glucose are present in the broth as the result of amylase activity. The Sumerians also must have known how to handle fire because, after saccharification of the barley starch, the broth was boiled to precipitate proteins and cell debris. The resulting liquid, naturally contaminated with yeast, was then fermented, whereas nowadays, yeast cultures are added.

The growing of grapes and wine making were already recorded in the Bible. The history of wine making is connected with Mediterranean cultures, and it spread throughout the Roman Empire. Later, the Christian faith along with its foundation of monasteries as well as European imperialism during the Middle Ages played important roles in the development of the culture of wine production. A key prerequisite for making a good wine is the quality of the grapes. However, the soil, the way the grapes are cultivated, and the regional climate are also important factors affecting the quality of wine. Once the grapes have a sufficient sugar content, they are harvested and collected in a vat, where they are crushed. The juice produced is then placed in wooden casks or stainless steel tanks. Fermentation is allowed to proceed using the yeast naturally present on the grapes or, as with beer, by addition of special yeast cultures. After a year or so, the wine is bottled and stored for a period of time, depending on the kind of wine.

Bioethanol in the form of pulque, mead, beer, wine, and many other alcoholic beverages has accompanied mankind for millennia. A century ago, the application of bioethanol emerged as an important commodity in the chemical industry and, especially now, as a fuel to at least partially replace oil-based gasoline. But it has to be kept in mind what was said in Chapter 14: a further increase of bioethanol production may not be at the expense of our forests and of sufficient food for the world's population. This will be discussed further in Chapter 17.

> It is dangerous, says Voltaire, to be right in matters
> where established men are wrong
>
> Georg-Christoph Lichtenberg

Chapter 17
Energy conservation from renewable resources

The term "energy production," frequently used in connection with microbes, is incorrect. According to the law of conservation of energy, energy cannot be produced in a closed system, but it can be converted from one form to another. To use a "microbial example," part of the energy present in glucose can be converted to the energy present in methane, then into heat when the methane is burned.

What mankind increasingly needs are energy carriers for provision of heat and electricity, for mobility, and for production of goods. The most suitable energy carriers are those with a high energy density. Liquids and gases, or electrical power, are favored for ease of transport. The volume of the energy carriers required globally in 2003 amounted to 15.3 billion tons of coal equivalents (TCE); by 2030 it is expected to increase by 52 percent to 23.3 billion TCE. The proportion of renewable energy carriers was 13 percent in 2003 and will probably reach 14 percent by 2030. In view of this relatively small percentage of renewable energy sources, it is apparent that coal, oil, gas, and nuclear materials will continue to be the predominant energy sources in the near future.

Isn't that a rather conservative point of view? Can't we expect a boost in the use of renewables?

Let's look into renewables without regarding the use of wind, water, and solar energy. Instead, let's concentrate on biological processes. Whenever the use of biomass is being considered, the question of energy density is of increasing importance due to transportation costs. Microorganisms are therefore put to work to convert biomass into liquid or gaseous energy carriers. For example, liquid carriers are ethanol and butanol; gaseous carriers are methane (biogas) and molecular hydrogen. The industrial ethanol production with baker's yeast, *Saccharomyces cerevisiae*, is fully established. In addition, genetically engineered strains of *Escherichia coli* have been developed (see Chapter 16). Butanol is of interest as an energy carrier and, even more so, as a basic chemical for industry (see Chapter 15). However, the major drawback of the acetone-butanol fermentation is the low butanol yield of only a few percent, which makes recovery of butanol quite costly.

Whereas the production of ethanol and butanol requires sophisticated technical equipment and engineered strains of yeast, *E. coli* and *C. acetobutylicum*, the

Discover the World of Microbes: Bacteria, Archaea, and Viruses, First Edition. Gerhard Gottschalk.
© 2012 Wiley-VCH Verlag GmbH & Co. KGaA. Published 2012 by Wiley-VCH Verlag GmbH & Co. KGaA.

production of biogas is comparatively simple. Large vessels called anaerobic digesters are filled with biomass, and the microbial degradation process begins. First, any oxygen present is consumed, then fermentations leading to products such as organic acids and alcohols take place. Eventually the substrates for methanogenesis emerge: acetate, methanol, H_2, and CO_2. Then the methanoarchaea take over and produce biogas. The biomass composition has an effect on how much methane there is in the biogas produced. Biogas produced from sugars and carbohydrates consists of approximately 50 percent methane and 50 percent carbon dioxide. When a certain amount of fat is present, the methane yield is higher than 50 percent. As already mentioned, industrial plants for biogas production are rather simple. They can be built to different scales, with small ones to provide biogas for households or large ones to process all the activated sludge continuously produced in sewage plants of large cities. India is famous for its millions of small biogas plants that provide energy for cooking, especially in rural areas. Bioenergy villages have been developed, but there are certain limitations. Bioenergy cities are an illusion because biomass collection and transportation would simply use too much energy in the form of vehicle fuels.

It has been calculated that half the biomass produced on agricultural areas would have to be channeled into the biofuel fermentation industry in order to meet the annual global need for fuel. There we face the problem already mentioned in a previous chapter. Biofuel production must be kept in balance with global food production and may not be carried out at the expense of long-lived biomass.

Ethanol, butanol, and biogas – these will be the major products of the fermentation industry in the future, but it still will be difficult to meet more than 10 percent of global fuel needs with this type of bioenergy. These processes and some new developments were summarized in a report by the American Academy for Microbiology (Microbial Energy Conversion, 2006).

What about biodiesel?

Products recovered from oil plants, such as rapeseed and soybean oils, cannot be used directly as biodiesel. Such oils consists of long-chain fatty acids that are linked to glycerol. A chemical process has to be performed by which methanol replaces the glycerol, so methanol is required for biodiesel production. Glycerol is not only a byproduct of this process but also an important commodity for biotechnology (see Chapter 28). The biodiesel produced is a valuable component of the so-called energy mix but, again, only a small percentage of petroleum-based fuels can be replaced by biodiesel from rapeseed or related plants.

What other biological systems can be used for generation of useful energy carriers?

One of the most interesting processes is using solar energy for cleavage of water to molecular hydrogen and molecular oxygen. In fact, that is what plants and cyanobacteria do (see Chapter 9). They are capable of cleaving water with the help of photosystem II. The molecular oxygen produced in this process is released. The two hydrogen atoms (2H) of water, however, are not converted into molecular hydrogen (H_2) but transferred to carrier molecules, which then serve as hydrogen donors for conversion of carbon dioxide into starch or other cellular constituents. In order to produce O_2 as well as H_2 from water in light-dependent reactions, additional enzyme systems are required that are able to evolve H_2 like the hydrogenases. If the components, photosystem II, ferredoxin as hydrogen carrier, and a suitable hydrogenase, were to be combined, then water could indeed be cleaved in a light-dependent reaction:

$$H_2O \xrightarrow[\text{Photosynthetic apparatus}]{\text{Light}} \boxed{\tfrac{1}{2} O_2} + 2H$$

$$2H + \text{Ferredoxin}_{ox} \rightarrow \text{Ferredoxin}_{red}$$

$$\text{Ferredoxin}_{red} \xrightarrow{\text{Hydrogenase}} \text{Ferredoxin}_{ox} + \boxed{H_2}$$

That's it!

Unfortunately, this works only in principle. Such a system would only function for a few minutes, mainly because the oxygen formed is so reactive that it would inactivate the whole system (see Chapter 5, "Oxygen is a nasty stuff"). The question is how plants manage to cope with this radical action of oxygen. They have developed a protective system in which the O_2 reacts with a target molecule called D1 protein, which is damaged in the process. A repair mechanism continually replaces damaged D1 protein. It's like having to change the spark plugs of a car every five minutes. Obviously, driving a car under these conditions would not be very convenient and, for similar reasons, the enzyme system explained above would not be suitable for H_2 production. An additional drawback is the oxygen sensitivity of most hydrogenases, which also would be inactivated. This whole field finds broad interest and the research required is nearly unlimited. The sunlight-driven cleavage of water into hydrogen and oxygen would eventually lead to a hydrogen-powered economy. In our opinion, this is the only process of global importance in which biological systems or chemical systems mimicking biological processes could be employed on a large scale for generation of energy carriers. The other energy carriers mentioned operate essentially on relatively narrow roads, whereas what we need is a broad new avenue. Together with endeavors using solar energy in photovoltaic plants or in plants using solar energy to heat liquids, these processes will put the sun at the hub of energy-conservation technologies. And that is exactly what we have to do.

> Hast thou not poured me out as milk, and curdled me like cheese?
>
> Job 10:10

Chapter 18
Cheese and vinegar

In many ways the existence of mankind is associated with milk, not only breast milk but also cow, sheep, goat, or camel milk products. These products, with an annual global consumption of around 670 million tons, are mainly produced by lactic acid fermentation. The social and cultural development of mankind has strongly been influenced by this fermentation, which facilitated the conservation and storage of food, a prerequisite for development of human settlements. Furthermore, lactic acid fermentation is part of the basic metabolism of muscles, without of course the participation of bacteria, and it is of vital importance in the intestinal tract of infants (see Chapter 10).

If raw milk is allowed to stand at room temperature, sour milk will be formed after about 10 hours. The naturally present population of lactic acid bacteria multiplies by fermenting the lactose present in the milk. The resulting lactic acid causes the principle milk protein, casein, to coagulate at a pH value of around 4.6. This reaction leads to curdled milk products such as curds and various types of cheese. Most cheeses, for example, Emmental, Gruyère, Cheddar, or Gouda, are produced in a slightly different process. First, the milk undergoes a mild lactic acid fermentation that does not coagulate the casein. Instead, coagulation is caused by addition of what is called lab-ferment, or rennin (chymosin). This enzyme from calf stomachs, now also produced by genetically engineered microorganisms, splits a certain bond in the casein molecule. This alters the solubility of casein, so it begins to coagulate at a pH value of 5.5. In the cheesery, the coagulated casein, or curd, is pressed to separate it from the liquid whey. The curd then undergoes a maturation process that finally results in the various types of cheese, such as goat's cheese (Figure 33).

And how is yogurt produced?

The production of yogurt also follows the same pattern but begins with milk thickened by heating. Then a mild lactic acid fermentation followed by soft coagulation gives a product with the consistency typical of yogurt. The lactic acid bacteria used today to make such products include *Streptococcus thermophilus*, *Lactobacillus bulgaricus*, and *Streptococcus cremoris*. Dairies and cheeseries no longer rely on the milk's natural population of lactic acid bacteria; instead, so-called starter cultures

---

Discover the World of Microbes: Bacteria, Archaea, and Viruses, First Edition. Gerhard Gottschalk.
© 2012 Wiley-VCH Verlag GmbH & Co. KGaA. Published 2012 by Wiley-VCH Verlag GmbH & Co. KGaA.

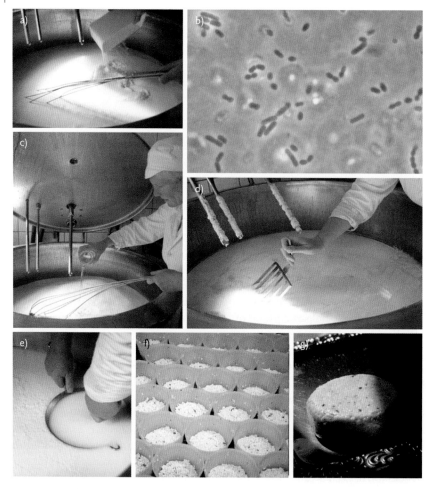

Figure 33 Making goat cheese: (a) the starter culture is added to the warm goat milk; (b) micrograph of the starter culture; (c) the enzyme chymosin (rennin) is added; (d) after coagulation of casein, a "cheese harp" is used to cut the mash so the whey can be drained; (e) the whey is sucked off; (f) the raw cheese is put into sieve forms; (g) cheese after ripening in a salt brine for three days. (Photographs: Anne Kemmling; cheese dairy in Landolfshausen, Lower Saxony, Germany.)

are used. These cultures are grown in special laboratories, tested for activity and safety, then made available to users. The use of these starter cultures allows such high-density bacterial growth in the milk that any undesirable bacteria are unable to compete against them.

Of course, milk is not the only substrate for lactic acid bacteria. Another product of these bacteria is sauerkraut, which is made from cabbage in a similar process, but leaves behind the cellulose fibers that cannot be fermented by lactic acid bacteria.

Where do the holes come from in some kinds of cheese?

These holes are actually bubbles in uncut cheese. Towards the end of the coagulation step, a special culture of propionic acid bacteria is added to the raw cheese. Propionic acid production makes a decisive contribution to the aroma and flavor of such cheeses. The fermentation carried out by propionic acid bacteria proceeds according to the following equation:

3 lactic acid → 2 propionic acid + 1 acetic acid + carbon dioxide

As the raw cheese thickens, the carbon dioxide is unable to escape, so it is trapped in the form of the bubbles mentioned above.

Not only Job referred to curdled milk in a simile, but also Homer, who wrote in the fifth song of the Iliad,

> "Just as fig juice added quickly to white milk clots it at once as it's stirred, that's how fast headstrong Ares healed."

In Germany, "alles Essig" (it's all vinegar) is an old expression meaning that something went wrong or the results are not as expected. Nowadays, this expression unfortunately has largely been replaced by more drastic ones. Many wine drinkers have had the experience of opening the bottle, expecting to inhale the wine's wonderful bouquet, only to realize after the first sip that "it's vinegar!" That there is a connection between wine turning to vinegar and its exposure to oxygen was observed long ago.

Wine connoisseurs may notice crystals that often collect at the bottom of wine bottles, known as "wine diamonds." They consist of the potassium/hydrogen salt of tartaric acid, potassium hydrogen tartrate. A related compound, sodium ammonium tartrate, is of historical importance. While inspecting crystal samples of this salt, Louis Pasteur noticed that they came in two asymmetric forms. These forms are now called D- and L-forms, or R- and S-forms. The theory behind this observation was first developed by Jacobus Henricus van't Hoff (1852–1911), who described the tetrahedron model of carbon, having realized that a carbon atom with different substituents can exist in two stereoisomeric forms. It is interesting that nature mainly consists of one type of stereoisomer. Amino acids, the constituents of proteins synthesized in the ribosomal protein factories, all belong to the L series. However, in bacteria there are a few amino acids of the D series that have special functions. For example, the cell wall of bacteria contains D-alanine, whereas only L-alanine is found in proteins.

After this excursion into stereoisomers, we will now focus on vinegar, which has been valued since ancient times for its flavor and as a food preservative. In order to make the best out of early observations how wine turns to vinegar, wine and other alcoholic solutions were purposely exposed to oxygen. The Orléans process was thus developed, in which alcoholic solutions were exposed to air in large pans or vats. Within a few days, a mat termed as *Mycoderma aceti*, formed on the surface. In the 1870s, this layer was found to contain bacteria, which since then have been called acetic acid bacteria. Today we know that one species of these bacteria, *Acetobacter xylinum*, is able to form cellulose threads that help form this mat-like layer. Bacteria are able to make all kinds of polymeric compounds, but it is quite rare that bacterial species are capable of making cellulose.

Acetic acid bacteria, so to speak, settle in the environment they "prefer," at the liquid surface of the pans and vats mentioned above. There, they have access to the nutrient ethanol from below and to oxygen from above, which they require for oxidation of ethanol to acetic acid. An expert would describe this process as incomplete oxidation. The alcohol is not completely respired to CO_2; instead, respiration stops at the level of acetic acid. Some acetic acid bacteria initially produce acetic acid, but once the alcohol has been consumed or a high concentration of acetic acid is reached, the latter is again taken up and completely respired to CO_2. Of course, this is not what the producers of vinegar have in mind, so they have to interrupt the process once the concentration of acetic acid has reached its optimum, around 10 percent (volume/volume).

The process of incomplete oxidation is also of ecological interest. Acetic acid bacteria live in a niche characterized by an excess of nutrients. They draw on nearly unlimited resources and, in a sense, avoid the cumbersome process of completely metabolizing the nutrients. Instead, they oxidize them incompletely. Such niches are found on the surface of juices or fruit. Relatives of the ethanol-oxidizing acetic acid bacteria often settle in such an environment. They follow the same strategy, for example, by oxidizing glucose to gluconic acid or glycerol to dihydroxyacetone; the latter is an active component of self-bronzers. The synthesis of vitamin C also includes a reaction step carried out by incomplete oxidizers. Hermann Sahm (Juelich, Germany) has intensively studied these processes, which he tells us about:

> "When today more than 100 000 tons of vitamin C per year can be produced cheaply; this is because of the bacterium *Gluconobacter oxydans*, which belongs to the group of aerobic acetic acid bacteria. *Gluconobacter* has a large number of membrane-associated enzymes that efficiently metabolize various sugars and sugar alcohols by incomplete oxidation. This is the reason why this bacterium has already been used since 1934 for oxidation of the sugar alcohol D-sorbitol to L-sorbose for vitamin C production. After this biotechnological-chemical vitamin C synthesis has successfully been carried out for decades, efforts are being made to develop a purely biotechnological procedure for the production of vitamin C."

The classical process is outlined in the Study Guide. Here, we have asked an expert, Karl Sanford (Palo Alto, California, USA), to summarize the development of an innovative biological process for vitamin C synthesis:

> "Modern-day biotechnology provides a means of gathering all of the relevant genes required to convert sugars, such as glucose and fructose that come from renewable feedstocks, e.g., cornstarch, into ascorbic acid within

a single microorganism. By having all the enzymatic steps in a single microorganism, the environmental sustainability and economic features are improved. Fewer manufacturing steps mean that fewer chemicals, less solvents, and less energy are used to make each pound of ascorbic acid.

The history for this process development is a long one, starting initially with Genentech and Lubrizol in the 1980s. The technology was subsequently licensed to Genencor, who then organized a consortium of companies and the United States Department of Energy's Argonne National Laboratory to work on improving the economics and process designs. This substantial effort started in the mid 1990s and ended in the early part of this century. This work was funded in part by the Advanced Technology Program of the United States Department of Commerce as a vanguard to technologies for conversion of renewable feedstock to chemicals using continuous biocatalysis. Much of this work anticipated the subsequent developments in biofuel production from renewable feedstock using sustainable manufacturing processes and the broader industrial or white biotechnology commercial developments. A focus of this work was the development of continuous biocatalytic reactors for the production of chemicals from renewable feedstocks with ascorbic acid as a first product.

Ascorbic acid was selling for between 15 and 20 US dollars per kilogram in the mid-1990s. This pricing was used as a benchmark to design a process that would be cost-competitive or, even better, cost-advantaged relative to the state-of-the-art at that time. However, globalization of the chemical industry was beginning; within a few years, the price had collapsed to just a few US dollars per kilogram and there were more than twenty new ascorbic acid producers. Most of these new entrants were in China, using 'old' technology with a different economic model. Although the Genencor process became a very low-cost process to make ascorbic acid as well as other types of related sugar/acid products such as erythorbic acid, the less-appealing commercial opportunity has kept this process from being put into large-scale manufacturing. With recent prices of ascorbic acid rebounding into the range between 15 and 20 US dollars per kilogram, maybe the time is ripe for reconsideration. Certainly, there is a world-class ascorbic acid process based on exemplary metabolic pathway engineering in a *Pantoea citrea* cell factory."

This is just one example for many biotechnological processes that have been designed and worked out but for economic reasons cannot yet be put into industrial reality.

Now back to vinegar. Of course, its production did not stop at the level of the Orléans process. A faster vinegar process was developed in the 1820s, in which the alcoholic solution passes through a reactor loosely filled with beechwood chips. From below, air flows upward past the alcoholic solution that trickles down through the reactor. Before long, acetic acid bacteria settle on the beechwood chips,

Figure 34 A glance into the vinegar plant of the company "Essig Kuehne KG" in Hamburg (Germany). The huge wooden reactors are made of oak. (Courtesy of the company.)

where they then fulfill their task just as they did in the *Mycoderma aceti* mats mentioned above. All modern processes for production of vinegar are more or less advanced versions of this so-called Schuetzenbach process. In Figure 34 you can see such a plant for vinegar production operated by the "Essig Kuehne" company in Hamburg, Germany, until a few years ago: beautiful wooden vats filled with beechwood chips. This form of biotechnological practice is even pleasing to the eye. Submerged fermentation processes have also been developed. These processes use large, aerated bioreactors of stainless steel in which alcoholic solutions, acetic acid bacteria, and air bubbles are continuously kept in contact by stirring.

We should mention once more the fermenting microorganisms that live in the absence of oxygen and those that are incomplete oxidizers, especially their diverse habitats. When sugary solutions such as fruit juice or milk are left to stand, the dissolved oxygen is quickly respired by aerobically living bacteria. As soon as the oxygen is gone, fermentation processes begin. These lead to formation of large amounts of lactic acid, alcohol, or other fermentation products. A different habitat develops at the interface between the fermented solution and the air. Such a habitat is often dominated by incomplete oxidizers.

> All things are poison and nothing is without poison,
> Only the dose permits something not to be poisonous.
>
> *Paracelsus*

Chapter 19
The periodic table of bioelements

There were coastal pastures in Australia that appeared to be suitable for raising sheep. However, the sheep grazing there did not thrive as expected: they all developed the Australian "coast disease." Chemical analysis revealed that these pasture soils were extremely low in cobalt salts. When small amounts of cobalt salts were added to the soil, the sheep gained weight and developed normally. What is the connection? The one central compound in nature that contains cobalt is vitamin B_{12}. Plants don't require vitamin B_{12}, but sheep and cattle do. Actually, cobalt is needed primarily for vitamin B_{12} synthesis by the microorganisms in the rumen of these animals. Bacteria and archaea are the only organisms capable of synthesizing vitamin B_{12}. From a structural standpoint, it is a beautiful compound with a cobalt atom right in the middle of a "net" of carbon, nitrogen, and hydrogen atoms (Figure 35). The methane-producing archaea in the rumen require a derivative of vitamin B_{12} to perform certain reaction steps in their metabolism. Humans also need vitamin B_{12}, which we ultimately get from microbial sources. A highly specific protein factor in our stomach and intestine binds the traces of vitamin B_{12} in our food, then transports the vitamin to sites in our body where it is involved in essential metabolic conversions. If this specific protein factor is defective, vitamin B_{12} cannot be taken up, and deficiency symptoms such as pernicious anemia may develop.

Cobalt and nickel are often mentioned in the same breath. It's remarkable that both metals, as the positively charged ions Co^{2+} and Ni^{2+}, are located at the center of complex compounds such as vitamin B_{12} and coenzyme F_{430}.

That's right. In microorganisms, vitamin B_{12} is more common than coenzyme F_{430}, which plays a key role in methanogenesis and was described in Chapter 13. Other groups of compounds should be mentioned in this context: the chlorophylls, which are structurally related to vitamin B_{12} and coenzyme F_{430} but contain magnesium (Mg^{2+}) as their central atom, as well as the cytochromes and hemoglobin, with iron (Fe^{2+}) at their center.

Magnesium and nickel ions have additional functions in the cells. Certain protein complexes such as the ribosomes are held together by Mg^{2+}. Phosphates such as ATP exist primarily as magnesium salts. Urease, which splits urea into

Discover the World of Microbes: Bacteria, Archaea, and Viruses, First Edition. Gerhard Gottschalk.
© 2012 Wiley-VCH Verlag GmbH & Co. KGaA. Published 2012 by Wiley-VCH Verlag GmbH & Co. KGaA.

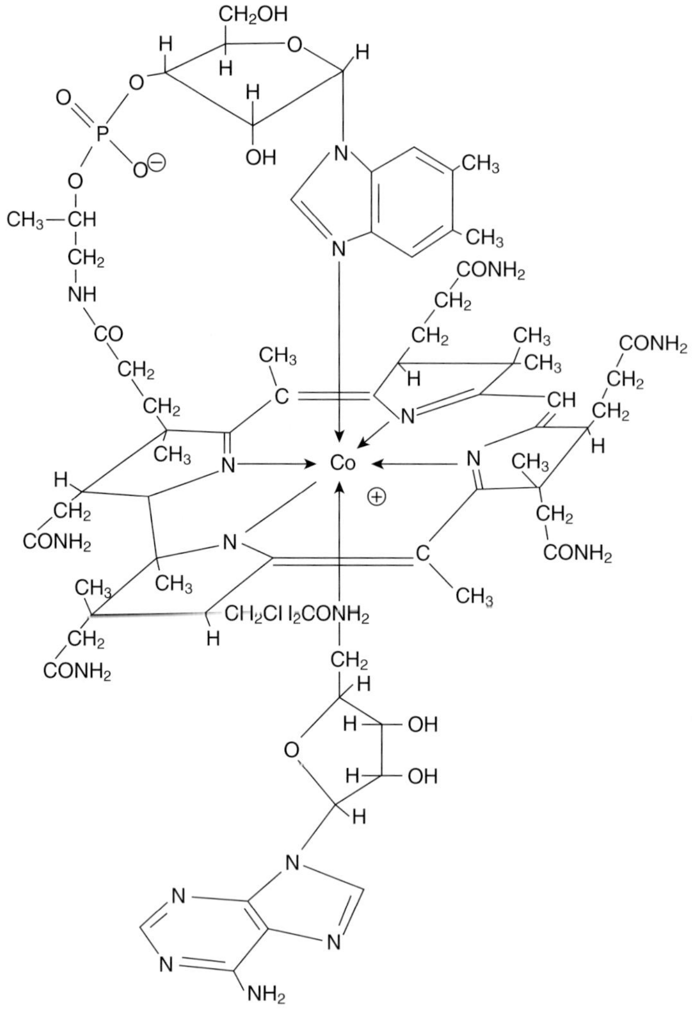

Figure 35 Formula of coenzyme B_{12}. This is one of the biologically active forms of B_{12}. In vitamin B_{12} the deoxyadenosine-residue (bottom part of the molecule) is replaced by a CN- or an OH-group. B_{12}-compounds can only be synthesized by a number of bacterial and archaeal species. (H.P.C. Hogenkamp, Vitamin B_{12} Coenzymes. *Ann. Acad. Sci* 112, 552–564 [1969].)

ammonia and carbon dioxide, was the first enzyme discovered to contain nickel. Furthermore, some of the hydrogenases mentioned previously also contain nickel.

Quite a few discoveries of metal requirements were more or less accidental, such as the nickel dependence of methanogenesis (see Chapter 13). Botanists found that manganese salts are essential for oxygenic photosynthesis. Later it was shown that manganese is directly involved in the production of oxygen, so this metal plays an especially key role on our planet! Its incorporation into one of the most important protein complexes in nature was one of the prerequisites for development of aerobic life.

The periodic table of elements created by Dimitri Ivanovich Mendeleyev (1834–1907) in 1869 is essentially the alphabet of chemists. It begins with H for hydrogen and, for a long time, it ended with U for uranium, which at the time had the highest atomic number in the table, 92. This number has increased impressively in the past 60 years to over 115. With the help of approaches truly reminiscent of war, bombardments with neutrons and other particles have enabled scientists to produce more and more heavy elements. The life expectancy of these unstable elements is admittedly often shorter than the seconds of happiness enjoyed by their discoverers. In other words, the rapid decay of these so-called transuranic elements makes them irrelevant to the biological world.

Just to continue with military terminology, the bacteria have used every trick in the book to successfully exploit the elements of the periodic table in their battles for nutrients (Figure 36). As previously mentioned, the presence of magnesium in chlorophylls opened a whole new horizon with light as a source of energy. The incorporation of manganese into a protein system led organisms into the new world of oxygen evolution during photosynthesis. Molybdenum succeeded in luring atmospheric nitrogen into the living world. Methanogenesis was only made possible by coenzyme F_{430} and vitamin B_{12}. This list could go on almost indefinitely. Bacteria were able to incorporate zinc into an active alcohol dehydrogenase, an enzyme system from which we still benefit today. Even tungsten was able to catch the attention of microbes, as Jan Andreesen (Halle, Germany) explains,

> "So far, tungsten is the heaviest bioelement. As an element of the sixth group, it is chemically and physically very similar to molybdenum. Tungstates, the salts of tungsten, are toxic to classical molybdenum enzymes, such as nitrogenase (enzyme for nitrogen fixation), nitrate reductase (generates toxic nitrite from nitrate), or sulfite oxidase (an important detoxification enzyme in the liver). The toxic effect is caused by prevention of molybdate uptake by the cells or molybdate incorporation into respective

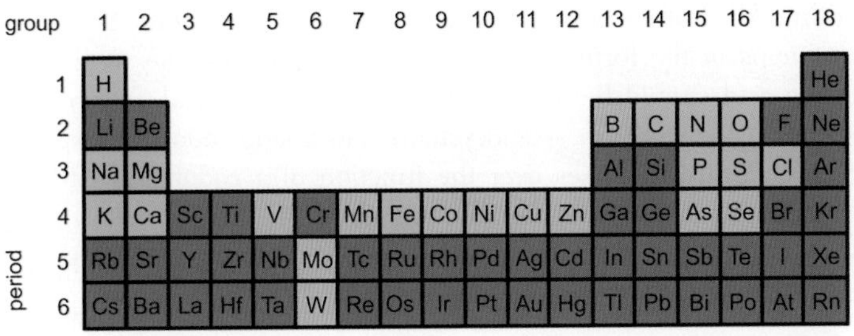

Figure 36 The bioelements in the periodic table. The macrobioelements are shown in green, the micro-bioelements, in yellow. Of the remaining elements, a few such as silicon play a role in certain organisms. (L.P. Wackett et al., Appl. Environ. Microbiol. 70, 647–655 [2004], modified.)

enzymes. By testing for this antagonism between molybdate and tungstate, a variety of molybdenum-containing enzymes were discovered. The test was simple. When growing bacterial cells were treated with molybdate, a certain enzyme activity could be detected, but when molybdate was replaced by tungstate, this activity was no longer detectable. Hence, it was a surprise in 1971 when Lars Ljungdahl (Athens, Georgia, USA) and I conducted an experiment with an anaerobic bacterium called *Clostridium thermoaceticum* (now *Moorella thermoacetica*). This bacterium produces an enzyme called formate dehydrogenase that oxidizes formate, the salt of formic acid, to CO_2 and even the reverse reaction. When we carried out the above test, the results were just the opposite. We measured enzyme activity in the presence of tungstate but virtually none in its absence. Later, it could be shown that the formate dehydrogenase of this bacterium contains tungstate.

Since the availability of chemical elements in nature decreases with their atomic weight, bacteria have to "work hard" to find the tungstate they need. That's why they possess two highly specific systems for tungstate uptake (Tup and Wtp). Both systems are specific enough to guide tungstate but not molybdate into the interior of the bacterial cell. At first sight it seems that the preference for tungstate over molybdate in some enzyme systems is a quirk of nature, but there may be a rationale behind it."

Iron plays a central role in all organisms. Bacteria love metals that are capable of rapid valence changes. For example, bivalent iron (Fe^{2+}) can easily be oxidized to trivalent Fe^{3+} (rust). This process is reversible, so iron-containing proteins such as cytochromes and iron-sulfur proteins are components of electron transport chains in which they accept and transfer electrons to other electron carriers. Phosphate and sulfate, as well as the salts of calcium, sodium, and potassium, all fulfill important functions in microorganisms. Bacteria are also highly interested in selenium, which in some proteins replaces sulfur at certain positions, thereby altering the catalytic properties of these proteins. As a result, a number of bacterial species require selenium for growth. Like these bacteria, humans can develop a selenium deficiency. Symptoms in humans include hair loss, myocardial muscle weakness, or diminished protection from reactive oxygen compounds. Selenium is present in proteins in the form of the amino acid selenocysteine. Without a doubt, this amino acid should be listed as the twenty-first natural amino acid. There is no specific code word for selenocysteine, but a stop codon in a special messenger RNA environment takes over the function of a codon for a t-RNA loaded with selenocysteine (▶Study Guide). All these fascinating details were discovered by research teams headed by August Boeck (Munich, Germany) and Terry Stadtman (Bethesda, Maryland, USA).

How can we visualize the role of metals in enzymes?

Much like a spider lurking in its web, metal ions such as Mg^{2+}, Fe^{2+}, Ni^{2+}, or Co^{2+} "sit" in proteins or special chemical structures, for example, vitamin B_{12} or cytochromes. The substrates of the enzymes are "prey" to these metals. However, they are not gobbled up when they get near these metal ions. Instead, the substrates are oxidized or reduced or otherwise modified, depending on the catalytic properties developed by a particular enzyme during evolution. Let's take catalase as a simple example. This enzyme, already mentioned in Chapter 5, contains iron:

$$2\,H_2O_2 \xrightarrow{\text{Fe-containing catalase}} 2\,H_2O + O_2$$

hydrogen peroxide → water + oxygen

Aqueous solutions of hydrogen peroxide are quite stable, but when a small amount of catalase is added, hydrogen peroxide decomposes instantaneously and oxygen is evolved.

It's fascinating that metals play such an important role in biology. On the other hand, we hear so much about soil contaminated with heavy metals. How do bacteria get by in such an environment?

Bacteria are unable to live in soil with extremely high cadmium levels, for example. However, bacteria are able to develop resistance to much higher concentrations of heavy metals than animals can tolerate. Hans Guenter Schlegel (Goettingen University, Germany) tells us about the nickel-resistant bacteria he isolated in New Caledonia, a group of islands northeast of Australia,

> "How did we become interested in studying heavy-metal resistance in the first place, in particular the resistance of bacteria to nickel? I began to isolate so-called 'knallgas' bacteria in 1951 as postdoctoral researcher. These bacteria require H_2, CO_2, and O_2 as nutrients, but they only grew when I added soil extract to the nutrient broth. Richard Bartha, who was at Goettingen University in the early 1960s then went on to Rutgers University (New Brunswick, USA), was the one who discovered that soil extract could be replaced by nickel salts. Later, Baerbel and Cornelius Friedrich (now Humboldt University, Berlin, and Dortmund University, Germany) showed that nickel is a constituent of the H_2-activating hydrogenase enzyme system in these bacteria. Actually, this should have been realized much earlier. In the course of evolution, why should nature not have made use of nickel's wonderful catalytic capabilities to activate H_2, thus serving as a reducing agent in microbial metabolism? In the meantime, nickel has been recognized as a common bioelement and a component of many enzyme systems.
>
> A nickel-salt concentration range between 1 mg and 10 mg per liter growth medium meets the nickel requirement for bacterial growth. However, the presence of nickel ions in much higher concentrations (between 50 mg and 2 g per liter) is inhibitory to microbial growth or even lethal. Nickel is one of the heavy metals that is toxic to microorganisms as well as to higher organisms such as humans. We found that most bacteria are unable to grow in heavily contaminated soil or close to industrial plants unless they

possess protective systems that confer resistance. This resistance is not achieved by blocking heavy-metal uptake into the cell. Instead, bacteria have developed specific efflux mechanisms to pump toxic ions out of the cell as quickly as possible. In contaminated ecosystems, the genes that code for these efflux pumps are located on plasmids, not on the bacterial chromosome where the 'housekeeping genes' required for basic cell maintenance are located. The efflux pump genes are located on plasmids because these pumps are only needed from time to time when the organisms are confronted with high nickel concentrations.

In 1989 I traveled to New Caledonia, whose islands primarily consist of iron-, manganese-, and nickel-containing rock. Interestingly, around 165 nickel-accumulating plants grow on New Caledonia. A sensational representative of this flora is the nickel tree, *Sebertia acuminata*. Its leaves contain more than 1 percent nickel and its blue-green latex, nearly 25 percent. For hundreds of thousands of years, the topsoil there has been permeated by the nickel from decomposing leaves of these trees. From the soil beneath these trees we isolated bacteria resistant to nickel concentrations of 1.5 g or even 3 g per liter. What surprised us most was the location of the genes conferring this resistance. They were not located on plasmids but on the bacterial chromosome. During the long course of natural selection, these genes became chromosomal housekeeping genes because their products, the efflux pumps, were needed continuously. New Caledonia was truly worth the week I spent there."

Nickel is a good example of how the effect of a metal may depend on its concentration. Nickel is essential in low concentrations but very toxic in high concentrations. The importance of the resistance-conferring efflux pumps described by Hans Guenter Schlegel can be illustrated by the following example: if your boat springs a leak, you have two alternatives: plug up the hole to stop the leak or bail out the water as fast as it comes in. In the case of heavy metals, bacteria only have the latter option: they have to bail them out. The nature and role of plasmids in bacteria mentioned above will be discussed in Chapter 23.

> It is not the strongest of the species that survives,
> nor the most intelligent that survives.
> It is the one that is the most adaptable to change.
>
> Charles Darwin

Chapter 20
Bacterial sex life

Aren't you exaggerating?

Perhaps, but there is something to it in the sense that genetic information is acquired. Before we go further into this, let's deal with the question of what the actual driving forces are behind bacterial evolution. Mainly, there are two.

One of these is the creation of genes containing information for new or altered biological activities. Here, the underlying reactions are mutations, gene duplications, and the disassembly, reordering, and reassembly of gene fragments. Of course, it is hard to imagine that these reactions sufficiently explain the whole scope and speed of microbial evolution. However, we shouldn't forget the enormous and almost inconceivably large number of microbial cells. There are billions and billions of cells, so even extremely rare mutations have the possibility of occurrence, perhaps resulting in a microbial species with a new characteristic. When we try to come up with something similar in the laboratory, we don't handle the microbes with kid gloves. At a time when appropriate methods for systematic genetic manipulation of bacteria were not yet available, irradiation with ultraviolet (UV) light was a method of choice for triggering mutations. Here's the approach: after UV-light exposure long and intensive enough to kill 99.9 percent of the bacteria in a suspension, the remaining 0.1 percent are screened for desired mutations. Only one cell in every thousand is still viable. However, if one trillion cells were present at the beginning, there will still be one billion left, including many mutants, perhaps even the mutants of interest.

It is similar in nature. Bacterial DNA can be permanently altered by defective replication. Radiation or chemicals can also damage the DNA. A number of mutations are silent, so they do not alter the phenotype, the characteristic features of the bacterium. In addition, bacteria possess very efficient DNA repair systems by which damaged DNA can be repaired. Mutations may also be lethal and lead to cell death, but every now and then a mutative event occurs that leads to a selective advantage. Enhanced fitness of such an altered bacterium will enable it to prevail, eventually changing the whole population.

Discover the World of Microbes: Bacteria, Archaea, and Viruses, First Edition. Gerhard Gottschalk.
© 2012 Wiley-VCH Verlag GmbH & Co. KGaA. Published 2012 by Wiley-VCH Verlag GmbH & Co. KGaA.

Couldn't you give an example from the laboratory?

We will choose an experiment conducted by Hans Kulla (1949–1993) of the Swiss Federal Institute of Technology (ETH / Zurich, Switzerland). It had to do with the reduction of an azo dye. Azo dyes are primarily produced by organic chemists. These dyes contain an N=N double bond with chemical residues on both sides and have the general formula R–N=N–R. There are bacteria that form an azo reductase by which simple azo dyes are cleaved into two parts, as is the case with dicarboxylic azobenzene (DCAB). However, more complicated azo dyes such as carboxy-Orange II cannot be cleaved by bacteria, so they are unable to utilize it. Hans Kulla prepared a bacterial culture maintained under continuous growth conditions. In such a culture, the volume of the continuous inflow of fresh medium corresponds to the continuous outflow of bacteria and consumed medium. The growth medium contained the simple azo compound DCAB, which could be

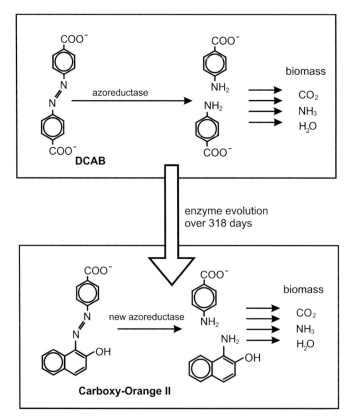

Figure 37 Emergence of an azoreductase by which, in addition to dicarboxybenzene (DCAB), a more complex azo compound (carboxy-Orange II) also is reduced. This type of enzyme first emerged in a continuously growing bacterial culture after 318 days. (H.G. Kulla in: L. Leisinger et al., Xenobiotics, p. 387–399, Academic Press, 1981, modified.)

utilized by the bacteria and, in addition, carboxy-Orange II. Incubation was continuous as generation upon generation of bacteria grew in the reactor. Despite UV irradiation to trigger mutations, nothing happened for months and months. Finally, after 318 days, the color of the culture medium changed, indicating that the more complicated azo dye had been metabolized. The bacteria then grew much more happily because they had succeeded in making a new substrate accessible (Figure 37). Through mutation the azo reductase, or rather the gene for its formation, had been altered. This resulted in an azo reductase with a new feature that had never existed before on our planet.

Here's a second example for the evolutionary potential of bacteria. Those were the days when beta(β)-lactam antibiotics, such as ampicillin, or methicillin, had been developed and germs had not yet become resistant to them. Some bacteria, however, already possessed the enzyme β-lactamase, which cleaves the β-lactam ring of penicillin and thus counteracts the action of this antibiotic (see Chapter 22). It was no problem for the scientists, who then developed antibiotics such as ampicillin, methicillin, and many others unaffected by β-lactamase. It was only a question of time until bacteria, through mutation and selection, caught up with one antibiotic after another. In the meantime, there are more than 200 different types of β-lactamase, all compiled in a genealogical tree of the various β-lactamases. Enzymes of this type restrict more and more the application of β-lactam antibiotics. This extremely alarming process forces a race between creative research for the development of ever-new antibiotics and the evolutionary potential of bacteria. Let's have the expert Timothy Palzkill (Houston, Texas, USA) tell us how the β-lactamase has changed in the last few years,

> The evolution of the TEM-1 and SHV-1 β-lactamases provides an example of evolution of altered enzymatic function by acquisition of point mutations. The TEM-1 β-lactamase is the most common plasmid-encoded β-lactamase in Gram-negative bacteria. The SHV-1 β-lactamase is 68 percent identical to TEM-1 and is also commonly found in Gram-negative bacteria. The TEM-1 β-lactamase was first discovered in the early 1960s, not long after the widespread introduction of ampicillin as a treatment for Gram-negative infections. The TEM-1 β-lactamase hydrolyzes ampicillin very efficiently to provide high levels of bacterial resistance to this drug. The corresponding gene was initially found in *E. coli* but it spread rapidly under the selective pressure of β-lactam antibiotic therapy to other Enterobacteriaceae (bacteria such as Salmonella).
>
> The cephalosporin β-lactam antibiotics such as cefotaxime and ceftazidime were introduced in the early 1980s, in part because they are hydrolyzed poorly by TEM-1 β-lactamase. In addition, the combination of an inhibitor of β-lactamase, such as clavulanic acid, with a β-lactam antibiotic such as amoxicillin was introduced as a therapy to avoid the action of β-lactamases. These therapeutic options were effective but placed selective pressure on the bacteria containing TEM-1 β-lactamase to become resistant. In the

1980s came the first reports of transferable resistance to cephalosporins and β-lactam antibiotic–β-lactamase inhibitor combinations. In many cases, the cause of this resistance was due to TEM-1 β-lactamase enzymes that had amino acid substitutions near the active site that resulted from the acquisition of point mutations in the gene encoding the TEM-1-β-lactamase gene. These TEM-1 variants were able to hydrolyze cephalosporins with increased efficiency and could therefore provide bacteria with resistance. In addition, TEM-1 variants with amino acid substitutions that allowed the enzyme to avoid the action of inhibitors were discovered. Thus, TEM-1 β-lactamase variants have evolved resistance to these therapies due to the acquisition of point mutations that confer a selective advantage to the bacteria during antibiotic treatment. A similar process has been observed with the SHV-1 β-lactamase. The new TEM-1 and SHV-1 variants are named by increasing number, such as TEM-3, TEM-4, SHV-3, and SHV-4, as new variants arise. This process has continued to the present so that at present the variants are up to TEM-178 and SHV-133 (see http://www.lahey.org/Studies/). The evolution of the TEM-1 and SHV-1 β-lactamases provides a clear example of how bacteria can quickly respond to selective pressure by the accumulation of mutations.

Isn't that a gloomy picture painted by Tim Palzkill? So there are two types of β-lactamases that are 68 percent identical, the TEM and the SHV family. The former was already discovered in the early 1960s and has spread worldwide, appearing in hospitals and in the environment. Point mutations have enabled the TEM lactamases to destroy the ever-new β-lactam antibiotics being employed for patient treatment. Altogether, there are 311 variants of the β-lactamase and this number will increase continuously. This variability is amazing. It's as if a bouquet of 311 differently colored roses were to arise from a red and a yellow one, and this within a relatively short period of time.

The reader should keep this in mind to better understand some of the alarming reports appearing in the press, such as the dangers arising from methicillin-resistant *Staphylococcus aureus*, MRSA (see Chapter 23).

But what does this have to do with bacterial sex life?

We will come back to this question after discussing the second driving force of microbial evolution. Successful mutations may be rare, but the resulting superior genes can be spread by proliferation of their microbial carriers. In addition, mechanisms have evolved in bacteria by which genetic information in the form of DNA is guided from outside the cell into its interior. Many but not all bacterial species have developed a system by which naked DNA fragments are taken up from the immediate environment and screened for genes whose products may be of use. This process is called transformation.

The situation in soil can be compared to that at a car junkyard: here, the lysed cells along with fragments of cell walls, proteins, and DNA fragments; there, rusting car bodies, old tires, dented mufflers, and grimy engines. In soil, still-living bacteria may take up DNA fragments and use the information encoded within the DNA (Figure 38a). At the junkyard, dealers may look for engines and other reusable parts.

Transformation was discovered by Frederick Griffith (1877–1941) in 1928. He worked with a bacterium that causes pneumonia, *Streptococcus pneumoniae*, often

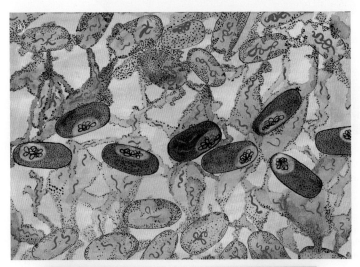

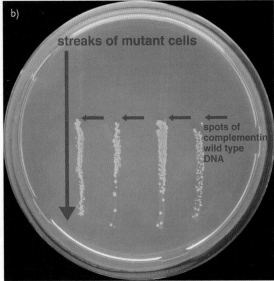

Figure 38 Transformation, an important mechanism for horizontal gene transfer. (a) Uptake of DNA fragments (red) by bacteria in soil. A few living organisms between dead and lysed cells (watercolor and gouache, Anne Kemmling, Goettingen, Germany). (b) Transformation experiment with a mutant of an *Acinetobacter* (continued)

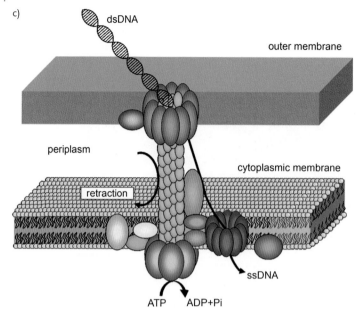

Figure 38 (continued) species that can no longer grow on p-hydroxybenzoate. The mutant lacks the gene for p-hydroxybenzoate hydroxylase. The lacking or defective gene is replaced by transformation with wild-type DNA. For this purpose, wild-type DNA is spotted on agar at points indicated by red arrows. The only carbon and energy source present in this agar is p-hydroxybenzoate. Mutant cells are now streaked out from the upper part of the Petri dish through the wild-type DNA spots downwards. Growth is only possible after contact of the cells with wild-type DNA (Beate Averhoff, Frankfurt/Main, Germany; original experiment: Nicolas Ornston, Beate Averhoff). (c) Model of the apparatus for DNA uptake by transformation in Gram-negative bacteria. The pilus consisting of helically arranged proteins is depicted in green. Double-stranded DNA (ds-DNA) is threaded into the transformation apparatus and pulled into the cell with concomitant hydrolysis of one strand. Single-stranded DNA then arrives in the cell. The whole process is driven by ATP hydrolysis. The space between the cell wall (a major part of it is the outer membrane) and the cytoplasmic membrane is called periplasm. (Model: Beate Averhoff, Frankfurt/Main, Germany.)

called pneumococcus. This microorganism occurs in two forms. One has a smooth surface because it is surrounded by a carbohydrate capsule, called the S strain. The second form, the R strain, has a rough surface but not a capsule. Only the S strain exhibits virulence.

Now we will describe the experiments carried out by Griffith, beginning with the controls. Cells of the R strain injected into mice had no effect. The same result was obtained when heat-killed S-strain cells were injected. But living S-strain cells killed the mice within a short period of time. Now comes the experiment: when R-strain and heat-killed S-strain cells were injected together, the mice became sick and living S-strain cells could be isolated from body fluids of these mice. This surprising result could only be so interpreted that the information for capsule formation and thereby the virulence had been transferred from the dead S cells to the living R cells. An explanation for this phenomenon was offered by Oswald Avery (1877–1955) and colleagues in 1944. They isolated DNA from the S strain and mixed this DNA with living R cells. Mice injected with this mixture became sick and living S-strain cells could be isolated. The conclusion was that DNA fragments had apparently been taken up by the R cells and that these fragments contained the information for capsule formation. This, by the way, was the first experiment to show conclusively that DNA is the carrier of genetic information.

As already mentioned, not all microbial species are capable of performing transformation. A certain mechanism is required that cannot be produced by all species. In addition, transformation-positive organisms such as members of the genus *Bacillus* have to be in a certain physiological state for DNA uptake, called competence. We have asked Beate Averhoff (Frankfurt, Germany) to elaborate on the transformation process.

> "Transformation is indeed nothing occurring by accident; it is genetically encoded. Transformation systems are highly complex. They are dynamic transport engines consisting of many special gene products. These engines require energy to develop a tractile force by which bound DNA is pulled into the cells. Also, the regulation of transformation activity is further evidence for its entirely biological nature. It is a transient phenomenon and the components of the transformation apparatus are only synthesized in response to certain signals that are generated in the environment of the cells. Such signals are, for instance, certain shortages of nutrients. Figure 38c gives an impression of how complex the DNA uptake system is. The double-stranded DNA is guided through the ring structure of the outer membrane. Then one strand is degraded and the remaining one is pulled into the cells' interior by the action of proteins such as comEA and comEC."

Not only Griffith's classical approach but also other experiments are suitable for demonstrating convincingly the occurrence of transformation. Let's look at an experiment carried out by Beate Averhoff (Frankfurt, Germany) and Nicholas Ornston (New Haven, Massachusetts, USA) and depicted in Figure 38b. Wild-type DNA from bacteria of the genus *Acinetobacter* is spotted on agar as indicated by the four red arrows. Then cells of a mutant of this organism that had lost the ability to grow on p-hydroxybenzoate are streaked out. The four streaks run downward from the upper rim of the dish through the wild-type DNA spots. The mutant cannot grow on this agar because the only carbon and energy source in the agar is p-hydroxybenzoate. However, cells of this mutant are able to grow after contact with wild-type DNA. This finding can be so explained that DNA fragments containing the genes for p-hydroxybenzoate degradation have been taken up by transformation, incorporated into the chromosome, then expressed.

This is a convincing experiment, an example for a process taking place on Earth millions of times every second. The occurrence of transformation is a fact, although the average person finds it difficult to accept the existence of such processes in nature.

Another process involving DNA transfer is conjugation, which finally brings us to bacterial sex life. It was discovered by Joshua Lederberg (1925–2008) and Edward Lawrie Tatum (1908–1975) and is an excellent example of how well-conceived experiments and far-reaching interpretation of the results have lead to important discoveries. Lederberg and Tatum worked with mutant strains of *E. coli*. They had isolated mutants incapable of synthesizing certain nutrients, so that these nutrients were required for growth. One strain required methionine (M) and biotin (B); the other one, threonine (T) and leucine (L). Such defects can generally be repaired by reversion of the mutations. Reversion rates for a single mutation are in the order of 10^{-6}, one in a million, so one of 10^6 cells will regain the status of the wild type. However, in the case of double mutants, two independent reversions of the mutations have to occur, so the rate would be 10^{-12}. Such an event is extremely rare. Lederberg and Tatum mixed the two double mutants and, much to their surprise, observed the occurrence of the wild-type phenotype at the unexpected rate of 10^{-5}. The M^-B^- strain must have somehow acquired the missing genes from the T^-L^- strain and/or vice versa, a result that could not be explained at the time. Like Frederick Griffith, mentioned above, Joshua Lederberg and Edward Tatum did a milestone experiment in microbiology. They made the following observations:

1) Direct contact between the two double mutants was required for the occurrence of the wild-type-like cells.

2) DNA was apparently transferred from one strain to another, and this transfer proceeded in one direction, with one strain serving as donor (F^+) and the other one as acceptor (F^-) (Figure 39).

Sometimes the donor strain (F^+) is called the male strain; the female strain (F^-) is then the acceptor. The process has been termed parasexual, but now it is generally called conjugation.

Now I understand the title of this chapter, but how can DNA be transferred from one cell to another?

The donor strain contains a genetic element that is essential for gene transfer, the so-called F plasmid, where F stands for fertility. Plasmids are small DNA rings in the bacterial cell that are separate from the chromosome. They contain genetic information that is not essential for basic bacterial life but allows microbial populations to survive under certain stress conditions, for example, the presence of an antibiotic (to be discussed in Chapters 22 and 23). The F plasmid has two impor-

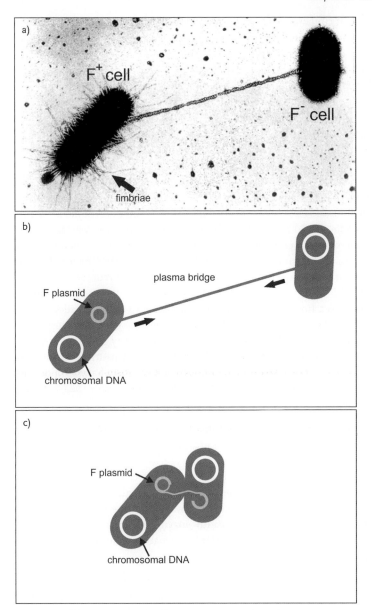

Figure 39 DNA transfer in E. coli by conjugation. (a) Electron micrograph, the F$^+$ cell carries fimbriae, and it produces the pilus by which contact is made with the F$^-$cell (according to Charles C. Binton Jr. cited in: R.Y. Stanier, J.L. Ingraham, M.L. Wheelis, P.R. Painter. The Microbial World, Prentice-Hall, NY, USA, 1986). (b) Schematic presentation of a), showing the chromosome of both F cells and the F plasmid in the F$^+$cell (diagram: Anne Kemmling, Goettingen, Germany). (c) One strand of the F plasmid migrates via a plasma bridge to the F$^-$ cell. The single-stranded parts of the plasmid in both cells are immediately replenished to double strands. (Diagram: Anne Kemmling, Goettingen, Germany.)

tant properties. It contains genes that facilitate the contacting and approach of the F^+ and F^- cells and the subsequent formation of a plasma bridge between the two cells. Furthermore, the F plasmid encodes the genetic information to convert its double-stranded DNA into single strands, to open one of these strands, and to transport this linear strand into the F^- cell (Figure 39c). The missing complementary strand is synthesized in both donor and acceptor cells, so that both cells are then in possession of the F plasmid.

This process explains the spread of F^+ cells, but not the results obtained by Lederberg and Tatum. The F plasmid has an additional important property: it is able to integrate into the bacterial chromosome at a certain position, thereby becoming part of the circular chromosome. The *E. coli* chromosome consists of about five million base pairs, and right in the middle sits the F plasmid with its one hundred thousand base pairs. Surprisingly, the gene products of the F plasmid are able to handle the 50-times-larger chromosome in the same way as its own DNA. First, the bacterial chromosome is opened within the integrated plasmid. Then the linear single strand of plasmid DNA with bacterial DNA attached travels into the F^- cell. At some point, the genes for the synthesis of threonine and leucine will arrive in the F^- cell.

We now come to a further mechanism, the homologous recombination, which explains the results of Lederberg and Tatum. The healthy genes arriving in the F^- cell are "recognized" by the defective T- and L-genes. They come into contact, and the defective genes are replaced by the healthy ones, just like replacing a defective circuit board with a new one. Let's briefly summarize what was said about transformation and conjugation:

1) A number of microbial species are naturally equipped with a system for uptake of free DNA. Of what use is this to bacteria? Let's imagine a rotting sugar beet, where trillions of bacteria grow at the expense of sugars, pectins, and other substances. When the nutrients are used up, the site becomes a "microbial cemetery." Most of the bacteria undergo cell lysis and release cell debris, proteins, and DNA, but a few living cells are still present. The ability to take up DNA by transformation now proves advantageous: mutations can be repaired and genes encoding useful enzyme systems can be acquired. Thus, transformation is an important mechanism for horizontal gene transfer.

2) Genetic material can be transferred from one microbial cell to the other with the help of conjugative plasmids. Conjugation takes on a particular importance when these plasmids not only encode basic genes but also genes conferring resistance to antibiotics. These resistances can then spread like an avalanche. As a result, the composition of the microbial population in a certain habitat might change dramatically. Microorganisms resistant to a certain antibiotic take over, whereas the sensitive ones disappear. The spread of β-lactamase-containing microorganisms in the world is a sad example of this.

I am impressed, but these mechanisms sound a bit artificial or constructed, as if they were concocted by scientists.

This reluctance is understandable, but bear in mind that many of the pioneers involved in such research – Griffith, Lederberg, Tatum – were awarded the Nobel Prize for their discoveries. In addition, the results of thousands of experiments performed over the years could otherwise not be explained without the existence of transformation and conjugation.

> For God's sake, send some other messenger
>
> William Shakespeare, The Comedy of Errors

Chapter 21
Bacteria can also catch viruses

Every so often, we all catch a virus – actually, it's a viral infection (see Chapter 30). Surprisingly, even bacteria can catch viruses. The viruses that attack bacteria are called bacteriophages, which literally means "bacteria eaters." Bacteriophages were discovered by the British microbiologist Frederick Twort (1877–1950) and the French microbiologist Felix d'Hérelle (1873–1949). They noticed that round, transparent areas developed on agar medium covered with a bacterial lawn, which looks a bit like vanilla pudding. These transparent areas were called plaques (Figure 40). When a bit of the material was taken from one of the plaques and transferred to another bacterial lawn, plaques also developed. This experiment and others showed that, obviously, the material in plaques was infectious. Not until the invention of the electron microscope years later was it possible to actually see the bacteriophages present in plaque material.

Bacteriophages, the most complex viruses known, have specialized in attacking bacteria. A good example is the T4 bacteriophage, which attacks *Escherichia coli*. Judged by its structural appearance, T4 looks like it has been designed and built in a workshop for medical equipment or in the prop department of a science fiction movie. Look at this phage as shown in Figure 41. If a creature like this were two meters or six feet tall, it would be quite frightening. The genetic information of this bacteriophage is contained in a large, linear, double-stranded DNA molecule located in the "head." The spectacular spring-like contractile tail surrounds an internal tube. With the help of the spikes and base plate, the bacteriophage scans the cell surface of *E. coli*, looking for a specific docking protein, much like a space shuttle locating a docking port of the International Space Station, ISS. What happens next is hard to believe: The tail contracts, and the internal tube is pushed through both cell wall and cytoplasmic membrane of the bacterium, much like a hypodermic needle, right into the cytoplasm. The viral DNA in the head of the bacteriophage is then injected into the bacterial cell. The viral DNA now controls all further activities of the cell. Within 60 minutes, the genetic information characteristic for this bacterium is destroyed and the components for around 100 bacteriophages are synthesized. The phage components self-assemble, then comes the final command of the phage DNA, to synthesize lysozyme. This enzyme, which is also present in our tears, hydrolyzes the bacterial cell wall and opens the "doors" to release the bacteriophages. They immediately start looking for other *E. coli* cells

---

Discover the World of Microbes: Bacteria, Archaea, and Viruses, First Edition. Gerhard Gottschalk.
© 2012 Wiley-VCH Verlag GmbH & Co. KGaA. Published 2012 by Wiley-VCH Verlag GmbH & Co. KGaA.

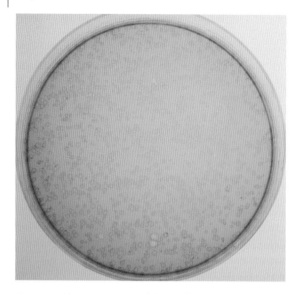

Figure 40 Plaques on an *E. coli* lawn. The transparent spots arise when bacteriophages enter *E. coli* cells, where they proliferate and lyse the cells. (Photograph: Svetlana Ber, Goettingen, Germany)

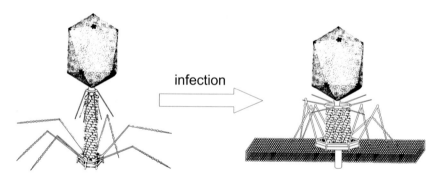

Figure 41 Model of the bacteriophage T_4. Left, the "attack" position. Constituents are: the hexagonal head made of protein and filled with DNA, the collar with six whiskers, the contractile sheath, the baseplate and six wedges. Right, the phage during infection. Contraction of the sheath allows injection of DNA through the internal tube. The size of the phage is about 200 nm. (Drawing: Anne Kemmling, Goettingen, Germany, after F.A. Eiserling. "Structure of the T_4 Vision." In: Ch.K. Mathews, E.M. Kutter, G. Mosig, P.B. Berget, *Bacteriophage T_4*, ASM, Washington, D.C., 1983.)

to attack. The round zones of cell destruction on agar plates, the plaques formed by bacteriophage activities, are transparent because less light is dispersed by tiny viruses and lysed cells than by intact bacteria.

Does a bacteriophage infection always kill bacteria?

The T4 bacteriophage is practically 100 percent lytic, so infection of E. coli with intact T4 DNA always leads to bacterial lysis and death. However, not all bacteriophages work this way. Another well-known E. coli-specific bacteriophage is called lambda. It consists of a head and a noncontractile tail. Its DNA contains only 50 genes, whereas the very complex T4 has 100 genes. Lambda can behave just like T4: infection is followed by viral proliferation, then lysis of the bacterial host cell. In many cases, however, lambda may reside in the bacterial cell as a sort of tenant. Lambda DNA becomes circular after it arrives in the E. coli cell, where it then persists much like the F plasmids mentioned earlier (Chapter 20). Sequence homology enables the small lambda-DNA ring and the large circular bacterial chromosome to form a complex (Figure 42). Next, both rings open and interfuse to form a figure eight. In this way, phage DNA is integrated into the bacterial chromosome. Such integrated bacteriophages are called temperate. Temperate-phage DNA is replicated along with that of the bacterial chromosome, so any bacterial offspring are temperate-phage carriers. Events such as exposure to toxins or deterioration of bacterial living conditions may activate regulatory signals and lead to excision of lambda-phage DNA, separating it from the bacterial chromosome. The lambda phage then becomes lytic like the T4 phage.

The lambda bacteriophage has been and still is an important tool in gene technology. In the early development of gene technology, a special property of this bacteriophage was used to introduce foreign DNA into E. coli cells. During assembly of lambda-phage particles, the head is filled with phage DNA. However, there is no mechanism by which foreign DNA can be excluded by the head-filling

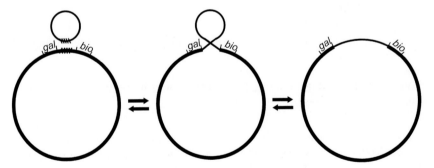

Figure 42 Integration of phage DNA into the E. coli chromosome. The two circles, the E. coli DNA (large) and the phage DNA (small) associate in a region of homology, between the chromosomal regions *gal* and *bio*. Then both rings open up and interfuse to form a figure eight. Unfolding of this structure gives a larger circular DNA molecule carrying the integrated DNA molecule. (Drawing: Anne Kemmling, Goettingen, Germany.)

machinery, which merely requires the presence of two lambda-specific regions. Between these regions, some of the lambda-phage DNA may have been excised and replaced by DNA encoding for nonlambda protein, much like the cuckoo that lays its eggs in the nests of other birds. Such altered phages retain their infectious potential, however. After infecting a bacterial host, they become temperate phages. When the host DNA is replicated along with the "smuggled" DNA, the information is transcribed and translated into proteins. Obviously, it's the end of the line for such phages because some of the information required for viral replication is missing, so there can be no active descendents.

This is another process that would be considered by most people as a delusion of molecular genetics experts.

Perhaps these processes will be more plausible when we discuss restriction enzymes in Chapter 25.

Bacteriophages were major objects of research in the 1940s and 1950s, the early days of molecular genetics. To give an example, Max Delbrueck (1906–1981) and Salvador Luria (1912–1991) (Nobel laureates in 1969) did experiments on the emergence of phage-resistant *E. coli* cells. The results led them to conclude that mutations are not induced but occur spontaneously. In an elegant analysis they showed that phage-resistant mutants arise irrespective of the presence of phages. The phages were simply required in the experiments to demonstrate the presence of such mutants. To give another example: during growth of *E. coli*, a variety of mutants is generated, among them are some resistant against the antibiotic Streptomycin. They appear spontaneously, but for the demonstration of their presence, Streptomycin has to be added to an agar medium. Growth of *E. coli* will be largely inhibited, but a few colonies developing from Streptomycin-resistant mutants will appear.

Only the T4 and lambda bacteriophages have been presented here. In *E. coli*, more than 20 different types of phages are known. When we consider the great variety of bacterial species, it's easy to imagine a world of invisible bacterial viruses as numerous and as versatile as the bacteria. Like higher organisms, bacteria also have developed mechanisms to prevent phage infections. A common way for them to become phage resistant is to modify the cell-wall protein required for phage docking or discontinue its synthesis. The phages are no longer able to locate potential target sites, so they wander aimlessly over the cell surface – sooner or later they will be degraded.

> Instead of casting out the contaminated culture (...),
> I made some investigations
>
> *Alexander Fleming*

Chapter 22
Antibiotics: from microorganisms against microorganisms

Mankind has been afflicted by epidemics for thousands of years, including infectious diseases that have raged throughout the world, such as the Black Death (plague) and cholera. The most lethal plague occurred in Europe between 1347 and 1355, when around 30 million people died, according to estimates by the Vatican. We may also recall that Boccaccio's *The Decameron* describes the amusements of several members of Florentine high society who have fled to a remote area of Tuscany in order to avoid exposure to plague victims and rats. Other infectious diseases, including tuberculosis, tetanus, gas gangrene, and all sorts of diarrhea, have been a real scourge of humans. Before strategies could even be developed to combat these diseases, the causative agents first had to be identified. It was not until 1883 that Robert Koch (1843–1910) showed that a bacterium, now called *Bacillus anthracis*, was responsible for anthrax in cattle. Based on his experimental studies on anthrax, Koch formulated his famous postulates, sometimes also referred to as Henle-Koch postulates: The pathogen must be present in all animals suffering from the disease, but absent in healthy animals; it must be possible to isolate the pathogen in pure culture; upon injection into healthy animals, it must induce the disease.

Independent of these findings, the vaccination against smallpox had already been introduced years earlier by Edward Jenner (1749–1823). Hygienic measures in hospitals and wound dressings to avoid exposure to the air were established by Max von Pettenkofer (1818–1901) and Lord Joseph Lister (1827–1912). The first pioneer to focus on combating infectious bacteria was Emil von Behring (1854–1917). Together with Shibasaburo Kitasato (1853–1931), he published the epoch-making article on the development of immunity against diphtheria and tetanus in animals. This made it possible to treat humans who were suffering from diphtheria or tetanus by using serum from infected animals. Von Behring and Kitasato carried out their decisive experiments while they were working as assistants in Robert Koch's laboratory in Berlin. They needed an animal for serum production, so Robert Koch allowed them to use an old castrated ram that otherwise would have been slaughtered to save feed costs. This ram made history as the first animal to serve as a supplier of serum for immunotherapy. Emil von Behring thus became the founder of immunotherapy, and in 1901 he was awarded the first Nobel Prize in Medicine for his pioneering discoveries.

Discover the World of Microbes: Bacteria, Archaea, and Viruses, First Edition. Gerhard Gottschalk.
© 2012 Wiley-VCH Verlag GmbH & Co. KGaA. Published 2012 by Wiley-VCH Verlag GmbH & Co. KGaA.

Chapter 22 Antibiotics: from microorganisms against microorganisms

So far this has nothing to do with the title of the chapter.

You're right, but let's first look at the development of antimicrobials. The next era was dominated by the concept of a chemotherapy using antibacterial chemical agents. It all began with salvarsan, a compound containing arsenic that proved very effective in combating syphilis. It was developed by Paul Ehrlich (1854–1915) and Sahashiro Hata (1873–1938), coming onto the market in 1910.

Salvarsan is a red compound, and it was Paul Ehrlich's notion that certain dyes capable of staining bacterial cells might also be effective as chemotherapeutics. At the time, chemotherapy was essentially a treatment with dyes. The discovery of Prontosil, the first chemotherapeutic on the market, is along the same line. The name of Gerhard Domagk (1895–1964) is associated with its discovery. He was awarded the Nobel Prize in 1939 but was not allowed to accept the prize until after the end of World War II. The first promising compound Domagk had in his hands was a compound labeled D4145, with which he reportedly treated his daughter after she developed a severe *Streptococcus* infection in her arm. The treatment was successful. Prontosil (below) was derived from this D4145, but had a higher solubility.

$$H_2N-\langle\bigcirc\rangle-N=N-\langle\bigcirc\rangle-SO_2NH_2$$
$$\qquad\quad\;\;\;|$$
$$\qquad\quad\;\;NH_2$$

For a number of years this antibacterial drug was used worldwide. Prontosil was even awarded the "Grand Prix" at the Paris World Exposition in 1937. Later, Jacques Trefouël (1897–1977), Federico Nitti (1903–1947), and Daniel Bovet (1907–1992) from the Pasteur Institute in Paris discovered that the active compound was not actually Prontosil but sulfanilamide (para-amino benzenesulfonamide). This compound is formed from Prontosil inside the body by reductive cleavage of the azo bond (see also Chapter 20). Thus, sulfanilamide became the first real antibiotic and the mother compound of the sulfonamides. When Prontosil and related compounds finally reached the market, word of their effectiveness had already spread and efforts to isolate the active compound were already underway.

There are many things for visitors to see in London, but for microbiologists or people interested in the history of science the one place to go is the Fleming Museum at St. Mary's Hospital. There the laboratory of Alexander Fleming (1881–1955) has been restored and you can even see where penicillin was discovered in 1928. One day Fleming had returned to the lab and had seen some Petri dishes in which colonies of *Staphylococci* were growing. An unwelcome colony of fungus had grown close to the edge of one of the Petri dishes and had apparently inhibited bacterial growth. In the immediate vicinity of the fungus there was no growth at all; somewhat further away, small bacterial colonies could be seen. Even further away, bacterial growth was normal (Figure 43). Fleming identified the fungus as a *Penicillium* mold and coined the name "penicillin" for the inhibitory agent. Not being a chemist, he was unable to isolate and characterize the compound respon-

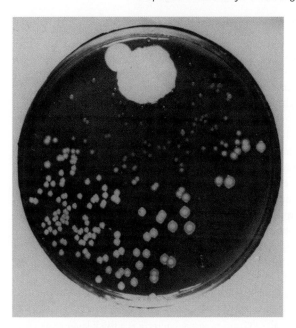

Figure 43 Alexander Fleming's own photograph of a penicillin mold. The staphylococci around the mold were obviously destroyed. (Courtesy of the Alexander Fleming Museum, London, UK.)

sible for this inhibitory effect. However, the biochemists Howard W. Florey (1898–1968) and Ernst B. Chain (1906–1979) followed up on his discovery. Florey was from Adelaide (Australia) and Chain, from Berlin (Germany), where he had graduated with a degree in chemistry in 1930. Both headed a team of British scientists to isolate penicillin. They had it in their hands by 1940 and were able to demonstrate its inhibitory effect on bacteria in higher organisms. The fundamental importance of penicillin had thus been recognized. An enormous research and development program was initiated in the United Kingdom and the United States, with the result that penicillin was already available in 1944 as the first antibiotic for treatment of wounded Allied soldiers. This began a new era in the continuing battle against microorganisms, an era that continues at present and will continue into the future.

Who coined the word "antibiotic" and how do antibiotics work?

The word antibiotic was coined by Selman Abraham Waksman (1888–1973). He was born in the Ukraine and emigrated to the United States in 1910. He began working at the Department of Microbiology of Rutgers University in 1930. His special interests were soil microorganisms and the products they secrete. His major discovery was the antibiotic streptomycin, which – together with Albert Schatz (1920–2005) – he isolated from bacteria called *Streptomyces griseus*. With this discovery, the filamentous microorganisms of the genus *Streptomyces* came into focus as an important group of antibiotic producers. Waksman defined an antibiotic as a compound that is produced by microorganisms and capable of inhibiting or killing other microorganisms, even in low concentrations.

To answer the second part of the question – how antibiotics work – we'll need to digress a bit. Let's start with penicillin, which is said to be effective against Gram-positive bacteria. The staphylococci present on Fleming's original Petri dish were indeed Gram-positive. Obviously, there also have to be Gram-negative microorganisms as well. Going back to Chapter 2, we saw that a bacterial cell is surrounded by a cytoplasmic membrane that is not the outermost structure of bacteria. In turn, this is surrounded by the cell wall. The functions of these structures may be compared with an inner tube and a tire of a bicycle, or with the corresponding parts of a football. The cytoplasmic membrane is a tight barrier that keeps all the nutrients, coenzymes, and metabolites inside the cell. However, since the concentration of all these compounds is generally much higher inside the cell than outside, a considerable osmotic pressure is exerted against the membrane, which would burst like an inner tube without a tire if it weren't surrounded by a cell wall. The latter gives rigidity and shape to the bacterial cells. For instance, bacteria can be round like a ball (*Staphylococcus aureus*) or shaped like a sausage (*Escherichia coli, Bacillus anthracis*). The cell wall of Gram-positive bacteria is a thick layer consisting of murein, a huge macromolecule. Of course, bacteria have somewhat of a problem during growth. As the cell increases in size, the cell wall needs to expand as well. There is therefore a mechanism by which some of the murein bonds are cleaved, then new murein components are inserted and new bonds are formed. It's like a pair of trousers that have became too tight and have to be altered by inserting new material. This is a very critical phase for microbial cells during which they are vulnerable for a given time. This is when an antibiotic such as penicillin can strike, inhibiting reconstitution of the bonds that have been cleaved to extend the murein layer. As a result, the cells burst. These facts reveal a basic disadvantage of penicillin: it only works on growing cells, but not on resting cells. The latter simply have no "weak spots" where penicillin can go into action.

How are Gram-negative bacteria different?

The Gram-negative bacteria are different because they have a relatively thin murein layer, which in turn is surrounded by a second layer called the outer membrane. This membrane is very important for Gram-negatives because it can keep toxic compounds, for example, out of the cell. Many compounds are unable to penetrate this outer membrane and interact with cell components inside the membrane. Penicillin is kept outside by the outer membrane, so it is unable to reach its target and thereby exert its antibiotic effect in Gram-negatives.

The differentiation between Gram-positive and Gram-negative bacteria goes back to the Danish bacteriologist Christian Gram (1853–1938), who worked with dyes that could stain bacteria. He developed a dye combination that could firmly stain Gram-positive bacteria and not be washed out. These dyes could also stain Gram-negatives but the stain was not retained by the cells and could easily be washed out. Gram, of course, could not explain the observed differences in staining behavior. An explanation had to wait until the composition of bacterial cell walls had been unraveled.

Which antibiotics are effective against Gram-negative bacteria? Does this mean that antibiotics of the penicillin type have no effect at all on Gram-negative bacteria?

It would be fatal if this were true. Penicillin is the prototype of the so-called beta-lactam antibiotics. When we look at the formula of penicillin G in Figure 44, we see a four-sided ring in the center of the molecule, the characteristic beta-lactam ring. If this ring is hydrolyzed and thereby opened, penicillin loses its antibiotic activity. This is exactly the role of beta-lactamase, which was discussed in Chapter 20. There is also another site where the structure of penicillin G can be altered. Researchers have learned to gently remove part of the molecule, shown in red in Figure 44. This is done with an enzyme that has no effect on the very sensitive beta-lactam ring. The resulting compound is called 6-amino-penicillanic acid. Chemists can then replace, at their pleasure, the red residue with others, for example, the ones shown at the bottom of Figure 44. The resulting products are the antibiotics ampicillin and methicillin. In this way, a great variety of second-generation antibiotics, so to speak, could be produced. Some of these, including ampicillin and methicillin, also work against Gram-negative bacteria, indicating that they are able to penetrate the outer membrane and reach their action sites.

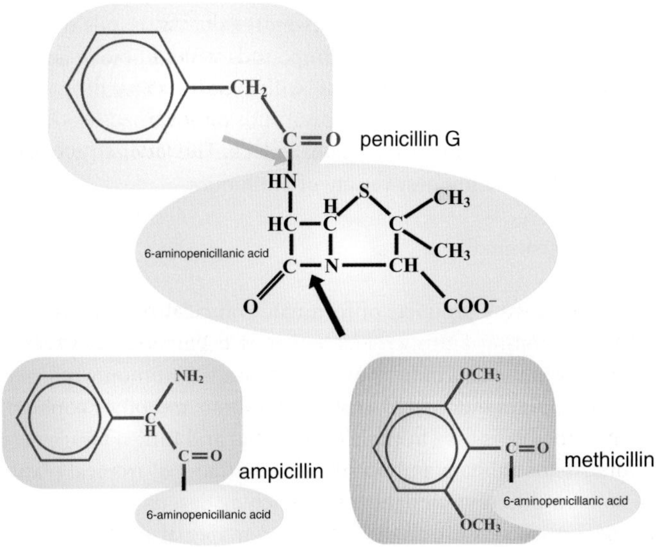

Figure 44 Structure of penicillin G. Enzymatic cleavage of penicillin G (green arrow) yields 6-aminopenicillanic acid, which can be linked to residues to give ampicillin or methicillin. Hydrolysis of penicillin by (beta-) β-lactamase (black arrow) leads to products with no antibiotic activity. (Diagram: Anne Kemmling, Goettingen, Germany.)

Penicillin and its derivatives are not the only beta-lactam-antibiotics. Another important group is called the cephalosporins, which are produced by fungi of the genus *Cephalosporium*. Many of these antibiotics affect Gram-positive as well as Gram-negative bacteria.

Are antibiotics also harmful to humans?

A good antibiotic should be nontoxic and may not interfere with our metabolic functions. The best targets for antibiotics are certain functions or structures found only in bacteria but not in humans. This is the great advantage of the beta-lactam antibiotics: their target, murein, is simply not found in higher organisms. As long as a person is not allergic to penicillin, he may even be treated with a daily dosage of this compound in gram amounts.

Do all antibiotics interfere with cell wall synthesis?

Antibiotics can also interfere with other bacterial functions or growth processes. For example, the bacterial protein factories are very important targets. The ribosomes of bacteria are lighter than those of higher organisms because the proteins that make up the bacterial ribosomes are different. There is a large group of antibiotics that interfere with bacterial protein synthesis, including streptomycin and the tetracyclines gentamycin, erythromycin, and clindamycin, to name just a few. Many of these antibiotics are produced by strains of *Streptomyces*, the filamentous soil microorganisms already studied by Selman Waksman. Thousands of *Streptomyces* species have been isolated in academic and industrial laboratories and analyzed for products having an antibiotic effect. Hundreds of thousands of compounds have been isolated, and around 18 000 compounds with antibiotic activity have been tested in detail. Several hundred of these, with a market value of approximately 60 billion US dollars per year, have been put into production. The major producers of antibiotics are the fungi *Penicillium* and *Cephalosporium*, whereas *Streptomyces* bacteria produce the greatest variety of antibiotics.

How are new antibiotics discovered?

Whereas the first antibiotics were products of fortunate coincidence, a systematic search for compounds exhibiting antimicrobial activities began soon thereafter. The procedure used in academic and industrial laboratories is commonly referred to as screening, which employs automated laboratory robots and other sophisticated equipment. The principle is easy to explain. Bacterial and fungal species are first isolated from all sorts of habitats, for example, the surfaces of tropical plants or various types of soils in different climate zones, such as deserts, podsols, and permafrost. Thousands of isolates are then cultured in a special nutrient broth. After growth, this broth is analyzed with agar diffusion techniques to recognize the presence of antibacterial compounds. A Petri dish with agar medium is first inoculated with a test microorganism. Little holes are punched in the agar, then

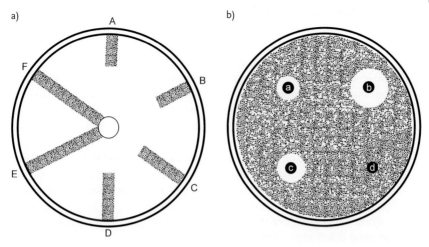

Figure 45 Antibiotic susceptibility tests. (a) Different strains are streaked on an agar plate, then the antibiotic to be tested is placed in the central hole. Strains E and F are insensitive, strains A and B are most sensitive. (b) Disks impregnated with different antibiotics are placed on an agar plate seeded with a certain bacterial species and incubated overnight. Result: the antibiotic on disk b exhibits the highest inhibitory effect. There is no inhibition by the antibiotic on disk d. (Drawing: Anne Kemmling, Goettingen, Germany.)

filled with solutions of the substances to be tested. The test organism grows on the plate to form a dense microbial lawn. If an inhibitory principle has diffused from the holes into the agar, growth will be prevented. The size of the halo formed is a measure of the concentration or the effectiveness of the compound present (Figure 45). Likewise, the specificity of a certain compound against various microbial species can be tested as shown in Figure 45b. Various bacterial cultures are streaked out onto an agar plate with a central hole containing the compound solution to be tested. Strains unaffected by the compound will grow all the way to the hole, whereas other strains may only be able to grow at some distance from the center hole.

These are just simple tests that have been modified and automated over the years to allow the assay of thousands of microbial products. This type of screening has been carried out for decades in many laboratories, so it is quite a rare occasion when a completely new compound or a new principle is discovered by which microorganisms are affected. Let's ask one of the experts, David Hopwood from the John Innes Center in Norwich (UK), what we might expect in the future from *Streptomyces* and other organisms as antibiotic producers:

> "The procedure of isolating microorganisms from natural habitats and screening them for antibiotic production was very successful in the 1950s and 1960s and many important antibiotics were found by this approach. These included tetracycline and erythromycin as antibacterials and

candicidin and amphotericin as antifungals. In retrospect this period of antibiotic discovery came to be called the "Golden Age" because it was relatively easy to screen thousands of streptomycetes and have a reasonable chance of finding a useful antibiotic that had not been seen before. However, after the Golden Age the task became much more difficult as the same antibiotics kept being discovered. By the early 1990s nearly all the pharmaceutical companies concluded that all the good compounds had been found and gave up the search, switching instead to screening large collections of synthetic chemicals made by robotic synthesis – combinatorial chemistry. This approach was a failure: the compounds had not evolved over millions of years to interact with living cells, like the natural antibiotics, and did not show antibiotic activity or, if they did, they were highly toxic to the mammalian host. Luckily along came genomics at this opportune moment. When the genomes of the first two streptomycetes to be sequenced were studied it was apparent that each of them had the innate capacity to make many potentially interesting compounds that had not been recognized when the bacteria were grown under standard screening conditions. The gene sets for these compounds were "sleeping" under standard laboratory conditions but presumably could be "woken up" under special conditions in the natural habitat when the compounds would be adaptive to the producing organism. The challenge now is to find general methods to wake up such sleeping gene sets and so make available a whole gamut of potentially interesting natural compounds that have not previously been tested. They are likely to include compounds with different modes of action as antibiotics compared with the compounds already discovered, which are made in relatively large quantities under a whole variety of growth conditions. Thus there is a good chance that they will overcome the resistance that has developed amongst pathogens to the antibiotics that have already been used widely."

The approaches discussed by David Hopwood will be of growing importance in the future. Clearly, mankind has an ongoing need for new substances with antibiotic activity due to the increasing spread of resistances, which will be discussed in the next chapter.

Why do microorganisms produce antibiotics?

First of all, it must be pointed out that not all microorganisms produce antibiotics. Major producers have already been mentioned, the filamentous fungi of the genera *Penicillium* and *Cephalosporium* as well as the bacteria of the genus *Streptomyces* and a number of representatives of the genus *Bacillus*. These organisms all undergo morphological changes related to the formation of resting forms like spores, during which the vegetative cells are broken down and the resulting nutrients are reused to make the spores. This makes them vulnerable at this phase to competing organisms that may hijack the nutrients, so it is reasonable that antimicrobial compounds are formed and secreted while such differentiation processes are taking place. These processes are able to proceed under the protection of a cloud of such compounds. This, of course, requires that the organisms secreting a certain antibiotic are also resistant to it.

The discovery of antibiotics had a great impact on the pharmaceutical industry. Penicillin production by British and American companies began in 1942. Streptomycin, discovered by Selman Waksman and Albert Schatz as a product of *Streptomyces griseus*, was first manufactured by Merck & Co. in 1946 and was on the market in 1947. It was the first antibiotic used to combat tuberculosis. Later it was replaced by other drugs because some patients suffered from side effects such as ototoxicity causing deafness. It is no exaggeration to conclude that penicillin and streptomycin launched a whole branch of the pharmaceutical industry that is now of global importance.

> This is the day appointed for the combat
> and ready are the appellant and the defendant
>
> William Shakespeare, Henry VI

Chapter 23
Plasmids and resistances

One type of plasmid has already been introduced, the F-plasmid. Joshua Lederberg (1925–2008) and others noticed in their work on bacterial conjugation that the information for this process was not localized on the bacterial chromosome but on the so-called F-factor, which was then designated "plasmid." In 1961 it was shown by Stanley Falkow and colleagues (Stanford, USA) that plasmids are small, circular DNA molecules that are present in bacterial cells in addition to the chromosome. These plasmids can be made visible as shown in Figure 46.

Why did nature come up with the idea of storing genetic information outside of the chromosomes?

Let's first look briefly at an analogy with personal computers or laptops on one side and CD-ROMs on the other. CD-ROMs have the advantage of storing information without taking up space on the hard drive. As long as this information is not needed, it can be stored as CDs on a shelf. If the situation changes, the information or data are quickly available. The situation with plasmids is similar. Happily growing bacteria usually do not need them. However, when a stress situation comes up, cells that more or less by chance still contain a useful plasmid are in demand. The whole arsenal of DNA uptake (conjugation or transformation) mechanisms is then activated because the availability of plasmid-encoded genetic information may decide over the life or death of these bacteria, over growing happily or wasting away.

This sounds a bit theoretical. What kind of stress situations do you have in mind?

Let's take a situation in which bacteria can only survive, for instance, by becoming resistant to heavy metals. The principle was described by Hans-Guenter Schlegel in Chapter 19. In this case, those plasmids conferring resistance are armed with information required to synthesize efflux pumps and to bring them into position. These pumps are rapidly synthesized and installed into the cytoplasmic membrane so that the intercellular heavy-metal concentration can be kept low. We learned about another resistance mechanism in Chapter 20. Bacteria produce plasmid-encoded beta-lactamases in order to cleave and thereby inactivate beta-lactam

---

Discover the World of Microbes: Bacteria, Archaea, and Viruses, First Edition. Gerhard Gottschalk.
© 2012 Wiley-VCH Verlag GmbH & Co. KGaA. Published 2012 by Wiley-VCH Verlag GmbH & Co. KGaA.

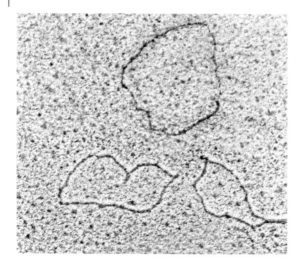

Figure 46 Electron micrograph of a plasmid (pBR322). The diameter of the upper ring is approximately 0.25 µm. (Photograph: Michael Hoppert, Goettingen, Germany.)

antibiotics. It was also discussed how beta-lactamases altered by mutation and selection are able to inactivate new types of beta-lactam antibiotics that have come onto the market to treat infections. There are also additional mechanisms conferring resistance to certain antibiotics, which in some cases are inactivated by phosphorylation or acetylation as soon as they arrive inside the cells. The protein synthesis factories, the ribosomes, are the targets of a number of antibiotics, including streptomycin. Bacteria have gained resistance to these antibiotics by modifying the target. The protein that normally interacts with streptomycin, for example, is so modified that the antibiotic is no longer able to bind, resulting in the development of resistance.

These are just a few examples of how antibiotic resistances develop. There has been an additional fatal development. Not only is there a permanent race between the application of new antibiotics and the appearance of new resistances, but also bacteria are able to collect resistances on their plasmids, like trophies. Due to the rapid propagation of bacteria and the conjugative events resulting in plasmid transfer, the accumulated resistances may spread like an avalanche. Therefore, bacteria containing such plasmids can be isolated everywhere. Just one example: the research group of Michael Teuber from the Swiss Federal Institute of Technology/ETH Zurich (Switzerland) isolated a strain of *Enterococcus faecalis* from raw sausage with plasmid-encoded resistances against tetracycline, lincomycine, chloramphenicol, and erythromycin. Harmless microbial populations that are normally common in man and animals are transformed more and more into enemy legions that no longer can be controlled effectively. One example, the plasmid PSMS-130 of a pathogenic strain of *Escherichia. coli*, is depicted in Figure 47. The resistance genes against eight antibiotics are united on this plasmid. In

Chapter 23 Plasmids and resistances

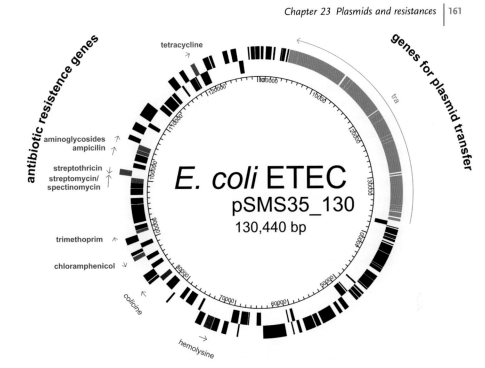

Figure 47 A plasmid of enterotoxinogenic *E. coli* strain, an ETEC strain carrying various resistance genes. The double ring depicted consists of 130 440 base pairs. The tra region (in green) contains the genes for plasmid transfer by conjugation. Genes in red confer resistance against eight different antibiotics. Also present are genes encoding a toxin (colicin) and a virulence factor (hemolysin). (Diagram: Elzbieta Brzuszkiewicz, Goettingen, Germany.)

addition, genes for a toxin (colicine) and a hemolysin (destroys erythrocytes) are present. What a potential threat is present on just this one type of resistance plasmid!

What about MRSA? There was something about it in the newspaper yesterday.

MRSA is the abbreviation for methicillin-resistant *Staphylococcus aureus*. Since methicillin is no longer commonly used as an antibiotic, MRSA now also refers to multiresistant *Staphylococcus aureus*. The danger associated with MRSA is twofold: First, the mechanism of resistance rests on the fact that a certain protein on the bacterial cell surface is so modified that antibiotics such as penicillin, methicillin, or oxacillin are no longer bound. As a result, cell-wall synthesis cannot be prevented. Secondly, the information for this modified binding protein is not encoded on a plasmid but on the bacterial chromosome and is therefore stable. This makes MRSA a real threat inside and outside of hospitals. Thousands of people die each year because such infections cannot be treated effectively with antibiotics.

Is there anything good about plasmids?

Indeed, a number of plasmids are very useful. The genetic information for the degradation of compounds that are only available from time to time is encoded on plasmids. As examples, let's take the degradation of naphthalene, toluene, salicylic acid, or the components of crude oil. These compounds are not found everywhere in nature, so it is unnecessary for bacteria to always have the genes for degradation enzymes available. These genes are outsourced to plasmids. Among the huge populations of bacteria in soil and in bodies of water, there are always a few containing plasmids with the genetic information required in case these populations are confronted with any of the above-mentioned compounds. Tanker and oil-platform accidents are negative examples of conditions under which bacteria carrying plasmids with genetic information for degradation of oil components are real champions. But we shouldn't forget the examples of nickel-resistant bacteria in New Caledonia (Chapter 19) and MRSA. If the challenge for bacteria is an ongoing situation, the genetic defense machinery will shift from plasmids to the chromosome. The fact that bacteria have the necessary tools for such shifts makes it even more difficult to combat them during infections.

I am trying to imagine how life on Earth would be without plasmids. Of course, horizontal gene transfer would be less efficient. And what about gene transfer into certain plants? At least the spread of resistances would be greatly reduced.

Speculations of this kind are not very useful. You should view the plasmids from the standpoint of the so-called selfishness of DNA. In his famous book, *The Selfish Gene*, Richard Dawkins writes about the existence of numerous rebellious DNA fragments. We only notice them if they contain information for self-replication and if they are in an environment that provides the machinery and resources for replication. Such an environment is the microbial cell. We should consider the microbe–plasmid systems from a somewhat different point of view. Plasmids keep bacteria in order to survive and to propagate. Those plasmids encoding resistances have a better chance of survival because their bacterial hosts are not destroyed so easily.

But where did resistance genes against antibiotics come from?

We have asked Julian Davies (Vancouver, Canada):

> "This is a complex question. One of the most striking examples of the influence of human activity on the biosphere is the development of antibiotic-resistant bacteria following the introduction of therapeutic antibiotics in 1941. Millions of metric tons of potent antimicrobials have been produced and released into the environment in the last 60 years. Plasmids carrying antibiotic resistance genes were first reported in Japan in the mid-1950s and were greeted with skepticism by scientists and doctors in the

west. How could multidrug resistance possibly be infectious? Plasmid-encoded transmissible drug resistance rapidly became a major worldwide threat to infectious disease treatment and the pharmaceutical industry has been unable to keep up with new resistance mechanisms. The origins of plasmids and their resistance genes pose a question of great evolutionary importance: it is generally assumed that they are environmental in origin; in fact a bacterial penicillinase was identified in 1940, before penicillin was in clinical use! In 1974 it was shown that soil actinomycetes possess enzymes that inactivate the antibiotics that they produce. It is assumed that these are for "self-protection" of the producer, but there is no convincing proof of this supposition. Antibiotic-producing organisms have biochemical mechanisms of drug inactivation, export systems and target inactivation protection mechanisms. These are related in mechanism to those found in pathogenic organisms in the clinic. Do all resistance mechanisms arise from producing strains in the environment? Recent metagenomic studies have shown that antibiotic resistance is widespread in nature; this has been termed the "resistome". Interestingly, resistance genes have been detected in bacterial strains that do not produce antibiotics. For example, resistance to vancomycin is widespread. Thus, resistance and production do not necessarily go together. The human gut microbiome is also a reservoir of many resistance genes but their relationship with disease bacteria is not known. Are the resistance genes in hospitals the same as those found in soils and other environments that have not been exposed to the human use of antibiotics? They work in the same way but the genes differ in their DNA sequences and gene regulation signals. The connection between the resistome and the clinic is not yet complete! It would require significant genetic "tailoring" to convert an environmental streptomycin phosphotransferase gene to that found in a *Shigella* or *Staphylococcus* pathogen.

The ecology of antibiotics and their resistance genes is poorly understood, especially since it is difficult to detect the presence of antibiotics in native soils and water sources. A better understanding of the roles that these wonderful small molecules play in nature might provide information on how resistance develops and perhaps lead to better ways of finding much-needed new antibiotics!"

So, in all our discussions on resistance, it has to be taken into account that the production of antibiotics by microorganisms is intrinsically tied to resistance of the producer. From there, resistance will spread, and the genetic "tailoring" mentioned may then lead to the complex picture drawn by Julian Davies. But we must go on with the isolation and design of new antibiotics.

> In nature's infinite book of secrecy
> a little I can read
>
> William Shakespeare, Antony and Cleopatra

Chapter 24
Agrobacterium tumefaciens, a genetic engineer *par excellence*

Actually, the official name of this bacterium is *Rhizobium radiobacter*, but let's stick to the more common name. You could say that *A. tumefaciens* causes plant cancer. Whenever these bacteria manage to enter the roots of plants through lesions or wounds, they initiate development of the crown gall disease. Galls consisting of tumorous tissue develop when *A. tumefaciens* stimulates formation of the phytohormones auxin and cytokinin, which trigger growth and proliferation of the tumors. Incidentally, the American phytopathologists Erwin Smith (1854–1927) and Charles Townsend identified *A. tumefaciens* as the microorganism causing the crown gall disease. Something very exciting goes on inside these galls. The plant suddenly begins to produce substances called opines, unusual derivatives of amino acids. Nopaline, for example, consists of the amino acid arginine and 2-oxoglutarate, a precursor of the amino acid glutamate. The latter plays a very important role in the intermediary metabolism of practically all organisms. The interesting thing about these opines is that *A. tumefaciens,* but not the plant itself, can use them as nutrients. *A. tumefaciens* seems to have human-like traits. It intrudes into someone else's territory, that of the plant; makes itself at home in the gall, and forces the host to serve daily meals in the form of opines.

Needless to say, researchers were intrigued by this phenomenon. They studied various strains of *A. tumefaciens* and found that some of them had lost the ability to infect plants. The Belgian geneticists Jozef Schell (1935–2003) and Marc van Montagu (Ghent, Belgium) recognized the underlying mechanisms that were involved. They thus became the pioneers of green gene technology. In 1975, together with their colleagues, they published a paper in *Nature* entitled "Acquisition of tumor-inducing ability by non-oncogenic agrobacteria as a result of plasmid-transfer." The researchers had been able to recognize the difference between noninfectious and infectious strains of *A. tumefaciens*: only the infectious strains contained a large plasmid, the Ti plasmid. This plasmid is able to do something that researchers would love to do themselves. It contains a number of genes in a so-called T-DNA region, which can be excised from the Ti plasmid and transferred into the plant cell, somehow managing to overcome all sorts of barriers to reach the plant nucleus and become integrated into a plant chromosome (Figure 48). The T region can be compared to a freight or goods train with locomotives at both ends and freight cars/wagons in the middle. In the case of the Ti plasmid, the

Discover the World of Microbes: Bacteria, Archaea, and Viruses, First Edition. Gerhard Gottschalk.
© 2012 Wiley-VCH Verlag GmbH & Co. KGaA. Published 2012 by Wiley-VCH Verlag GmbH & Co. KGaA.

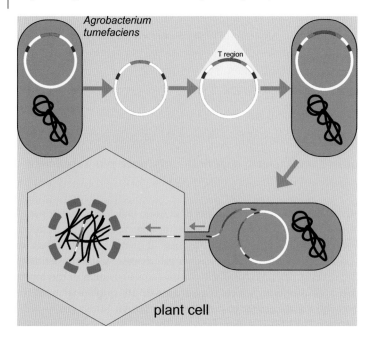

Figure 48 Gene transfer by means of the Ti-plasmid of *Agrobacterium tumefaciens*. Not the whole plasmid but the T-region is transferred into the nucleus. The signals for transfer of the T-region are located in the areas marked in black. Foreign genes (in red) are inserted into the green area. These genes arrive in the nucleus, where they are inserted into one of the chromosomes. The steps depicted are: isolation of the Ti-plasmid, insertion of genes to be transferred, transformation of *A. tumefaciens*, and transfer of the T-region. (Drawing: Anne Kemmling, Goettingen, Germany.)

freight consists of genes for synthesis of the phytohormones auxin and cytokinin as well as genes for synthesis of opine nutrients for *A. tumefaciens*. Knowing these facts, the alert reader will immediately recognize the basic concept behind the experimental approach in green gene technology. Between the two locomotives, the cars loaded with freight simply have been replaced by cars loaded with genes for synthesis of desired products. For example, the freight train delivers genes required for synthesis of a certain protein, which will be synthesized by the plant under appropriate conditions.

The first successful experiments of this kind were carried out in 1983 at the University of Ghent (Belgium) and at the Max Planck Institute for Plant Breeding Research (Cologne, Germany). The genetic information conferring resistance to the antibiotic chloramphenicol was successfully transferred into plants. This was the key experiment of green gene technology, which in the past decades has gone through a breathtaking development and has become the most important approach for development of crop plants in agriculture. Essentially we owe this to the bacterium *Agrobacterium tumefaciens* and, of course, to the ingenious researchers.

What is the latest state-of-the-art?

Let's begin with Bt corn as an example, chosen because it involves still another bacterium, *Bacillus thuringensis* (Bt). It owes its name to the microbiologist Ernst Berliner (1880-1957), who reisolated and described it in 1911. He came across *B. thuringensis* while he was studying the cause of a disease of butterfly larvae, caterpillars. During spore formation (see also Chapter 15), *B. thuringensis* produces in its interior a toxin crystal. When this toxin enters the gut of insects, the intestinal wall is paralyzed and becomes porous. Gut content, including bacteria, is released into the body cavity and the insect finally dies. Higher organisms are insensitive to Bt toxin, primarily because of differences in the pH value of the intestinal tract and the composition of the intestinal wall.

The Bt toxin can be isolated easily. It has already been in use for years in insect traps. However, this is not the application we want to discuss here. The sequence of the toxin gene has been known for quite some time, so Ti-plasmid technology made it relatively easy to insert this gene into plant cells, where it was then expressed. The resulting plant, called Bt corn, is protected in this way from the corn borer, which otherwise causes severe damage resulting in crop losses of nearly 30%. Cotton can also be protected in a similar way.

Worldwide, Bt corn and Bt cotton are grown on an area of around 100 million hectares (nearly 250 million acres), which amounts to approximately 8% of the total agricultural area. These Bt crops represent the technology of the future. It should be pointed out, however, that the introduction of such crops has to be accompanied by risk assessment including not only human health aspects but also any possible effects on plants, insects, and other animals.

Two new developments originating in the "green workshop" should be mentioned. These demonstrate the enormous potential of directed modification of plants. The first example is commonly called the gene potato. Potato starch consists of two components, amylose and amylopectin, both polymers of glucose. In amylopectin, the glucose chains are highly branched and thereby responsible for the adhesive properties of starch. It was obviously tempting to develop a potato that produces only amylopectin but no amylose. Such a potato would be a source of starch with better adhesive properties. This type of gene potato is now available. We now come to the second example, golden rice. The use of this rice would help prevent vitamin A deficiencies in certain parts of the world. Golden rice produces beta-carotene (provitamin A) on the basis of genes from daffodils and from the soil bacterium *Erwinia uredovora*. The genes were introduced into rice plants by *A. tumefaciens*, which seems to have become the workhorse for genetic engineering of plants.

At present, green gene technology is in the process of revolutionizing agriculture, just like the introduction of synthetic fertilizers 150 years ago.

> Stand and unfold yourself
>
> William Shakespeare, Hamlet

Chapter 25
Eco R1 and PCR – molecular biology at its finest

More of those abbreviations! What do they mean this time? The first one sounds like a satellite, and what in the world is a PCR?

Let's start with *Eco*R1. In Chapter 21, we saw how the lambda phage was integrated into the chromosome of *E. coli*: "... both rings open and interfuse to form a figure eight." Of course, the controlled opening of a DNA ring is no easy task. The term "integrase" has been coined to describe the highly complex system of enzymes required for this process. This is only one of the ways a circular chromosome can be opened. The second, very commonly used way to cut DNA is to use enzymes called restriction endonucleases. They were discovered in the 1960s by Werner Arber, who in 1978 was awarded the Nobel Prize together with Daniel Nathans and Hamilton Smith for this groundbreaking discovery.

*Eco*R1, a type of enzyme that is used for cutting DNA, is present in *E. coli*. All these enzymes have a name indicating the organism of origin: *Bam*H1 does the cutting job in *Bacillus amyloliquefaciens* and *Sma*1, in *Serratia marcescens*. Because of the importance of restriction enzymes in molecular biology and gene technology, researchers have looked for even more of them. So far, several hundred have been described. Here we will concentrate on *Eco*R1, where *Eco* stands for *E. coli*, R for a resistance plasmid present in the strain (see Chapter 23), and 1 for the first enzyme of this type from this strain. We have asked Werner Arber (Basel, Switzerland) to describe how restriction enzymes were first discovered:

> "The host range of bacteriophages is restricted for a number of reasons. One reason is host-controlled modification. This phenomenon, first described in the 1950s, is known today as microbial restriction modification. Earlier work on bacteriophages revealed that there was a strong restriction of phage reproduction when the phage was transferred from its original host to a closely related bacterial strain. However, it became apparent that the few descendants of these phages had adapted to the new host but were strongly restricted in reproduction when transferred back to the original host. This restriction and adaptation to a new host was observed in several consecutive back-and-forth transfers of phages between two hosts. Therefore, it was unlikely that genetic variations were the reason for

the observed modification because hereditary factors would have been more stable.

We came across this known but unexplained phenomenon in 1960 when we studied radiation damage to microorganisms. This biosafety research was done within programs to promote the peaceful use of atomic energy. In our project, we studied the effect of irradiation (ultraviolet light, X-rays, and gamma-rays) on bacteria and bacteriophages. Soon it became clear that the DNA of irradiated bacteriophages is degraded to acid-soluble fragments after the phages have infected their normal bacterial host. Interestingly enough, we could demonstrate that a relatively rapid degradation of phage DNA to similar fragments also occurs when phages are transferred to a closely related bacterial host. Therefore, the reason for the modification of microbial restriction, known since the 1950s, apparently was DNA degradation. As a result, we focused our research on bacterial restriction modification systems.

When we started to concentrate on these systems, we already had bacterial strains that accepted phages grown previously on other bacterial strains. By comparing these strains with the original ones that were restrictive towards phages, experimental evidence was obtained that host-controlled modification of phage DNA must involve an epigenetic modification of the DNA. We could then demonstrate that this modification was a sequence-specific methylation of nucleotides (mostly adenine or cytosine). These results in the mid-1960s led to a better understanding of the phenomenon of bacterial restriction modification systems. The modification has to involve an enzymatic, host-encoded sequence-specific DNA methylation, and restriction has to be attributed to an endonuclease that is able to differentiate between the methylation pattern of the host bacterium and the absence of this pattern on restricted phage DNA. In the latter case, restriction enzymes would initiate DNA degradation. This would explain the original observations.

In the light of these results, it seemed unlikely that restriction and modification were limited to infectious viruses. In experiments on bacterial conjugation and on transformation with purified microbial DNA, it could clearly be shown that restriction modification systems are also active in this case, but they are not as completely effective as with phage DNA. An acceptable methylation pattern will be imposed on a certain percentage of the foreign DNA early enough to prevent its degradation. From our current point of view, restriction modification enzymes were important in biological evolution because they reduced horizontal gene transfer to a meaningful and useful frequency.

Around 1970, some research teams succeeded in isolating restriction and modification enzymes from various bacterial strains. Biochemical studies with these enzymes fully confirmed the model outlined above on the functionality of restriction modification systems. The studies revealed that one type of restriction enzymes found in nature cuts foreign DNA endonucleolytically at strain-specific recognition sites in the absence of methyl groups (type II enzymes). In contrast, type I enzymes cut the unmodified DNA outside of the recognition sequences but also only in the absence of sequence-specific DNA methylation.

Soon it was realized that type II restriction endonucleases were very useful tools primarily for molecular studies of genomes of various organisms. With the help of restriction enzymes, enormously long DNA threads could be cut into handy fragments. Even in the 1970s, this allowed the directed generation and propagation of DNA fragments to be used for DNA-sequence analysis and functional studies. The door for genomics had been opened."

This is an important historical document describing an epochal discovery. Indeed, the door for genomics really had been opened. DNA could then be cut in a controlled way, and the resulting fragments could be employed to carry out what is now called genetic engineering, to be described later. A few additional comments might be helpful. In Chapter 21 it was mentioned that host specificity of some bacteriophages may depend on the presence of receptor molecules on the surface of bacterial cells. The starting point of Werner Arber's investigation was a much narrower host specificity. A bacteriophage from bacterial strain A was able to enter the closely related bacterial strain B but was not propagated there. This phenomenon is called restriction. Later it was found that the phage DNA in strain B had been fragmented and could no longer produce intact phages. This phenomenon, however, was not an all-or-nothing situation. A few of the phages produced descendants. It was a rocky road before researchers arrived at the concept that the host organism labels its own DNA with a certain pattern, you could say with a barcode, to prevent its own DNA from being degraded by its own restriction endonucleases. However, foreign DNA is not labeled, so it will be degraded. The few phage descendants that recovered in Werner Arber's experiments were the result of phage DNA acquiring a host-specific label just in time to avoid degradation. This label consists of a certain profile of methylated A or C bases of the DNA. In this case it is called an epigenetic modification of the DNA, in which the genes but not their sequence have been altered, by methylation for example.

*Eco*R1 recognizes the base sequence GAATTC on the DNA, where it attaches. Then it hydrolyzes the double strand so that "sticky ends" are produced (Figure 49). The two strands have been cut, but the ends produced are sticky or cohesive. These two ends, under appropriate conditions, can be brought back into position by base pairing. Healing or repair is catalyzed by competent enzymes known as ligases, whereas the process itself is called ligation.

The basic principles of gene technology were developed on the basis of these insights. Plasmid DNA and a fragment containing the desired gene are both cut with *Eco*R1 (Figure 50). The plasmid DNA and the gene fragment locate each other with the help of their sticky ends, and a ligase helps to form a bond between the two. The plasmid, enlarged by the gene fragment, is then transported into *E. coli*

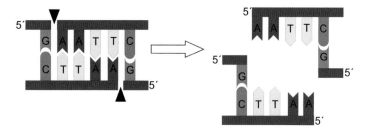

Figure 49 Cleavage of double-stranded DNA by the restriction endonuclease EcoR1. It cuts one strand between G–A (direction 5' to 3') and the other strand, shifted by four bases, between A–G (direction 5' to 3'). Sticky ends are produced that "find each other" under appropriate conditions. (Diagram: Anne Kemmling, Goettingen, Germany.)

cells by transformation (see Chapter 20), where it proliferates together with the bacterial cells. There the encoded information will be transcribed and finally translated into a certain protein. When the gene fragment contains the gene for proinsulin, for example, then proinsulin will be produced by E. coli cells.

At least on paper, we have just cloned a gene by generating a recombinant E. coli strain and using it for production of a recombinant enzyme, in this case, proinsulin (see Chapter 28). These discoveries and developments have laid the foundation for the gene technology industry, which has rapidly become an important part of the world economy.

We finally come to that other abbreviation, PCR, which stands for polymerase chain reaction. It was invented by Kary Mullis (Nobel Prize in 1993), who experimentally verified his concept of this reaction, thereby initiating a further, rather stormy, development of this method. Together with the restriction modification enzymes, PCR revolutionized all of molecular bioscience. It's probably no exaggeration to assume that millions of PCRs are run every day in thousands of laboratories around the world.

Let's take a look at the results of a PCR by means of an agarose gel (Figure 51). In this type of gel, DNA fragments can be separated when voltage is applied. The bands on the left- and right-hand sides of the gel represent DNA fragments of varying lengths. The shortest fragments, near the bottom, have migrated faster and farther than the longer ones. Lane 2 contains chromosomal DNA that has only migrated a short distance because of its high molecular weight it. A certain gene has been amplified by PCR; and the results can be seen in lanes 3–6. After 17 PCR cycles, a band is barely visible in lane 3. However, a nice fat band of DNA appears in lane 6 after 25 cycles. Apparently the amount of DNA has been enhanced enormously by PCR. This is fantastic because amplification doesn't alter the length of the gene fragment nor the base sequence or order. Within only two hours, an amount of DNA barely visible on the gel can be amplified by a factor of several million. In this way, large amounts of DNA can be produced for use in laboratory experiments.

In order to understand PCR, we have to jump back to DNA replication, in Chapter 2. In the discussion there, one important detail was left out. If we were

Chapter 25 Eco R1 and PCR – molecular biology at its finest

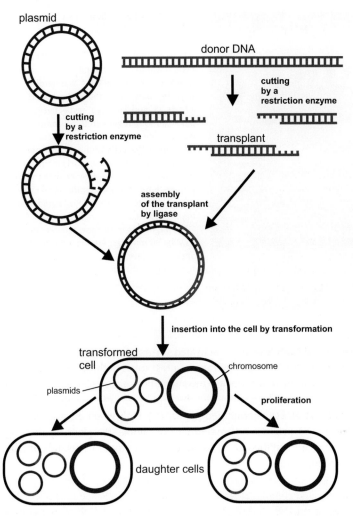

Figure 50 Cloning of a gene. DNA with the desired gene as well as a plasmid are cut with EcoR1. The fragment with the desired gene and the "opened", linearized plasmid, are incubated together under conditions under which the two fragments associate because of their sticky ends. The two gaps between the fragments are closed by addition of the enzyme ligase. The resulting plasmid containing the desired gene is introduced into a bacterial host by transformation. With proliferation of the bacteria, the plasmid is also proliferated. The desired gene can then be isolated from the cells. Under certain conditions, the gene can be expressed so that the encoded protein is produced. (Diagram: Anne Kemmling, Goettingen, Germany, on the basis of Figure 38 in *Das Spiel* by M. Eigen and R. Winkler, Piper Publishers, Munich and Zurich, 1981.)

Chapter 25 Eco R1 and PCR—molecular biology at its finest

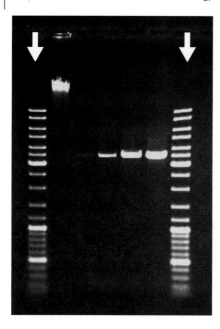

Figure 51 Demonstration of a PCR product by agarose gel electrophoresis. Left and right lane (lanes 1 and 7): the so-called DNA ladder, DNA fragments from 10 000 base pairs down to several hundred base pairs; lane 2, chromosomal DNA that hardly migrates; lane 3 to 6, the PCR products after 17, 20, 23, and 25 cycles, respectively. A 3500 base-pair-long region on the chromosome was amplified using specific primers. Separation (melting) of the two chromosomal strands was achieved by incubation at 96 °C for two minutes. After cooling to 56 °C, primers bind and DNA synthesis takes place. After 4 minutes the sample is heated again, then recooled, and so on. Separation of the DNA fragments in agarose gel was done at 85 V for 90 minutes. Arrows indicate the application level of the samples at the beginning of the run. (Experiment: Frauke-Dorothee Meyer, Goettingen, Germany.)

to mix single-stranded DNA with the activated forms of A, T, C, and G as DNA building blocks and with DNA-polymerase in a test tube, nothing would happen. That's because the so-called primers are missing, short pieces of DNA around 20 building blocks in length. The sequence of the primer must fit exactly to the start region of the single strand to be amplified (Figure 52). Only when the primer binds to the single strand to produce a tiny segment of double-strandedness does the DNA polymerase recognize the task to be done. Then it goes to work to complete the double-strandedness.

Which components and conditions are needed to carry out PCR?

The reaction mixture contains DNA with the gene to be amplified, with let's say a length of 2000 base pairs. Primer for the start region is added, which, of course, is only possible when the sequence of the start region is known. Furthermore, activated building blocks and thermostable DNA-polymerase are added. The enzyme could be, for example, the *Taq* polymerase of *Thermus aquaticus* mentioned in Chapter 6. The mixture is heated and kept at 96 °C for 20 seconds. The double-stranded DNA separates into two single strands, a process called DNA melting. It is then cooled down to 56 °C. The primer molecules bind to the single-stranded gene, and the DNA polymerase does its job. The two single strands yield two double strands, and in the next cycle, four single strands are generated, and these form in turn four double strands. In the cycle that follows, eight single strands are converted into eight double strands, and so on. It's easy to calculate: one DNA fragment can yield over one billion fragments in 30 cycles, and this in

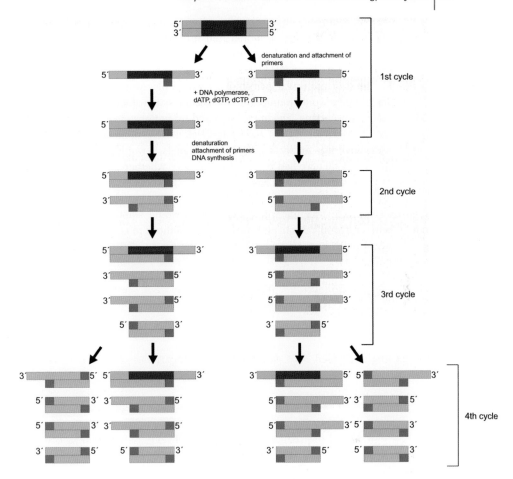

Figure 52 Principle of PCR. Cycle 1: melting of the double-stranded DNA is followed by primer binding and DNA synthesis; cycle 2: after melting and primer binding, DNA is synthesized. After cycle 3, two double-stranded products have the correct length, and eight after the fourth cycle. After 20 or so cycles, the fragments with an overhang are practically negligible. (Diagram: Anne Kemmling, Goettingen, Germany.)

just a few hours. Obviously, thermostable DNA polymerases, such as the enzyme from *T. aquaticus*, are particularly suited for many reaction cycles because they even remain active during the high-temperature step required to "melt" the double-stranded DNA.

What are the specific uses for PCR?

We'll answer this question with a few lines from the autobiography of Kary Mullis. There he describes how the idea for PCR hit him at mile marker 46.58 while driving along State Highway 128 in Northern California. He wrote,

> "The procedure would be valuable in diagnosing genetic diseases by looking into a person's genes. It would find infectious diseases by detecting the genes of pathogens that were difficult or impossible to culture. PCR would solve murders from DNA samples in trace materials—semen, blood, hair. The field of molecular paleobiology would blossom because of PCR. Its practitioners would inquire into the specifics of evolution from the DNA in ancient specimens. The branchings and migrations of early man would be revealed from fossil DNA and its descendant DNA in modern humans. And when DNA was finally found on other planets, it would be PCR that would tell us whether we had been there before or whether life on other planets was unrelated to us and had its own separate roots."

It is impossible to imagine where biosciences would stand today without the methods described in this chapter, without the methods to cut, copy, and amplify genes. And finally, bacteria and archaea are the major players in providing the tools required.

> Union gives strength
>
> *Aesop*

Chapter 26
Interbacterial relationships

Let's begin with two experiments. The first one was originally conducted by the botanist Wilhelm Pfeffer (1845–1920; Leipzig, Germany) and taken up again in the 1960s by Julius Adler (Madison, Wisconsin, USA). A fine-bore capillary tube is lowered into a mineral salt solution containing cells of *Escherichia coli* (Figure 53). Then a solution of an amino acid (aspartic acid) is introduced into the capillary. Only minutes after the aspartic acid begins to diffuse into the bacterial suspension from the capillary, most of the bacteria accumulate in and around the opening of the capillary.

The second experiment has to do with luminescent bacteria and is often carried out in courses in practical microbiology. In one such protocol, instructions for students begins with a quote by the naturalist Placidus Heinrich (1758–1825), who in 1815 had bought a codfish at the fish market in Regensburg (Germany):

> "The fish was three feet long, quite fat, and, like all shipped fish, it had already been gutted in the Netherlands. As soon as it arrived there, it was scaled, soaked in fresh water for a day, then hung in a food cellar at approximately 12 °C. The next evening, the eye sockets and some areas of the fish's back were visibly glowing already. The light only appeared on the surface of the fish, occasionally as bright spots or stripes. This phenomenon was even more beautiful the next evening. The luminescence continued to spread before reaching an apparent maximum on the third day, when the light was very bright, exceeding the light of white phosphorous. The fluorescent substance could be wiped off with the hand or scraped off with a knife. At this point, the fish was washed in the dark and prepared for the next day's meal."

What can we say except "bon appétit!" To do this experiment we need a fresh, iced sea fish, preferably a freshly gutted herring. The fish should be put in an open glass bowl or a Petri dish and partly covered with water of seawater salinity (3.4 percent salt). The fish should be examined in the dark after one or two days between 10 °C and 15 °C (50 °F and 60 °F). The little shining dots that are visible are bacterial colonies. They are picked off the fish with a needle, transferred to a nutrient salt solution (30 g/l salt plus 5 g peptone, 0.5 g glycerol, and 1 g yeast

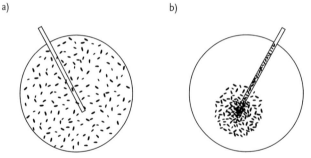

Figure 53 Capillary test for demonstration of chemotaxis. (a) Bacteria are equally distributed in a flat vessel. (b) An aspartate solution is added through a capillary. Within a few minutes, all of the bacteria present in the vessel have gathered around the tip of the capillary, with some even swimming into the capillary. (Drawing: Anne Kemmling, Goettingen, Germany.)

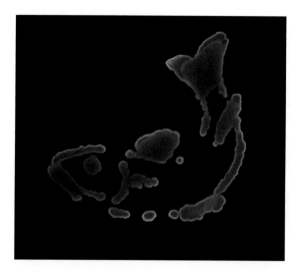

Figure 54 Bioluminescent bacteria on a Petri dish. The organisms (*Vibrio fischeri*) were grown on a saline complex medium. The photo was taken in complete darkness; exposure time: 15 seconds. (Photograph: Anne Kemmling, Goettingen, Germany.)

extract), and incubated overnight. Petri dishes containing agar and the above solution are prepared. Drops of the grown culture are spread on the surfaces of agar plates and incubated at 12 °C for one or two days. It is quite impressive to inspect these plates in the dark. The colonies really do glow, and you can even use them to make little glow-in-the-dark drawings on the plates, such as the fish in Figure 54. If you look at the luminescent colonies closely, you may notice that the large colonies glow very intensely, whereas the tiny colonies don't glow at all. We will come back to this later.

Those are interesting experiments, but I can't detect anything interbacterial about them.

You're right, as least when it comes to the first experiment where each of the bacteria is fighting to get to the food (aspartic acid). But how do they do it? Well, they do it by chemotaxis. In our case, the aspartic acid acts as an attractant for the bacteria, which are in a mineral salt solution without any other source of energy and carbon such as glucose. We know the mechanism by which microbes tend to migrate toward food. It's not at all like a sailing regatta in which the skipper keeps his eye on the turn buoy and sails straight toward it, wind permitting. *E. coli* cells are able to move around because they have flagella, like a galley equipped with oars. However, they can't deliberately swim toward a food source, which in our case is inside the capillary.

That's interesting. Do all bacteria have flagella?

No, not all of them. Nearly 50 percent of the bacterial species are motile because they have flagella. A few examples of immotile bacteria are *Staphylococcus aureus*, already mentioned in connection with MRSA; archaea such as *Thermoproteus tenax*; *Corynebacterium glutamicum*, to be introduced in Chapter 28; and *Mycobacterium tuberculosis*, to be discussed in Chapter 29. Examples of motile microbes are *Escherichia coli* and *Salmonella typhimurium*, both peritrichous, with flagella extending in all directions. Others are polarly flagellated, good examples being *Chromatium okenii* (Figure 8a) or sulfate-reducing bacteria such as *Desulfovibrio vulgaris*. Flagella have a helical or spiral shape, and they are assembled by transporting the building blocks through the inside of the growing flagellum up to its tip, like building a chimney by transporting the bricks upward through the inside. Flagella are rigid, and their rotation is driven by a membrane-inserted motor resembling the rotor of ATP synthetase (see Chapter 8). Individual cells can swim forward, backward, sideways, up, or down.

Now we'll come back to chemotaxis. A bacterium swims straight ahead when its flagella rotate in a counterclockwise direction. Swimming is interrupted every now and then by a modulating system that stops the flagellum's motor and switches to a clockwise rotation. This causes confusion, and the cells tumble for a moment instead of swimming. Then order returns and swimming resumes in a random but different direction, depending on the spatial orientation of the cell when tumbling stopped. This is where regulation of tumbling frequency comes in. When the cell is moving toward the food source, the capillary in our case, it senses the increasing concentration of the attractant. As a result, the usual tumbling frequency is reduced. The cell is able to swim for longer periods of time in the direction of the food without being interrupted by tumbling. On the other hand, when the bacterium happens to swim away from the food, the usual tumbling frequency resumes. The cell has to pause more often to tumble and to change direction, until the food has been detected. This mechanism allows the bacteria to swim further in the "right" direction, so they eventually end up accumulating around the tip of the capillary, the source of food.

These are probably the simplest sensory abilities found in organisms. The cells measure how the concentration of aspartic acid changes over time. This rate regulates the tumbling frequency. You may wonder how cells react when suspended in a solution with a uniform nutrient concentration. This system is also prepared for such situations: the chemotactic response simply becomes attenuated. There are proteins in the membrane called MCPs (methyl-accepting chemotaxis proteins). These are signal-transmitting proteins that bind the attractant, aspartic acid in our case. By way of a cascade of signals, the MCPs pass on the information to the motors of the flagella. When the nutrient concentration remains more or less constant, the whole system more or less relaxes. The MCPs become insensitive and the chemotactic response weakens, so the cells become sedate and swim around more slowly (▶Study Guide).

There are also other, related phenomena. A number of bacterial species have developed a comparable sensory response toward oxygen, called aerotaxis. Phototaxis is also very impressive. Organisms such as *Thiospirillum jenense* have a bundle of flagella at one end that acts like a ship's propeller when the organism is swimming toward the light. However, when this organism enters a dark area, the bundle swings around and pulls the cells back into the light.

Aren't these mechanisms of response by bacteria to their environment quite impressive? The principles of tactical behavior, widely used today in modern technologies, had already been realized ages ago by bacteria: measuring the change of a parameter over time, transfer of signals, and responsive action to emerging situations. Let's now jump back to the glowing fish. The bacteria responsible for this phenomenon are *Photobacterium phosphoreum* or *Vibrio fischeri*. They contain an enzyme system known as luciferase, which catalyzes an oxygen-dependent oxidation process linked to the emission of light, bioluminescence. The microorganisms mentioned and related species even bring some light into deeper parts of the ocean. They exist as symbionts in the luminous organs of deep-sea fish, helping them find mating partners or attract prey in the pitch-black ocean.

Which conclusions can be drawn from the fish experiment? For one thing, luminescence apparently depends on the actual bacterial cell density: When such bacteria form a dense population, they glow, but when the bacterial concentration is low, as in tiny colonies, they don't. This phenomenon is known as quorum sensing, and its significance in nature cannot be overestimated. Quorum sensing makes it possible for cells to determine whether a quorum, a certain threshold concentration of cells, has been exceeded at a given location. The underlying mechanism of this phenomenon is straightforward: all of the cells involved secrete a given signaling compound. In many cases, including ours, this compound is an N-acyl homoserine lactone (AHL). When few cells are present, the AHL concentration surrounding the cells is low. A high cell concentration results in a correspondingly high AHL concentration capable of exerting feedback on the intermediary metabolism of the cells. As a result, switches are thrown that eventually lead to the luciferase reaction and the ensuing emission of light.

Glowing is nice, but why is this quorum sensing so important?

Let's go into biofilms now. At first glance, one may tend to assume that most bacteria swim freely in oceans, rivers, and ponds, or they move around in a certain state of solitude in the liquid films found in soil or on surfaces. Of course, this nomadic way of life is important for colonization of new habitats and for rapid reproduction under favorable conditions. In recent years, however, it has become increasingly apparent that the majority of bacteria is present in the form of biofilms. These biofilms not only cover our teeth with dental plaque and lots of surfaces with slime but they also are responsible for formation of crusts on certain surfaces and the clogging of membranes, filters, and pipeline systems in industrial production facilities. These microbe–surface interactions in technical plants and devices are referred to as biofouling or biocorrosion processes. In the medical-care sector, especially in the fight against infections, biofilms play an extraordinarily dangerous role. Bacteria may grow on the surfaces of catheters, implanted stents, or other invasive devices, and thus contribute considerably to the infectious potential of microbes. This is especially problematic when we consider opportunistic germs whose species have developed a pathogenic lifestyle when they live in biofilms. As a typical representative of these microorganisms we will look at *Pseudomonas aeruginosa*.

The bacterium *P. aeruginosa* is very common in nature. Normally it is harmless because it does not produce any real toxins. However, it colonizes patients weakened by infections or suffering from cystic fibrosis or cancer, who are being treated in intensive-care units. By way of the respiratory tract, *P. aeruginosa* reaches the lungs. There it settles and produces an extracellular polysaccharide that forms a sticky biofilm in which the cells can grow and proliferate. The biofilm grows and grows, putting an increasing strain on the lungs, often with a fatal outcome. What causes the cells to embed themselves in such a matrix and produce virulence factors and enzymes for fat degradation under these conditions? The answer is quorum sensing, and the signaling molecules are AHLs as well. The team of Peter Greenberg (Seattle, USA) has found that at least 1 percent of the genes of *P. aeruginosa* can be switched on by quorum sensing. The genome of this microorganism is one-and-a-half times the size of the *E. coli* genome, which consists of 6.2 million building blocks in the form of base pairs. Of the 5570 genes of *E. coli*, quorum sensing controls the expression of about 5 percent of these, which are only translated into enzymes under biofilm conditions. We have asked Peter Greenberg for an assessment of the risks associated with *P. aeruginosa*:

> "*Pseudomonas aeruginosa* causes difficult-to-cure infections and often impossible-to-cure infections. For example, most patients with the genetic disease cystic fibrosis have lungs that are permanently infected with *P. aeruginosa*. Antibiotic therapy can beat back these infections but not eradicate them, and people with cystic fibrosis ultimately die from the infections. By aggressively using antibiotics that can eradicate other types of *P. aeruginosa* infections, we have extended the life expectancy of people with cystic

fibrosis. Now, half will live to see their thirty-fifth birthday, a great improvement over statistics from 20 years ago, when half were dead by their seventeenth birthday. But life for most of the years of a cystic fibrosis patient is difficult and having half a chance to live beyond 35 isn't good enough. The discovery that *Pseudomonas* infections involve coordination of group behavior and the discovery of some of the signals required for group activities has alerted us to the idea that we might be able to more effectively control cystic fibrosis infections and other infections caused by this bacterium and other quorum-sensing bacteria by targeting communication."

It is a goal of the research carried out by Peter Greenberg and others to disrupt the communication channels between the *P. aeruginosa* cells, thus preventing or at least impairing the formation of biofilms in patients' lungs. What they are trying to achieve is "antiquorum sensing", so that the cells revert to a solitary, nomadic lifestyle and are no longer aware of each other's presence.

Biofilm formation may also proceed in a similar way in other cases. First, a few cells of a certain species colonize a surface. The signals they emit may even be strong enough to attract other individuals. Then exo-polysaccharides, sticky starch-like substances, are produced. These form a matrix in which cells of other bacterial species also are incorporated. Growth of expanding mushroom-like structures have been observed. Finally, the thickness of the film increases. Many mobile bacteria develop such a liking for a sedentary life that they get rid of their flagella once they have settled in the biofilm, where they don't need them anymore. They have made themselves comfortable in an excellent habitat.

Anne Kemmling (Goettingen, Germany), who supplied many of the illustrations in this book, studies biofilms and is especially interested in the positive aspects of biofilms. She writes:

"When we look at several generations of bacterial species, almost all of them will be organized in biofilms at some stage of their lives. Even for bacteria that are otherwise mobile, having a home base once in a while has its advantages. When we talk about biofilms, we have all types of bacterial growth in mind. There are biofilms between a solid surface, such as a rock, and air, which we call subaereal biofilms (Figure 55, left). Subaquatic biofilms are widespread; we have all experienced these when we walk across a stony river bottom, which is often very slippery due to microbial growth (Figure 55, right). Biofilms in sewage plants are secret playgrounds for microbiologists. That's why first-year students of microbiology are taken there for an introduction to biological waste water treatment as well as biofilm formation. Several microbial species organized in biofilms help get rid of chlorinated hydrocarbons, toluene, oil, or insecticides in contaminated soil and water. But let us discuss the role of biofilms in nature a little more in detail. We already learned from Joachim Reitner, in Chapter 5, that bacteria in biofilms represent the oldest fossils on record. The strategy of protecting oneself, together with other microorganisms in a matrix, to withstand wind, tides, sun, or dessication can be considered very success-

Figure 55 Examples for subaerial (left) and subaquatic (right) biofilms. (Photograph: Anne Kemmling, Goettingen, Germany.)

ful. A more or less thick layer of cells also has the advantage that gradients can be built up, i.e. gradients of oxygen or nutrients, which allow microorganisms to find their proper ecological niches. Biofilms can be considered conceptual forerunners of the cell collaboration and communication achieved in genuine tissues.

The efficiency and the diversity of microbial interactions in biofilms make them most interesting to ecologists. Among the subjects explored is the question how biofilms enable the bacteria to survive under such adverse conditions as extreme aridity, high salt concentrations, low nutrient availability, or high pressure. Without being organized in biofilms, the bacteria would never be able to survive under these conditions. The interaction between biofilms and the environment is also of interest because rock-formation processes are affected. Biofilms protect the rock surface from weathering and, most important for arid sites, they increase the water-storage capacity of rocks. All this is not only beneficial to them but also to higher organisms."

There certainly is something to these interbacterial relationships. Positive and negative aspects are closely intertwined. On the one hand, biofouling incurs enormous costs. On the other hand, crust formation in deserts stabilizes the soil, preparing it in turn for settlement of higher organisms.

> The Red Queen said: It takes all the running
> you can do to keep in the same place
>
> Lewis Carroll, Through the Looking Glass

Chapter 27
From life as a nomad to life as an endosymbiont

In the previous chapter we saw that bacteria can live as single organisms in various habitats but may also tend to settle together on surfaces, creating biofilms. In addition, bacteria are even able to coexist with higher organisms by forming symbiotic relationships.

Symbiosis in a strict sense is an association between two different organisms for the benefit of both partners. Symbiotic nitrogen fixation is a somewhat borderline case because the rhizobial bacteria entering the roots of a plant are subsequently held captive, irreversibly transformed into nitrogen-fixing bacteroids. As individuals they don't gain a thing from this joint venture, but in the end they contribute bound nitrogen to the soil when the plant decays. This in turn benefits their free-living rhizobial colleagues and other microorganisms in the soil. Associations of the bacterial species *Frankia*, *Azospirillum*, and *Azoarcus* with plants, mentioned in Chapter 11, much better fit our definition of symbiosis.

At this point, we will take a short detour to the corals because of their beauty and importance and the numerous dangers they face. Corals live in clear, shallow waters that are low in nutrients, and they require photosynthesis for growth and reproduction. Corals are only able to perform photosynthesis with the help of other organisms, single-cell algae called *Zooxanthellae* that live inside the coral cells. This symbiosis is one of the most delicate relationships found in nature. Slight changes such as a rise in temperature may lead to coral bleaching and even death. For this reason it is feared that global warming, in addition to other severe environmental changes, will cause coral extermination. Bleaching of coral signals the loss of the symbiotic partner, and bacteria could even be the cause. Eugene Rosenberg (Tel Aviv, Israel) and his research team found that the *Zooxanthellae* in Mediterranean corals are attacked by a bacterium called *Vibrio shiloi* whenever the water temperature rises by a few degrees. We asked him to tell us about it:

> "For the last several decades coral reefs have been in a decline, largely due to emerging and re-emerging diseases. The largest environmental factors contributing to these diseases are pollution, over-fishing and rising seawater temperatures. On the global scale, coral bleaching is the most severe disease. Coral bleaching is the disruption of the symbiosis between the coral animal and its endosymbiotic algae, commonly referred to as

Zooxanthellae. As a result of the loss of the algae, the tissue becomes transparent and appears white because of the calcium carbonate backbone.

In two cases it has been demonstrated by applying Koch's postulates that coral bleaching is a result of bacterial infection: the coral *Oculina patagonica* in the Mediterranean Sea by *Vibrio shiloi* and *Pocillopora damicornis* in the Indian Ocean and Red Sea by *Vibrio corallilyticus*.

In both cases, increased temperature caused the pathogen to express virulence genes. The *V. shiloi* infection cycle has been studied extensively and it has been shown that a peptide toxin is produced which blocks photosynthesis of the algae. Recently, David Bourne and colleagues in Australia have reported that *Vibrio* appeared in large numbers in coral tissue just prior to a mass bleaching event on the Great Barrier Reef. The recovery of the corals when the temperature decreased correlated with the loss of the *Vibrio*.

At present there is a debate on whether mass coral bleaching is the result of bacterial activity or photoinhibition of the algae at high temperature and light intensity."

These findings, shocking enough in themselves, are documented in a photograph (Figure 56) that requires no further comment.

So bacteria were first nomads, then symbionts, and now endosymbionts?

On our way to the endosymbionts, let's first take a look at a few more host–microbe interactions. Pathogenic bacteria such as *Bacillus anthracis* or *Mycobacterium tuberculosis* are parasites (see Chapter 29). A number of bacteria have given up growing and proliferating as free-living organisms. They have become obligate parasites; they are able to divert preformed building blocks, amino acids, coenzymes, and nucleotides from their host cells – in some cases, even ATP. *Rickettsia prowazekii*

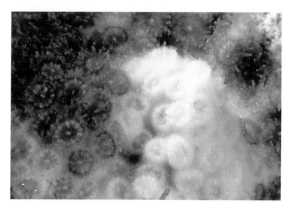

Figure 56 Partially bleached coral *Oculina patagonica*. (Photograph: Fine and Loya, 1994, provided by Eugene Rosenberg, Tel Aviv, Israel.)

is such a bacterium: It lives intracellularly and is the cause of typhus, also called typhus fever (not to be confused with typhoid fever). The genus Rickettsia is named after Howard Taylor Rickett (1871–1910), who discovered that Rocky Mountain spotted fever is caused by a species of this genus. During evolution, there has been a transition from certain obligate parasites to endosymbionts. A good example is *Buchnera aphidicola*, which lives in plant aphids and provides its host with certain aromatic amino acids (e.g., tryptophan). Bacterial endosymbionts of the *Buchnera* or *Carsonella* species contain very small genomes because they have given up many of the functions that normally make up a bacterial cell.

Could mitochondria and chloroplasts be the remnants of endosymbionts?

Yes. These ideas were first put forward by the German botanist Andreas F. W. Schimper (1856–1901) in the 1880s, then elaborated upon by the Russian biologist Constantin Mereschkowsky (1855–1921) and later taken up and further developed by Lynn Margulis (Amhurst, Massachusetts, USA).

Mitochondria (let's call them mitos for short) are, as mentioned earlier, the power plants of the cells. In their interior, reducing equivalents are generated that subsequently react with oxygen in the membrane. A proton motive force is generated, and the F_1F_0-ATPase as discussed in Chapter 8 takes advantage of this force to synthesize ATP.

But this surely is not enough to support the theory that mitos originate from endosymbionts.

When we look at other components of the mitos, two features are quite striking. They contain circular DNA, which can be thought of as a minichromosome, and their ribosomes are similar in size (70S) and structure to those from bacteria. In contrast, eukaryotic ribosomes are larger (80S) in general. Human mitochondrial DNA (mtDNA is the short form) contains only 17 000 base pairs and codes for 35 genes, including genes for ribosomal RNAs, some transfer RNAs, and components of the ATPase and the respiratory chain (Figure 57). By all indications, a lot of information has either been lost or else transferred to the nucleus. Despite this, the mitos have retained a certain degree of autonomy. Mitos are special: they carry genetic information that is not present in the DNA of human cells and they cannot be synthesized, so to speak, from scratch. Instead, they reproduce themselves by dividing, just like the bacteria do. Mitos are also in the egg cells of mothers, and that's how they are passed on to the next generation. It's funny that men, whose cells also contain mitos, don't pass them on to the next generation because sperm heads contain no mitos. Worldwide comparisons of mtDNA sequences in humans have made it possible to develop a migrational tree of mankind (or better, womankind). Researchers compared the base sequences of a hypervariable region of the mtDNA that lies between the base pairs 16 001 and 16 559 (Figure 57). An analysis of these sequences in humans from Australia, Europe, America, and other continents revealed not only some differences but also

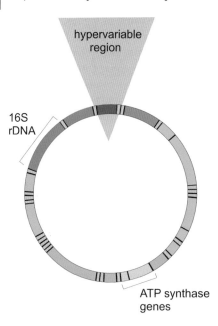

Figure 57 Human mitochondrial DNA. Indicated are: gene encoding 16S-rRNA, the hypervariable region, the transfer-RNA genes (yellow) and the genes for ATP synthase. (Diagram: Anne Kemmling, Goettingen, Germany.)

a basic pattern that is apparently the basis for all the others. This basic pattern is only found in the mitos of African women. That's why we speak of the "mitochondrial Eve," the common mother of all mankind who must have lived in Africa. Sequence analyses indicate that the migration of humans out of Africa and into other continents began less than 100 000 years ago.

As all cells are subject to aging, the mitos also grow "old" and become less efficient. After all, they are constantly exposed to radicals, the toxic reduction products of oxygen (see Chapter 5) generated in their close proximity. Since mitos are both the source and the target of such toxins, the mtDNA obviously is more liable to oxidative damage than the DNA enclosed and protected by the cell nucleus. As the mitos grow older, cumulative DNA damage leads to formation of nonfunctional proteins and to disturbances in the coupling of respiration and ATP synthesis. The resulting loss of efficiency and decline in cellular energy supply accelerates the aging processes even more.

Does anyone have an idea which bacteria could have migrated into the eukaryotic cells?

In 1998, evolutionary biologist Siv Andersson (Uppsala, Sweden) and her colleagues published a paper entitled "The genome sequence of *Rickettsia prowazekii* and the origin of mitochondria". *R. prowazekii* has already been mentioned as the microorganism causing typhus, which has been responsible for devastating epidemics in the past. We may recall that the decimation of Napoleon's army in Russia in 1912 was primarily due to typhus. Siv Andersson's team sequenced the genome of *R. prowazekii*, which consists of around 1.1 million base pairs. A comparison with known sequences revealed that the closest relatives of *R. prowazekii* are plant mitos as well as those of a protozoan called *Reclinomonas americana*. The DNA of human mitos is not suitable for such comparisons because it is much smaller and only contains a few genes, as we have seen. To a greater extent than the protozoans, human mitos have adapted to the task of energy conservation by reducing and simplifying their genetic information. The DNA not needed was simply lost, or it migrated into the nucleus. Especially the mt genes of said protozoans and those of *R. prowazekii* correspond to a high degree, leading to the conclusion that the precursor of *Rickettsia* once migrated into cells that were larger than bacteria. Such cells had a nucleus but still lived, more or less, like bacteria. Once inside the cell, the immigrants lived on waste materials in their hosts and respired with the oxygen present. They became more and more specialized in energy conservation and eventually evolved into mitos. In plants, the immigrants obviously could not have been *Rickettsia*, which lack a photosynthetic apparatus. Therefore, cyanobacteria must have been the ones to be engulfed by eukaryotic cells.

Again, bacteria have given us quite a lot. All higher organisms owe them their organelles for energy production: the mitochondria for respiration, the chloroplasts for photosynthesis, and the enzyme systems to combat ROS (reactive oxygen species).

> Time is the best appreciator of scientific work,
> and I know that an industrial discovery rarely
> bears all its fruits in the hands of its first inventor
>
> Louis Pasteur

Chapter 28
Bacteria as production factories

That sounds awfully technical – why don't you come right out and say biotechnology?

Biotechnology includes a whole spectrum of application areas, so we say it is multicolored. Red biotechnology refers to processes in which products are manufactured using animal cells. In green biotechnology, plant cells serve as production factories, as described in Chapter 24. We speak of white biotechnology when microbes are involved in the production process. The production of ethanol, lactic acid, vinegar, acetone, butanol, biogas, and antibiotics has already been discussed. Now, isn't that already an impressive collection of products?

Yes, but some of those processes seem old-fashioned, especially the ones for vinegar, cheese, or acetone and butanol.

We'll look at some of the high-tech processes a bit later. Although some biotechnological processes seem old-fashioned, they are still very important. Let's take glutamic acid for a start, an amino acid whose most common salt is sodium glutamate. You may recall that proteins consist of twenty different amino acids. Our body can only synthesize twelve of these, so the other eight are essential, including lysine, tryptophan, and methionine. We need to obtain them through our food because we would otherwise be unable to synthesize all of our own cell proteins. There's a big market for the production of these essential amino acids because they are added as supplements to numerous foods and animal feeds. Annual worldwide production of amino acids is in the range of 2.5 million tons with a market value of around eight billion US dollars.

Instead of discussing the synthesis of essential amino acids, we will turn to the synthesis of sodium glutamate. After all, its production makes up nearly 50 percent of the total amino acid production mentioned above.

Why do we need such large amounts of sodium glutamate if it's not even essential?

Sodium glutamate or, more precisely, the optically active L form, is an important flavor enhancer in foods. All those tasty gravies and sauces used to refine various dishes are classically made from meat extract prepared by cooking meat to obtain a delicious broth. Those in a hurry add water to a bouillon or stock cube. The effective ingredient in all of these is sodium glutamate, so why not produce it in large amounts and add it directly to foods? That's why glutamate has been so successful in enhancing food taste worldwide.

We are probably not exaggerating to say that fast food just wouldn't exist if Shuko Kinoshita and his coworkers had not isolated *Corynebacterium glutamicum* from soil samples in Japan in 1957. This bacterium is unsurpassed when it comes to producing glutamate. Ajinomoto means sodium glutamate in Japanese, and this is also the name of the production firm in which *Corynebacterium glutamicum* is at work in huge reaction vessels with a volume of 500 000 liters (125 000 gallons). Within two days of growth, several tons of bacterial cell mass have been produced, and much of the glucose added to the vessel has been converted to sodium glutamate, the final concentration being approximately 100 g/l. This product is purified and marketed. Naturally, you can't expect just any isolated strain of *Corynebacterium glutamicum* to be able to produce such large amounts of sodium glutamate from glucose. Researchers have discovered that the concentration of the vitamin biotin is critical for sodium glutamate production. Only when a certain concentration of biotin is present do the cells tend to secrete the amino acid. In other words, only under such conditions is the cytoplasmic membrane sufficiently permeable to glutamate. With *Corynebacterium glutamicum* we have a bacterium in our white biotechnology portfolio that certainly has changed the eating habits of mankind. Around 1.7 billion kilograms (1.7 million tons) of sodium glutamate are produced this way every year. The essential amino acid L-lysine comes second with an annual production of 0.7 million tons.

What about enzymes – are they also produced on a large scale, by the ton?

Hundreds of different enzymes are isolated for laboratory use. Think of all the DNA polymerase required for all those PCR reactions, or the lactate and glutamate dehydrogenases for assays in clinical chemistry. In such cases, several kilograms are sufficient to cover the annual demand worldwide. However, enzymes for industrial purposes need to be produced by the ton. Let's look at two examples, the first one being glucose isomerase. This enzyme is common in various bacterial species; and its actual function is to restructure (isomerize) sugars with five carbon atoms, converting xylose into xylulose, for example. Fortunately, this enzyme is not very choosy, so it even accepts a six-carbon sugar such as glucose as substrate, rearranging (isomerizing) it into fructose.

I never heard of this glucose isomerase enzyme. What's it for?

What would numerous foods and drinks taste like without their sweetness? Foods are sweetened by adding sugar such as saccharose, which comes from sugar cane or sugar beets. There are insufficient amounts of saccharose available to sweeten all the different jams and beverages, for instance. Therefore, a process called saccharification was developed to convert starch into glucose. Starch is actually a polyglucose consisting of many glucose molecules linked together. The chemical breakdown of starch into single glucose molecules is carried out on an industrial scale with the aid of enzymes, for example, alpha-amylase. The product is a glucose (dextrose) sirup that is not very sweet. Dextrose has a very refreshing taste but it lacks the sweetness of saccharose. Fructose, on the other hand, is 40% sweeter than saccharose, so glucose isomerase is used to convert glucose sirup into a glucose-fructose mixture. This mixture contains 60% glucose and 40% fructose, which reflects the equilibrium established by the action of glucose isomerase. When the table of "Nutritional Facts" on a food package says "fructose" or "liquid sweetener", we should thank glucose isomerase for this food's sweetness. Annually, 300 000 tons of this kind of liquid sweetener are produced in the United States.

The huge amounts of glucose isomerase required for liquid sweetener production are produced by special *Streptomyces* strains. These microorganisms are real friends of mankind because they are also some of the producers of antibiotics (see Chapter 22). Genetically modified production strains of *Streptomyces* are not in use, however, because glucose isomerase is part of our food chain. During the production process, glucose isomerase accumulates within the growing cells of *Streptomyces*, which then have to be destroyed to isolate the enzyme from the cell extract.

Other enzymes of industrial interest are secreted by the production organisms, making the enzymes easier to isolate and purify. The bacterial cell mass simply needs to be separated from the culture broth containing the enzymes, either by centrifugation or filtration. Bacteria have a good reason for secreting certain enzymes. When these organisms grow on natural products such as starch or protein, the uptake mechanisms of the bacteria are unable to cope with such large polymers because of their size. These nutrients first have to be degraded to smaller molecules outside the cells. After that, "handier" molecules such as glucose, maltose (consists of two glucose molecules), or amino acids can be taken up by the cells. For the purpose of extracellular degradation, the bacteria have to release the required enzymes into their surroundings, which is not easy. Various types of secretion systems are available for this task. In one case, the enzyme proteins to be secreted carry a distinct signal sequence, a number of amino acids at the tip of the protein chain. This sequence functions as a kind of barcode, which informs the secretion apparatus that this protein should be escorted out of the cell as a linear, threadlike chain of amino acids. Once outside the cell, the protein folds to form an active enzyme, for example, alpha-amylase or one of the proteases. The latter are also used for industrial purposes.

Where are the proteases used?

These enzymes are added to dishwashing or laundry detergents to remove proteins from dishes or clothing. Proteins are simply hydrolyzed to amino acids. Simple detergents are able to remove grease or fat from various materials but they are unable to remove protein completely. Therefore, proteases are indispensable helpers in our daily cleaning tasks. The proteases of choice belong to the subtilisins, especially those that are alkaline-stable and cooperate very effectively with detergents. These genetically modified proteases are produced primarily by

Bacillus species, that is, *B. licheniformis* or *B. subtilis*. According to Karl-Heinz Maurer (Duesseldorf, Germany), 900 tons of proteases were produced in the European Union in 2002. These 900 tons were then marketed as 20 000 tons of granulate or liquid products with a value of 350 million US dollars. The US market is described by Michael Rey of Novozymes in Davis, California (USA):

> "Market estimates place annual global revenues for enzymes at around three billion US dollars. This is a lot of enzymes! What do these enzymes do in detergents? Let's say I had a nice meal of pasta with meat sauce and chocolate ice cream for dessert. Of course, I spill some of my meal on my favorite blue-and-white striped shirt and have to wash it. The enzymes in the detergent work together on the stains: the proteases remove protein stains from the meat, the amylases remove starch-based stains from the pasta, the lipases remove the fatty stains from the olive oil, the cellulases whiten the fabric, and the mannanase removes guar gum present in the chocolate ice cream. Most people may be familiar with the use of enzymes in detergents but may not be aware of the positive impact this has made on the environment. Enzymes have helped to make detergents more environmentally friendly by reducing consumption of water and energy as well as the quantity of chemicals required to get clothes clean. Enzymes allow people to wash their laundry at half the normal temperature. By lowering the wash temperature, a 60 percent reduction in electricity can be achieved, resulting in a reduced carbon footprint for washing clothes. If everyone in Europe washed their laundry at 30 °C instead of 60 °C, the CO_2 reduction would equal the annual carbon emissions from 3 million cars and their clothes would still be clean. In countries like China and India, where handwashing of clothes is a common practice, these enzymes in laundry bars result in more effective clothes cleaning; saving time, effort and water for millions of people.
>
> Detergents contain surfactants, which create the foam seen in the wash water. These surfactants are fossil-fuel-based products and are influenced by the price of a barrel of oil. Enzymes can now replace some of these surfactants, providing a stable, sustainable, biobased solution."

Enzymes and proteins in general are one of the key areas of white biotechnology. The development of many of the relevant production processes is intrinsically tied to the appearance of biotech companies in the late 1970s.

One of the most important proteins produced by genetically engineered microorganisms is insulin, whose role in our body is well known. It regulates blood sugar levels and is released into the blood when this level begins to rise following a carbohydrate-rich meal. Insulin synthesis proceeds in the pancreas, in special cells called the islets of Langerhans. In humans with type I *diabetes mellitus* (commonly referred to as diabetes), islet degeneration due to autoimmune processes results in insufficient production of insulin. The incidence of type I diabetes in humans lies between one and five percent. This disease has been known for centuries. In the second half of the nineteenth century, there was mounting evidence that a functional disorder of the pancreas is responsible for diabetes. It was the French pathologist Gustave Edouard Laguesse (1861–1927) who identified in 1893 the "islands", first described by the German pathologist Paul Langerhans (1847–1888) as the sites of hormone production. The term insulin originates from *Insel*, the German word for island or islet. Isolation of insulin from cattle and the first treatment of diabetic patients with insulin began in Canada in the 1920s. For decades it was subsequently extracted from tons and tons of porcine (pig) pancreas because porcine insulin is very similar to human insulin. Hundreds of millions of people were treated with porcine insulin over the years and thus enabled to live a life without the serious consequences of high blood sugar.

What is insulin from a chemical standpoint?

Insulin is a protein consisting of two amino acid chains connected by sulfur bridges. In comparison with many other biologically active proteins, insulin is relatively small. The A-chain consists of 21 amino acids, and the B-chain, of 30 amino acids (Figure 58). The exact sequence of amino acids in both chains was determined by Frederick Sanger, for which he received the Nobel Prize in 1958. The total chemical synthesis of insulin was achieved by Helmut Zahn (1916–2004) (Aachen, Germany) in 1963. This was a remarkable feat and a magnificent confirmation of the amino acid sequence of insulin as determined by Sanger years earlier. However, it is much more expensive to synthesize insulin chemically than

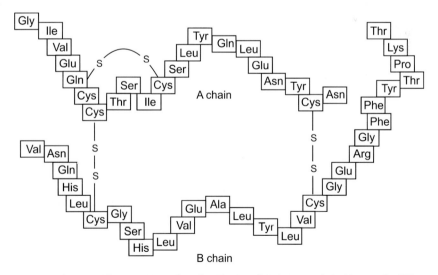

Figure 58 Amino acid composition of insulin. The A and B chains are linked by two disulfide bridges between cysteine residues (Cys). An additional disulfide bridge connects two cysteine residues within the A chain. (Diagram: Anne Kemmling, Goettingen, Germany.)

to isolate it from pancreas extracts, so this alternative is uneconomic. However, when genetic engineering emerged in the laboratories, a new era of insulin production began. In 1978, the gene of human insulin was cloned in *Escherichia coli* by Genentech scientists. Human insulin was Genentech's second recombinant protein, and it became the first one marketed by Eli Lilly in 1982. This was no easy task because insulin consists of two chains. Researchers had to take into account that the precursor of insulin formed in the islets of Langerhans, called proinsulin, is a single protein chain containing both of the insulin chains. Following its synthesis, proinsulin is cleaved at two positions to yield active insulin. The messenger RNA containing the genetic information for proinsulin synthesis was isolated. Then the two respective segments coding for the A-chain and the B-chain were each cloned into plasmids so that each chain was produced in a different *E. coli* strain. These chains were purified and, finally, linked by sulfur bridges. Insulin production processes were further developed by Eli Lilly and other companies, for example, Novo Nordisk and Hoechst (now Sanofi Aventis).

For the millions of diabetics worldwide who use reasonably priced human insulin, this is real progress. Tons of recombinant *E. coli* cells are grown and harvested. These cells are loaded with one of the precursors of insulin. Once the cells have been disrupted, these precursors are isolated and converted into the active hormone. Of course, researchers have tried to create new forms of insulin, which is relatively easy using current technologies. The nucleotide sequence is just altered in a certain position so that a particular amino acid in one of the chains is replaced by a different one. This modified insulin is then produced, and its function as a hormone for blood-sugar regulation is tested extensively. Modified insulins having a prolonged or an immediate effect were discovered in this way. There's no question that the use of human insulin from bacterial cultures has been the greatest success story of genetic engineering up to now.

Apart from bioethanol, are there other chemicals manufactured using bioengineered microorganisms?

Yes, and a good example is 1,3-propanediol:

$$HOH_2C-CH_2-CH_2OH$$

Propanediol has, so to speak, two hands in the form of two OH-groups, and these in turn can be linked on the right and on the left to other two-handed molecules such as terephthalic acid with its two COOH groups:

$$-OH_2C-CH_2-CH_2O-\underset{O}{\overset{\|}{C}}-\!\!\!\left\langle\bigcirc\right\rangle\!\!\!-\underset{O}{\overset{\|}{C}}-OH_2C-CH_2-CH_2O-$$

There are polymers formed that are in great demand as plastics. Many anaerobic bacteria, including *Clostridium*, *Citrobacter*, or *Lactobacillus* species, produce 1,3-propanediol from glycerol. Glycerol is also a 3-carbon compound like propanediol, but it contains an additional OH group on the central carbon atom. Two enzymes are required for removal of this OH group. The first one is glycerol dehydratase, which contains a certain form of vitamin B_{12} and acts to remove water from glycerol. This enzyme is a difficult partner in the production process because it is oxygen sensitive. The second enzyme is called propanediol dehydrogenase. Although there is a lot of glycerol on the market as a side product of rapeseed oil conversion with methanol to biodiesel, from an economical standpoint it was necessary to start with the glucose from starch saccharification. An account of the genetic modification of E. coli into an economical 1,3-propanediol-producing organism is given here by Gregory Whited from Genencor in Palo Alto, California (USA):

"Recently, E. coli has become a favorite production host for many bacterial production systems, including the production of commodity compounds, like 1,3-propanediol. Using the same metabolic pathways as E. coli, the baker's yeast Saccharomyces cerevisiae can make glycerol from glucose using two additional enzyme reactions that are not naturally found in E. coli. Glycerol itself can be converted into 1,3-propanediol by common organisms such as Klebsiella or Citrobacter in a fermentation process that employs two enzyme reactions that allow these bacteria to grow without oxygen, again not present in E. coli. When the two enzymes from Saccharomyces for making glycerol and the two enzymes from Klebsiella were combined genetically and expressed in E. coli, the recombinant E. coli made 1,3-propanediol from glucose. Although this process worked quite well, the recombinant E. coli still needed to be further modified genetically to function as a commercial production organism. This involved a number of steps such as 1) removing redundant metabolic pathways to restrict the flow of carbon into the pathway for 1,3-propanediol production and away from pathways leading to growth, 2) changing the primary way that glucose was transported into the cell to allow for a higher percentage of carbon to enter the production pathway 3) increasing the metabolic steps to reactivate and maintain the sensitive B_{12} cofactor, and 4) removing genes that could metabolize the product 1,3-propanediol to further acids or aldehydes that were unwanted. Altogether, the final production strain contained 6 bacterial genes from Klebsiella, 2 genes from Saccharomyces, and over 20 different modifications to the chromosome to delete, overexpress, or underexpress native genes and regulatory systems.

The biological production of 1,3-propanediol via this metabolically engineered E. coli strain has been commercial since 2006 when DuPont Tate & Lyle opened its first production plant in Loudon, Tennessee. This plant produces 100 million pounds [45 million kilograms] of 1,3-propanediol per year."

Looking at Figure 59, let's sum it all up: the genes for two enzymes were transferred from baker's yeast into E. coli. Together with its own gene portfolio, E. coli

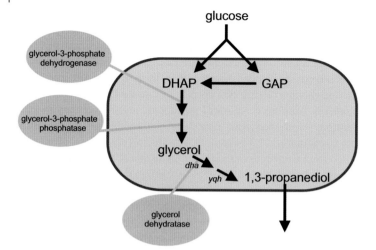

Figure 59 Construction of an E. coli strain able to synthesize 1,3-propanediol from glucose. Two genes have been introduced from the yeast *Saccharomyces cerevisiae*, which are able to convert dihydroxyacetone phosphate to glycerol. Furthermore, the B_{12}-containing glycerol dehydratase genes from *Klebsiella pneumonia* have been introduced. yqh is a dehydrogenase converting beta-hydroxypropionaldehyde into 1,3-propanediol. Certain pathways in the production strain have been blocked, as in the ethanol production strain depicted in Figure 32b. (Diagram: Petra Ehrenreich, Goettingen, Germany.)

was then able to convert glucose into glycerol. Two additional genes were taken from a close relative of *E. coli*, *Klebsiella*, and also put into *E. coli*. The result was a production organism. Something of interest was observed by the researchers when they attempted to optimize the process. As it turned out, *E. coli* didn't even need the dehydrogenase from *Klebsiella*. Propanediol yields were even higher when the *Klebsiella* gene was omitted. It had not been known that *E. coli* also has this dehydrogenase, which does the work even more efficiently than the *Klebsiella* dehydrogenase.. However, it was absolutely necessary to transfer the dehydratase genes. The high-performance organism that had been constructed could then be put to work making 1,3-propanediol in big tanks.

Finally, we come to a biopolymer that is typically bacterial and can be produced in large amounts. It is the "fat" of bacteria, poly-beta-hydroxybutyric acid with its easy-to-remember abbreviation PHB, along with its derivatives, generally called poly-beta-hydroxy fatty acids, the PHFs.

Like plants and animals that accumulate storage materials such as starch, glycogen, or fats, bacteria also "plan ahead" for hard times. Many microbes accumulate starch, but quite a number of them store PHB, which can accumulate inside the cells in large amounts. In time of need, PHB can be used as a nutrient to ensure survival of these bacteria.

Accumulation of PHB is especially pronounced in the bacteria known as knallgas bacteria. They are named for their ability to utilize a mixture of molecular hydrogen and molecular oxygen, which normally explodes with a big bang (Knall) when ignited, but not if the mixture is utilized by bacteria. They are capable of gently performing the reaction of H_2 and O_2 to water. The enzyme, hydrogenase, is able to extract electrons from hydrogen and transfers them via an electron-transport chain to oxygen, as already shown in Chapter 8.

In nature these bacteria are probably quite rare. Where would they even find knallgas, this mixture of $H_2 + O_2$?

Surprisingly, they are quite common. Knallgas bacteria can be isolated from practically any soil sample simply because H_2 is produced in deeper soil layers as a result of fermentation processes. This H_2 diffuses to upper soil layers where it meets O_2, and this is exactly the habitat of the knallgas bacteria.

The storage material PHB was discovered in the 1920s by the French chemist Maurice Lemoigne (1883–1967) in *Bacillus* species. There was a revival of interest in this material in the 1960s when it was discovered that the knallgas bacterium *Ralstonia eutropha* is able to accumulate PHB up to 80% of its cell mass. Now that's really overweight! PHB is present as intracellular granules that look much like filled flour sacks under the electron microscope (Figure 60).

PHB is a bioplastic material: it can be dissolved, precipitated, and used to make foils, bags, or even bottles. When we were discussing 1,3-propanediol, the two OH groups of the molecule were compared to two hands linked to terephthalic acid to form a polymer. In beta-hydroxybutyric acid, the two hands are an OH group and a carboxyl group. When activated, these molecules react with each other to form a polymer.

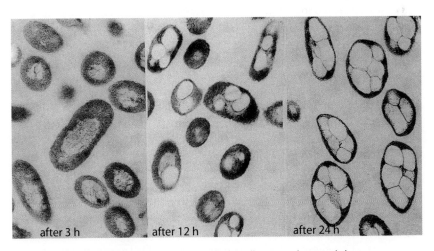

Figure 60 Electronmicrograph of *Ralstonia eutropha* cells accumulating poly-beta-hydroxybutyric acid (PHB) during growth. Pictures were taken after 3, 12, and 24 hours of growth. (Hans Guenter Schlegel *et al.* Nature 191, 463–465, 1961.)

$$-\text{OCH}-\text{CH}_2-\underset{\underset{O}{\|}}{\overset{\overset{CH_3}{|}}{C}}-\text{OCH}-\text{CH}_2-\underset{\underset{O}{\|}}{\overset{\overset{CH_3}{|}}{C}}-$$

Two discoveries have led us to expect that the PHFs soon will become a mass-market product to replace nonbiodegradable plastics. One of these developments is the production of copolymers consisting of beta-hydroxybutyric acid and beta-hydroxyvaleric acid (butyric acid consists of four carbon atoms, and valeric acid, of five). These copolymers have improved properties because they are not brittle like PHB. In addition, PHFs are thermoplastic, so they can be pressed into various forms. A whole array of biologically degradable, PHF-based biopolymers is now available. Only the price still restricts their production and their use. The second development involves genes that already were cloned in the 1980s from *Ralstonia eutropha* but can now be transferred to other microorganisms. Most importantly, they can be transferred to plants, which then begin to produce PHB. Should it be possible to induce plants to accumulate large amounts of PHFs, then the price will no longer stand in the way of a broad application of these biopolymers. The latest developments are summarized by Alexander Steinbuechel (Muenster, Germany):

> "Research on PHB and on PHFs has attracted me ever since I started my scientific career 25 years ago. The chemical structure of these molecules is relatively simple, but the metabolism as well as biogenesis and structure of the PHF granula is a class of its own. So it is possible to produce with bacteria a whole array of chemically different PHFs tailored for desired properties. Also the susceptibility for biological breakdown is determined by the chemical composition of PHFs. These polymers are close to their economical breakthrough: the company Metabolix (Clinton, Ohio, USA) started the production of PHFs on a 50 000-ton scale using genetically modified *E. coli* strains. The production of PHFs using transgenic plants has been delayed, especially in Europe, by discussions on the acceptance of products coming from green biotechnology."

Perhaps eyesores such as piles of packaging material on land or the accumulation of plastic bags in an otherwise beautiful ocean will soon belong to the past.

Finally, we come back to thermostable enzymes that were already discussed by Gregory Zeikus in Chapter 6. They are finding their way more and more into various applications, which Garabed Antranikian (Hamburg, Germany) tells us about:

> "Microorganisms living under extreme conditions have always fascinated me. Which adaptation mechanisms have they developed, and what possible fields of application are there for their robust enzymes? Even though we haven't yet elucidated all the secrets of these skilled survivors, I notice with pleasure that their enzymes (extremozymes) have become indispensable in many fields, and that they have revolutionized modern biotechnology. The

triumphal procession of PCR technology would not have been possible without the application (use) of thermostable DNA-polymerases. We owe *Taq* polymerase and its relatives from *Pyrococcus* and *Thermococcus* many breakthroughs in gene technology and in sequencing genomes from bacteria to humans. Genome sequencing, in turn, puts us in a position to explain the basic principles which permit survival under extreme conditions.

We come across extremozymes at home in detergents and washing up liquids. They make our washing spotless white and give jeans their washed-out look. Thermostable starch-processing enzymes give Coke its sweetness, and salt-loving microorganisms produce protective substances for our skin. But the real triumphal procession of enzymes and extremophile microorganisms is still impending: with the increasing use of renewable resources, stable biocatalysts will also become significantly relevant. The enormous diversity of extremophile microorganisms will provide us with novel enzymes to be used for the conversion of lignocellulose-containing biomass. In biorefineries of the future, extremophiles will play an important role."

Isn't it fantastic how bacteria, as natural wild-type strains or as part of a genetically engineered system, are able to contribute to our health and to life in general? Hormones, antibiotics, enzymes, solvents, fuels, and biopolymers – we owe all of these to the microbial world.

> The other lords, like lions wanting food,
> do rush upon us as their hungry prey
>
> William Shakespeare, Henry VI

Chapter 29
Plants, animals, and humans as food resources for bacteria

Isn't it macabre to think of us as a source of bacterial food?

We all have come to appreciate the role of bacteria in nature. They decompose all dead material and participate in the cycles of elements. Why should they stop when confronted by living organisms? Bacteria have developed strategies to invade animals and plants. Higher organisms are only able to survive by developing an effective counterstrategy. If we stop and look at counterstrategies for a moment, one of these is the protection offered by "bacteria-proof" surfaces such as the human skin or the surfaces of leaves and fruit in plants. How important this protection is becomes apparent when a skin wound gets infected or when insects manage to damage fruit or leaves.

Even when the outer surface remains intact, there are natural openings through which microbes can enter, including plant stomata (openings in leaves) or openings of the body: the mouth and nose as well as the respiratory, gastrointestinal, and urogenital tracts. However, these areas of the body are protected from microbes by a coating called epithelium. This coating is densely populated by bacteria, including natural, or indigenous, microbes that are tolerated (immune tolerance). Pathogens, however, that manage to penetrate the epithelial barrier are confronted with our immune system.

The innate immune response is immediate; it includes inflammatory processes and deployment of macrophages ("big eaters") to intercept an intruder such as a bacterium, which is then engulfed and digested. Similar to trophies, the macrophages subsequently display on their cell surfaces various components (antigens) of the digested intruder. This initiates the specific immune response, which deals with invaders escaping the first line of defense. The macrophages "show" these antigens to other cells of the immune system, the B cells and the T cells, which go into action. There are several types of T cells. The T helper cells coordinate the body's immune response by activating T killer cells that are then able to "recognize" and destroy the bacterial intruder directly, and by assisting the B cells in producing antigen-binding antibodies on the B-cell surface. The B cells then form two cell types, plasma cells and memory cells. In interaction with T helper cells, plasma cells produce and release specific antibodies that circulate throughout the body, functioning as watchdogs of the immune system. The long-lived memory cells are around for years, ready to proliferate when confronted with "their" antigen again.

Let's look at whooping cough, a bacterial infection caused by *Bordetella pertussis*. These bacteria colonize the epithelial cells of the upper respiratory tract and produce pertussis toxin, one of several virulence factors. The body's innate immune response to these bacteria is followed by the specific immune response described above. The memory cells produced in this case are still around when a second infection with *Bordetella* occurs. Therefore, the body's "rapid response team" is prepared, and the whole infection is quickly brought under control.

This immunological memory is helpful as long as the antigens, meaning the characteristic surface structures of bacteria or viruses, have not changed. Many viruses and bacterial species are genetically capable of changing their surface structures by mutation. This is especially the case for a number of *Salmonella* species or for the virus species that cause colds. In others, the antigenic variability is low, for example, in the polio virus or the bacterium that causes whooping cough.

We have all had bacterial and viral infections at one time or another. Once the symptoms appear, we know that our immune system is attempting to bring the infection more and more under control. However, massive infections of the respiratory tract or following extensive burns may end fatally without appropriate treatment. Immunodeficient patients that contract certain infections are especially at risk.

Which infections should I be worried about when the skin is broken or when there is dirt in a wound?

Such wounds will be quickly colonized by the pyogenic (pus-producing) organism *Staphylococcus aureus*. Gas gangrene, which is more common in deeper wounds, is caused by *Clostridium perfringens* and *Clostridium histolyticum*. This type of infection was life-threatening in earlier times, especially during World War I. Many wounded soldiers died from clostridial infections and the accompanying edema, an accumulation of fluid in the body. These clostridia are capable of degrading the structural protein of tissues, collagen. The resulting liquefied tissue is then subject to fermentation. Fortunately, this type of infection is now rare, and it can be kept under control by wound disinfection and antibiotic treatment.

Tetanus is an often deadly disease caused by *Clostridium tetani*, whose spores are commonly found in soil. In wounds with conditions favorable for spore germination of this microorganism, the resulting cells may produce the tetanus toxin. This neurotoxin is the second most toxic compound known, next to the botulinus toxin. The lethal dose is around 1 nanogram (one billionth of a gram, 10^{-9} g) per kilogram body weight. Fortunately, a highly effective vaccine has been available for the past 70 years. It is a tetanus toxoid produced by treating the toxin with formaldehyde. Immunization protection lasts for a period of six to ten years, so immunization should be renewed, especially after possible contact with these microbes after this period. Thanks to this immunization, tetanus is very rare in

Figure 61 Painting of a man suffering from tetanus. (Charles Bell, Royal College of Surgeons in Edinburgh, with permission.)

industrialized countries. Worldwide, however, it is still a big threat, with nearly 400 000 cases per year, including many newborns – of unvaccinated mothers – infected via the umbilical cord.

The tetanus toxin is an enzyme that cleaves a certain bond in a protein (VAMP) located at the endplates between nerves and muscles, thereby destroying a complex required for release of neurotransmitters. This causes a continuous muscle contraction affecting the whole body, usually beginning in the neck and jaw (Figure 61). That's why tetanus is commonly called lockjaw.

The genetic information for tetanus toxin synthesis is located on a plasmid. But we don't need to worry that this toxin-producing ability can spread to other bacterial species. Toxin production requires an additional factor encoded on the *C. tetani* chromosome. In other words, the genetic information on the plasmid alone is insufficient for toxin production. By the way, the tetanus protein is so toxic that even minute amounts can't be used for immunization. Our immune system simply is unable to respond to such a small dose, so it's good that immunization with the toxoid mentioned has proved effective.

Aren't insect bites and stings also gates of entry for bacteria?

Of course, not only for bacteria but also for the organism causing malaria, *Plasmodium falciparum*, which is transferred by *Anopheles* mosquitoes. *P. falciparum*, which has a nucleus, is not a bacterium but a protozoan. This organism proliferates in erythrocytes (red blood cells) and causes recurring high fever. Annually, several hundred million people are infected with this infectious disease, and nearly two million die.

The bubonic plague is a typical bacterial infection transmitted by insect bites. The microorganism responsible is *Yersinia pestis*, which is transferred by fleas from rats to humans. In medieval times outbreaks of the bubonic plague wiped out a considerable percentage of the human population. Fortunately, it is of little significance today. Another example is the tick-borne Lyme disease, or borreliosis, which has emerged in recent years. In the United States it is mainly caused by *Borrelia burgdorferi*. If not treated properly, early symptoms such as fever and a circular skin rash surrounding the bite will be followed by swelling and pain in the large joints, heart problems, and disorders of the nervous system. The circular rash is a symptom that does not always appear after infection with *B. burgdorferi*. If tick bites are known to have occurred, one should be on the alert for other symptoms such as fever, headache, and swelling of the joints.

We all have to breathe, but isn't it dangerous because of the bacteria in the air?

Yes, our respiratory tract is one of the most important gates of entrance for bacteria and viruses. Legionnaire's disease, known since 1976, is a good example of how people can be infected by continuously breathing in contaminated air. This disease affected 200 participants of the Pennsylvania State Convention of the American Legion in Philadelphia, and 30 of them succumbed to this disease. All participants had stayed at the same hotel, and the hotel staff was not affected. Finally, it was realized that the air-conditioning system for the hotel guests had something to do with the infection. During the night, the veterans had somehow been infected by inhaling contaminated air. It was a frightening situation all over the United States until the research group of Joseph McDade at the Center for Disease Control (CDC) in Atlanta, Georgia, clarified the situation. They isolated the previously unknown bacterium responsible for the disease, later to be named *Legionella pneumophila*. This microorganism lives and proliferates in its ameba host, which prefers warm water as a habitat. It is found not only in air-conditioning systems but also in warm-water systems in households and swimming pools. Legionnaire's disease is a good example of how we affect the microbial world around us by providing all sorts of new habitats, in this case warm water containing not only the bacteria but also protozoans and their nutrients. Under such conditions, the organisms may reach numbers sufficient for disease outbreaks. One of the conclusions drawn from this tragic event is that it is safer to store very hot water and mix it with cold water just before use. Legionnaire's disease also shows us that the appearance of new, pathogenic microorganisms is ongoing.

Now we'll go back more than 125 years. Visitors to Berlin usually walk down a wide avenue, "Unter den Linden," toward Brandenburg Gate, one of the main attractions of Berlin. It was also the site where the Berlin Wall and the division of Berlin were alarmingly visible until November, 1989. Why not take the parallel street, the "Dorotheenstrasse?" There, not far from Brandenburg Gate, you will reach a nice old building belonging to the Humboldt University. The front still shows some damage from World War II. There, next to one of the doorways, is a plaque with the following inscription:

> "In this building, Robert Koch delivered a lecture to the Physiological Society of Berlin on March 24, 1882, in which he announced his discovery of the organism causing tuberculosis."

This was one of the most outstanding scientific lectures of all times. Only a few days later, there was already a comment in the *New York Times*.

By using a special staining technique, Robert Koch (1843–1910) was able to demonstrate in tissues the presence of the organism causing tuberculosis. He used an old solution of methylene blue that had been standing in the laboratory and – as was found out later – had taken on an alkaline pH from the glass in which it was stored. Because of this, Koch had a hard time repeating his results with a fresh solution of methylene blue. The organism he discovered is now called *Mycobacterium tuberculosis*. Robert Koch concluded in his lecture that the bacteria in tissue from patients suffering from tuberculosis can be separated from this tissue and be grown in laboratory cultures. When these cultures are used to infect animals, the animals will also get tuberculosis. Thus, there is a connection between the development of tuberculosis and the presence of the bacteria that Koch had isolated.

As for tuberculosis, also called the "white plague," nine million people become infected every year, of which two million do not survive. There is still no potent vaccine available. There is one that was developed in the 1920s by the French microbiologists Albert Calmette (1863–1933) and Camille Guérin (1872–1961) from the Institut Pasteur in Paris. This BCG vaccine is based on the microbial strains causing bovine (cattle) tuberculosis. It is unfortunately not effective enough, so various research centers are intensely involved in the development of a new, more potent vaccine.

Mycobacterium tuberculosis (Figure 62) grows intercellularly, that is, it invades cells that are destroyed when the organism proliferates. There is something mysterious about tuberculosis bacteria: they may persist for years inside the cells in an apparently dormant state. A certain signal causes them to revert back to the active form – this is the outbreak of tuberculosis. Apparently the organisms contain a gene that codes for a regulator protein called PhoP. Whenever this protein is present in large amounts, the "sleeping" genes are turned off and the bacteria "awaken". These findings point the way toward development of a strategy to cure tuberculosis. If researchers find a way to keep the concentration of PhoP low, then the bacteria would stay dormant. We have asked Stefan Kaufmann (Berlin, Germany) to tell us about the various possibilities for prevention and treatment:

> "As already mentioned, approximately two million people die of tuberculosis every year. This situation has become increasingly critical because multiresistant or even extremely resistant organisms causing tuberculosis appear more and more frequently, which makes treatment even more difficult. The vaccine BCG protects small children from severe forms of tuberculosis, but the most common form of this disease, adult pulmonary tuberculosis, is not prevented by BCG. So it is urgent to have new drugs and vaccines available. We are trying to improve the BCG vaccine so that it stimulates a broader repertoire of T lymphocytes, which are involved in

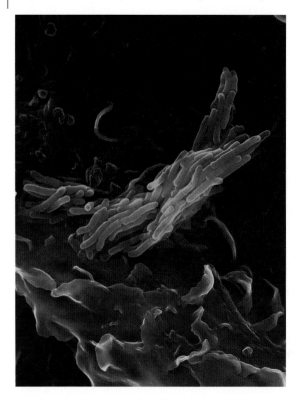

Figure 62 *Mycobacterium tuberculosis* in bone-marrow macrophages of a mouse. The bacterial cells are approximately 0.4 µm in diameter and 1.2 µm long. (Micrograph: Volker Brinkmann and Stefan H.E. Kaufmann, Max-Planck-Institute for Infection Biology, Berlin, Germany.)

the protection against tuberculosis. The new vaccine developed works in animals and has passed a first clinical trial successfully. The increasing resistance of *M. tuberculosis* to current drugs requires the development of new chemotherapeutics. So far, all drugs are effective against the metabolically active tuberculosis organisms. Drugs that also act on the organisms in their dormant state are urgently needed. The so-called dormant gene products are useful targets for the development of new drugs. Unfortunately, research on the control of tuberculosis has been neglected in the last few decades. Enormous efforts must be made to catch up and to effectively combat this scourge of humanity."

Let us hope that tuberculosis research receives the necessary attention and support.

If I ever get bronchitis or even pneumonia, can I safely assume that it's not caused by Legionella pneumophila or Mycobacterium tuberculosis?

You can be pretty sure because infections of the respiratory tract are primarily viral infections. Classical influenza viruses or corona viruses, which include the SARS virus, are the foremost causing agents. These viruses are responsible for colds, coughs, and inflammatory processes of the respiratory system. However, it gets dangerous when bacterial infections are superimposed upon viral infections. These superinfections are often caused by *Streptococcus pneumoniae, Chlamydia pneumoniae,* or *Mycoplasma pneumoniae.* The species names reveal which environment is preferred by the organisms: the *pneumon,* the Greek word for lungs. Colonization of the lung by *Pseudomonas aeruginosa* (Chapter 27) also may lead to infections and complications in patients. Much depends on the general health of the person infected. It's a shame that the elderly often die of pneumonia. In such cases, the immune system is unable to cope with a massive viral infection or with a combination of viral and bacterial infections, which the body no longer can keep under control. It is simply overwhelmed by slime and toxins.

We now direct our attention to a different bacterium that is capable of invading the respiratory tract, *Bacillus anthracis.* Shortly after September 11, 2001, five letters containing a white powder were mailed within the United States. Later, this powder was found to contain spores of *Bacillus anthracis,* the organism that causes anthrax. Among the addressees were Senators Thomas Daschle and Patrick Leahy. A total of eighteen people were infected, of which five died. *B. anthracis* is the organism with which Robert Koch first demonstrated the correlation between a specific disease and the causative organism (see Chapter 22). There are several ways for *B. anthracis* to invade its target organism. Of the three forms of anthrax disease (lung, skin, and intestine) the one infecting the lungs is especially dangerous. It develops when spores prepared for use in biological warfare get into the lungs. High fever, chills, and failure of lung and circulatory function may be the result. Only when it is identified at an early stage can lung anthrax be treated successfully with penicillin or other antibiotics, such as doxycycline. How dangerous anthrax really is was demonstrated in 1942 in a morbid experiment on the Scottish island of Gruinard, which was purposefully contaminated with *B. anthracis* spores. Even 20 years later, sheep brought to the island died within a short period of time. Finally, after 48 years, the whole surface of the island was disinfected chemically so that it could be made accessible to the public again.

Bacillus anthracis contains a chromosome and two plasmids, on which the genes required for virulence are located, including genes for capsule formation and several toxins.

Do anthrax toxins work the same way as tetanus toxin?

No, they are quite different. In anthrax, the bacterial capsule plays an important role in virulence. The capsule consists of polyglutamic acid, which envelops the bacterial cell and protects it from attack by the host cell. Following infection of the host cell, *B. anthracis* is able to proliferate with little interference from the host. The two toxins produced by this bacterium are essentially different combinations of the components EF, PA, and LF. Together, they cause edema (swelling caused by fluid buildup) and necrosis (tissue breakdown and cell death). When animals finally die of an infection with *B. anthracis*, up to ten million bacteria are present in one ml of their blood.

Great efforts have been made in recent years to develop new vaccines against anthrax by using a nonpathogenic strain of *B. anthracis*, which no longer has a capsule and produces only one of the three toxin components, PA.

We will continue to discuss toxins as we turn to infections of the gastrointestinal tract. In these cases, major sources of infection are contaminated water, contaminated food, and droplet infections, which are spread by sneezing or coughing, from man to man or from animals to man. *Salmonella* strains are often responsible for what is commonly called food poisoning, especially strains of *S. typhimurium*. Meat, fish, or eggs may contain so many *Salmonella* cells that the toxin present is sufficient to make people ill within hours. The toxin in this case is an exotoxin, a component of the bacterial cell wall, the extremely toxic lipoid A. It causes the macrophages to liberate large amounts of biomodulators (e.g., interleukins). The result may be high fever, circulatory problems, and organ damage. Although extremely rare, botulism fits into the discussion at this point. *Clostridum botulinum* is able to grow on high-protein foods. Under certain conditions, this bacterium produces a neurotoxin that is the most toxic compound known to man. The lethal dose is around 0.4 nanograms per kilogram body weight, a toxicity that exceeds our imagination. One kilogram or two pounds of the toxin would be enough to kill everyone on our planet. Nevertheless, it is also used in minute amounts to treat some forms of muscle spasticity, for example, to correct strabismus (cross-eye). Preparations of botulinum toxin have even been used successfully to smooth skin wrinkles. Fortunately, this toxin is not very stable, so it would not be useful for biological warfare. However, the spores of *B. anthracis* are extremely stable and capable of germinating in the lungs, where the organism can multiply and produce toxin. This unfortunately opens the door to possible misuse.

Enough of biological warfare. What causes, for example, diarrhea?

We'll consider two types of diarrhea, beginning with cholera, which is caused by *Vibrio cholerae*. This bacterium was first observed in 1854 by Filippo Pacini (1812–1883) and isolated by Robert Koch from patients suffering from cholera, in 1883. Actually, *V. cholerae* prefers marine habitats and proliferates in fish, mussels, and other marine animals. If it gets into humans by way of contaminated water, for instance, it colonizes the small intestine and produces cholera toxin, CTX. There have even been recent outbreaks of cholera, most recently in 2008 in Zimbabwe and in 2010 in Haiti.

What does CTX do in the human body?

Actually, it's quite ingenious. CTX consists of two subunits. Five of the B subunits first form a channel through the membrane by squeezing into the membrane structure of the host cell and forming a ring. Then an A subunit enters the cell through the ring, where it inhibits by substitution the activity of a regulatory G protein on the inside of the cell membrane. Experts would call it ADP-ribosylation. Various reactions follow, including a massive efflux of water and minerals from the cell. This loss completely upsets the body's balance of water and minerals.

Our second example is *Clostridium difficile*, an organism little known to the public. *Clostridium difficile* is normally a harmless inhabitant of the intestine. However, it has picked up a number of resistances against antibiotics. When a patient receives antibiotic treatment, the normal intestinal microflora is more or less destroyed. Under such conditions, *Clostridium difficile* eventually becomes dominant in the intestine. Diarrhea is caused by its two CD toxins, A and B, which bind to the cell membrane and are taken up by endocytosis. Within the cells, processes similar to the ones described for CTX (above) take place. As before, regulatory processes are disturbed. There is damage to the cells of the intestinal epithelium, loss of barrier function, and efflux of water are followed by inflammation.

Similar strategies are pursued by several bacterial species that also cause severe damage to the intestinal epithelium. The organisms causing plague (*Yersinia pestis*) or typhoid fever (*Salmonella typhi*) are among these. *Yersinia pestis*, for example, sends effector proteins into the host cell with the help of a transport apparatus, the type III secretion system (T3S). It is fantastic how this syringe-like protein structure grows out of the bacterial cell, punctures the eukaryotic host cell, and injects bacterial effector proteins into the host (Figure 63). These effector proteins then arrange uptake of the bacterium by the host cell, called phagocytosis, in which the bacterium is engulfed and internalized in a phagosome. The bacteria reside and proliferate in this compartment of the infected cell, which becomes increasingly weaker and is finally destroyed. The bacteria are released to continue their destructive work in other cells.

We should make a remark on our intestinal commensal, *Escherichia coli*. This bacterium is usually assumed to be harmless. It only makes up a small percentage of our intestinal flora as described in Chapter 10. However, there are *E. coli* strains that differ from the rest and have some really nasty properties. Some of them cause diarrhea, whereas others are responsible for infections of the urogenital tract. Many of those gastrointestinal upsets that may trouble us in some tropical countries are assumed to be caused by *Salmonella* species. Often this is not the case. Instead, pathogenic *E. coli* strains are responsible, including the ETECs (enterotoxigenic *E. coli*) or the EHECs (enterohemorrhagic *E. coli*). EHECs are especially dangerous because they express the Shiga-toxin which destroys erythrocytes. This pathotype was responsible for the *E. coli* outbreak in California and, subsequently,

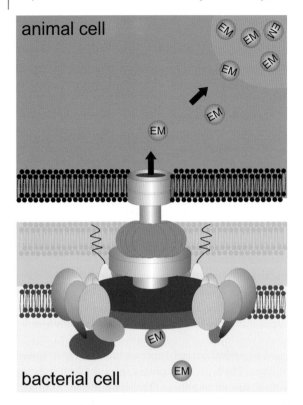

Figure 63 Model of T3S (Type III Secretion Systems). The apparatus is anchored in the bacterial cytoplasmic membrane. The "needle" of the apparatus penetrates the membrane of the animal cell. Effector molecules (EM) are released into the animal cell. (Drawing: Anne Kemmling, Goettingen, Germany.)

in many neighboring states in 2006. The 2011 outbreak in Germany was caused by an entero-aggregative *E. coli* (EAEC) which had picked up the genes for the production of the Shiga-toxin. Because of its aggregative abilities, it colonizes the large intestine and forms toxin-producing biofilms. It is the most dangerous *E. coli* pathotype known so far and may be called EAHEC (entero-aggregative hemorrhagic *E. coli*). Bladder infections are linked to the presence of UPECs (uropathogenic *E. coli*). How the bladder epithelium looks in such a case can be seen in Figure 64.

Why have the properties of E. coli changed so dramatically?

The genomes of these strains contain "pathogenicity islands" (PAIs), which were described in the 1980s by Joerg Hacker and Werner Goebel (Wuerzburg, Germany). PAIs are insertions in the genome that can make the host sick. These islands carry genes for erythrocyte-destroying hemolysins or genes for toxins. We have

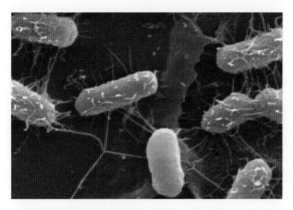

Figure 64   Uropathogenic strain 536 of *E. coli* on bladder epithelium. (Electron micrograph: Hilde Merkert, Wuerzburg, Germany.)

asked Joerg Hacker to describe the discovery and importance of the pathogenicity islands:

> "Pathogenicity islands are relatively large inserts in the genome of pathogens, and these inserts encode a number of morbid factors, the so-called virulence factors. In the 1980s we described that uropathogenic *Escherichia coli* strains would spontaneously lose genes encoding virulence factors. It then turned out that these genes resided on large genomic fragments, which we called pathogenicity islands (PAIs). They are characterized by a number of specific properties. They usually carry direct repeats at their ends, they are inserted in tRNA-genes, and they code for a specific integrase. This integrase is involved in the chromosomal excision but also in the integration of the PAIs. In the meantime, PAIs were found in more than 30 species of pathogenic microorganisms. Furthermore, it was shown that similar structures also occur in genomes of nonpathogenic microorganisms. These fragments are called genome islands, and the pathogenicity islands are just a subclass of them. Others are the fitness islands or the resistance islands. Because of these insights in genome sequences, we know today that genome islands play a significant role in the evolution of prokaryotes."

PAIs are dangerous, not only because of the genes encoding virulence factors but also because of their mobility. They resemble plasmids in a way.

Of all the organisms discussed so far, only the ones that cause plague and tuberculosis grow intracellularly. Are there any others?

Most pathogenic bacteria remain outside the host cells. They inject toxins, like *Vibrio cholerae*, or effector molecules, like *Yersinia pestis*, or they attack the cells with exotoxins. *Mycobacterium tuberculosis*, *Yersinia pestis*, and *Salmonella typhi* invade cells, but they remain surrounded by a membrane and they proliferate in the phagosomes of the cells. One of the few bacteria that multiply in the cytoplasm of a host cell is *Listeria monocytogenes*. How they manage to do this will be described by Werner Goebel (Munich, Germany):

> "Most of the readers of this book have probably never heard of *Listeria monocytogenes*. In a way, they should be glad because listeriosis, the disease caused by these human and animal pathogenic bacteria, can be life-threatening. The term "listeriosis" conceals the severity of the symptoms, including sepsis (blood poisoning) or encephalomeningitis. Fortunately, this pathogen is very effectively kept under control by our cellular immune system. Therefore, listeriosis is relatively rare, but the death rate of patients with listeriosis is around 25 percent!
>
> *L. monocytogenes* is a nightmare, especially for food producers, because these bacteria infect humans almost exclusively through consumption of contaminated food. Especially milk and meat products are sources of these microorganisms. To a large extent they owe their virulence to their ability to invade the cytoplasm of many cells in our body, to efficiently proliferate in this intracellular niche, and to subsequently attack neighboring cells without being liberated from the previously infected cell. This unusual accomplishment of the Listeria is achieved with the help of a toxin (listeriolysin), two phospholipases, and a surface protein called ActA. As already discussed, the intracellular bacteria *Salmonella typhi* and *Mycobacterium tuberculosis* reside and proliferate in specialized phagosomal compartments after invading host cells, From there, they have no way to escape without destroying the invaded host cells. *L. monocytogenes* is, however, able to escape from the primary phagosome soon after internalization. First, the listeriolysin drills holes in the membrane of the phagosomes. Then the two phospholipases cleave the phospholipids of the phagosome membrane, which is then solubilized. In this way, Listeria reaches the cytoplasm of the infected cell. There, something amazing takes place, which is catalyzed by the ActA protein: at one end of each *Listeria* cell, polymerization of the monomeric actin molecules inside the host cell takes place, forming a tail of actin fibers behind the bacterium. This actin tail serves Listeria as a means of motility. As the tail grows longer, the bacterium is propelled through the initial host cell and into a neighboring cell.
>
> A prerequisite for efficient proliferation of *L. monocytogenes* in the cytoplasm of infected cells is the close synchronization of bacterial metabolism with the metabolism of the host cell. Until now we do not understand much about these complex metabolic processes, but there are several new techniques that may help to solve this crucial problem in the pathogenesis of *L. monocytogenes* and other bacterial infections."

It's hard to believe that intracellular bacteria propel themselves forward by inducing a tail of actin fibers behind them.

It really is unbelievable, but this phenomenon has been shown in electron micrographs and in laboratory experiments, and it has even been filmed under a microscope.

All these pathogens—which of them are primarily responsible for infections in hospitals?

You have touched on a serious problem; we have asked an expert, Michael Gilmore (Cambridge, Massachusetts, USA) to tell us about it:

> "Although it varies depending on the hospital ward, overall about 5 percent of hospitalized patients develop infections AFTER being admitted for care. Most of these infections occur in patients who have indwelling intravenous lines, urinary catheters, ventilation tubes, or who have recently undergone surgery, especially involving the GI [gastrointestinal] tract. Because these infections occur in hospitals where large quantities of antibiotic are used, they are often caused by antibiotic-resistant microbes. Leading causes of hospital acquired infection are coagulase negative staphylococci (mainly *Staphylococcus epidermidis*), *S. aureus*, and the enterococci (mainly *Enterococcus faecalis* and *E. aecium*). These are all Gram-positive cocci that are often multidrug resistant and can be very difficult to treat. Gram-negative bacteria cause hospital acquired infections to a slightly lesser extent, and the leading offenders are *P. aeruginosa*, *E. coli*, and its relatives. The AIDS epidemic, which prior to the development of effective treatments resulted in severe immune compromise, was accompanied by large increases in opportunistic fungal infections. Since many of these patients were hospitalized, this significantly changed the landscape of hospital acquired infections. *Candida albicans*, *Aspergillus* species and other fungal organisms emerged among leading causes of infection. Although these infections have been dramatically reduced in AIDS patients because of treatment that preserves the integrity of the immune system, these infections are now found in patients receiving organ transplants, and in cancer patients (in whom the immune system is wiped out intentionally to prevent organ rejection, or as an unfortunate consequence of cancer chemotherapy). Recently, there have been two alarming developments related to hospital acquired infection: The first is development and spread of pan-resistant strains of the Gram-negative microbe *Acinetobacter baumannii*. These microbes are resistant to all of the common antibiotics and are extremely difficult to treat. A second major concern is the transmission of resistance to one of the last line drugs, vancomycin, from vancomycin-resistant enterococci (VRE) to methicillin-resistant *S. aureus* (MRSA), resulting in VRSA. Since 2002, VRSA

> infections have been reported at a rate of about one to two cases per year, mostly in foot wounds of diabetic patients. There are few drugs for effectively treating these infections. Since hospitals provide an environment where antibiotic use is concentrated, they are the breeding ground for the spread of antibiotic resistance. In a sense, hospitals are where Darwin meets Newton—for every ecological pressure we place on microbes, there is an equal and opposite reaction."

It is interesting to bring Darwin and Newton together in this context. It must mean that whenever we artificially vacate an otherwise inhabitable niche (e.g., by antibiotic treatment) sufficient variability in microbial life already exists for something to move in. And if that "something" is not already a good fit, natural mechanisms of variation, mostly mutations, allow for the settling of increasingly fitter occupants. This is our dilemma in combating antibiotic resistance.

Now we come to a bacterium that was discovered in 1983 by Barry Marshall and John Robin Warren (Perth, Australia), as a resident of our stomachs. This discovery failed to attract much attention for a number of years, but we know now that it can cause stomach ulcers and is possibly involved in the development of cancer. The organism's name is *Helicobacter pylori*. In 2005, these two Australian microbiologists were awarded the Nobel Prize for their discoveries.

When *H. pylori* gets into the stomach, it has to tolerate a bath in weak acid somewhat longer than the bacteria that make it from the stomach into the intestine. It has to tolerate the acid until it has settled into the gastric mucosa. There is one enzyme system produced by *Helicobacter pylori* that helps by cleaving urea into ammonia and CO_2, and immersing the cells in a cloud of ammonia. After getting established, *H. pylori* produces a number of factors, including a cytotoxin that promotes inflammation and causes damage to the mucosa. As a result, ulcers develop in nearly 10 percent of the individuals infected by this organism, several million people worldwide. These conditions also seem to favor the development of tumors.

H. pylori apparently is the bacterium that has most frequently infected mankind. Nearly 50 percent of the world population carries this organism in their stomachs. Investigations by Mark Achtmann (Berlin) and Sebastian Suhrbaum (Hanover, Germany) and their research teams, using samples from people in all continents, have produced really exciting results. These were based on comparisons of gene sequences of various isolates of *H. pylori*. Genetically, this organism is very variable, so many of us—in terms of the genome sequence—are inhabited by a unique strain of this organism. Evolution of the various strains isolated can be reconstructed, however, and the results are breath-taking. They largely correspond with the migration of *Homo sapiens* that was based on sequence comparisons of mitochondrial DNA (see Chapter 27). So *H. pylori* migrated out of Africa along with its host, *H. sapiens*, around 100 000 years ago. The genomes of strains from Australia, China, Africa, Europe, and America differ genetically to an extent comparable with the differences found in human mitochondrial DNA. It is fascinating how the use of modern methods can lead to exciting discoveries about bacteria.

Figure 65 Imprint of a soybean leaf onto a methanol-containing agar medium. The imprint has been produced by pink-colored colonies of methylobacteria. (Photograph: Julia Vorholt, Zurich, Switzerland.)

Bacteria obviously get into plants as well, but how do they do it?

The fact that plants are a good source of nutrients for bacteria was already mentioned at the beginning of this chapter. The surfaces of plants, especially the leaves, are occupied by bacteria, just like our skin (see Chapter 10). If a leaf is pressed onto an agar plate for a moment, colonies will grow in a leaf pattern. The imprint of a soybean leaf (Figure 65) is very impressive because the growth pattern of bacterial colonies even reveals details of the leaf veins. Globally, leaves represent an enormous habitat (estimate: $6.4 \times 10^8 \, km^2$) for bacteria, giving home to 10^{26} cells. Julia Vorholt (Zurich, Switzerland) explains these observations:

> "Plants accommodate a variety of different microorganisms on their leaves, which are adapted to a life at the interface between the atmosphere and the leaf surface, the so-called phyllosphere. Many of these inhabitants do not cause any damage to the plant. Such a group of organisms are the methylotrophic bacteria, which interestingly enough can be found on practically all leaf surfaces. They exhibit a special metabolism allowing them to grow on one-carbon compounds such as methanol. This compound (wood alcohol) is produced by plants. Methylotrophy, the ability of these bacteria to utilize methanol, enables them to successfully compete with other microorganisms. They prefer niches, for example, the base of leaf hairs or the junctions between adjacent epidermis cells, especially along the leaf veins and stomata.

> For life on the surface of leaves, several adaptations are required. One is protection from UV light. Therefore, members of the genus *Methylobacterium* produce carotinoids and are pink-colored. A number of proteins are also produced to facilitate survival of these bacteria under the rapidly changing conditions on leaf surfaces. It will require further intensive research to better understand the life of microorganisms in the phyllosphere, the interaction between microorganisms and plants, and their roles in the biogeochemical cycles, in particular of carbon and nitrogen."

That's fascinating – would you please explain again what methylotrophy means?

Methylotrophy is the ability of microorganisms to grow on reduced one carbon compounds containing methyl groups such as methanol (CH_3OH), methylamines such as trimethylamine ($(CH_3)_3$–N), or also methane (here the term methanotrophy is also common). Methylotrophic bacteria are highly specialized and many of them are unable to grow on more complex compounds. You might be wondering where the methanol in the leaves comes from. Plants contain methyl esters such as pectins. During synthesis and degradation of these methyl esters, methanol is an intermediate, and a certain percentage of it diffuses out through the leaf surface.

Now the phytopathogenic bacteria have to be discussed. At the beginning of this chapter, we said that they enter leaves via natural openings (stomata) or via wounds to the plant. Phytopathogenic bacteria follow various strategies. Organisms called *Erwinia* produce enzymes that degrade plant cell walls and make nutrients available to these bacteria but also to others. Subsequently, fermentation processes set in, which also involve fungi to a large extent.

More "sophisticated" bacteria such as *Xanthomonas campestris* or *Pseudomonas syringae* also make a living on plants. Ulla Bonas (Halle, Germany) writes about this:

> "As described for the organisms causing plague and depicted in Figure 63, most Gram-negative phytopathogenic bacteria develop a T3S system with which a long filamentous surface structure (type III pilus) is formed. It penetrates the plant cell wall and docks onto the cell membrane. Functioning like a syringe, between 20 and 40 different effector proteins are transported into the cytoplasm of the plant cells. The function of these effector molecules is the subject of intensive research; a few of the details are already known. There are effectors that weaken the defense reactions of the plant whereas others interact with plant signal transduction processes by a sort of molecular mimicry. Also, reactions in the cell nucleus are manipulated. A well-studied effector protein is AvrBs3 from a pathogenic variety of *Xanthomonas campestris* (pathovar *vesicatoria*), which is specific for paprika. This protein functions as a transcription factor. After it has entered the plant cytoplasm via the T3S-system, it is transported into the cell nucleus where it specifically binds to a number of genes. As a result, the plant reacts with cell enlargement and formation of pustules, which provide better growth

> conditions for the microorganisms. In resistant paprika plants, a so-called BS3-resistance gene is activated, which promotes suicide of the infected plant cell and thereby prevents further bacterial proliferation."

This example shows us that plant–microbe interactions are as complex and ingenious as human–microbe interactions. Of course, it seems not of high priority for us to learn more about phytopathogens, but actually it is, when we consider the severe damage to crops.

Another example for a clever approach to making a living on plants is the "ice-plus bacterium", *Pseudomonas syringae*. It expresses a protein that is inserted into the outer membrane of the cells and serves as a nucleating center for ice crystals at freezing temperatures. These crystals cause frost damage of the plants and allow *P. syringae* to invade. So-called ice-minus mutants were isolated by Stephen Lindow (Madison, Wisconsin, then Berkeley, California), and also generated by genetic engineering. Their effect on frost damage on a strawberry field was tested in 1987, an experiment of historical importance because this ice-minus bacterium was the first GMO (genetically modified microorganism) to be released to the environment.

This chapter has taken us all the way from meningitis to rotten strawberries, so what's the best way to finish it? We will conclude with a quotation from Albert Camus' *The Plague*, which describes how danger is lurking everywhere:

> "He knew that those jubilant in the crowd did not know but could have learned from books: that the plague bacillus never dies or disappears for good; that it can lie dormant for years and years in furniture and linen-chests; that it bides its time in bedrooms, cellars, trunks, and bookshelves; and that perhaps the day would come when, for the bane and the enlightenment of men, it would rouse up its rats again and send them forth to die in a happy city."

We know better now, so we are able to protect ourselves more effectively and to fight back as well.

> The (human) genome looks like a sea of
> reverse-transcribed DNA (from retroviruses)
> with a small admixture of genes
>
> David Baltimore

Chapter 30
Viruses, chemicals causing epidemics?

Viruses cause diseases, viruses are probably related to bacteriophages and to some small pathogenic bacteria, and some viruses look like floating mines with spikes or like snakes with big heads

Correct, except for viruses being related to pathogenic bacteria. In fact, it was already pointed out in Chapter 2 that in prevalent opinions viruses do not fulfill the criteria for living organisms. They lack an independent metabolism and can only proliferate by occupying the metabolic apparatus of host cells. The whole translation machinery is missing, and viral ribosomes simply do not exist.

But what about the title of this chapter?

This question was asked by Louis Villareal. We will come back to it. The viruses of bacteria, the bacteriophages, were already discussed in Chapter 20. Most animal and plant viruses are less complex than phage T4. They contain in their interior only one type of nucleic acid, either DNA or RNA, like the phages. The genetic material of the viruses is surrounded by a structure (capsid) consisting of protein molecules (capsomers), which are assembled to form a helix or icosahedron. The latter is a Platonic body bounded by 20 equilateral triangles. Many viruses are enwrapped in a phospholipid-containing envelope that they take along when leaving the host cell. Virus-specific proteins are embedded in this membrane-like envelope.

There are more than 10^{31} virus particles on Earth. They are all waiting for their chance to infect their prey and to proliferate. Together with phages they play a hugely positive role keeping biological systems in balance.

We humans inhale or swallow them, or they enter our bodies via the urogenital tract or with the help of ticks or mosquitos via the skin. In most cases, these viruses are neutralized by our immune system. Viruses exhibit a so-called tropism by which they specifically recognize receptors of certain cell types, where they bind and fuse with the cell membrane or are taken up by endocytosis. What follows is the uncoating of the nucleic acid, replication of the DNA or RNA, synthesis of messenger RNA and of viral proteins, assembly of virus particles, and their release. In many cases, a dead cell is left behind. One dead cell does no harm, but

---

Discover the World of Microbes: Bacteria, Archaea, and Viruses, First Edition. Gerhard Gottschalk.
© 2012 Wiley-VCH Verlag GmbH & Co. KGaA. Published 2012 by Wiley-VCH Verlag GmbH & Co. KGaA.

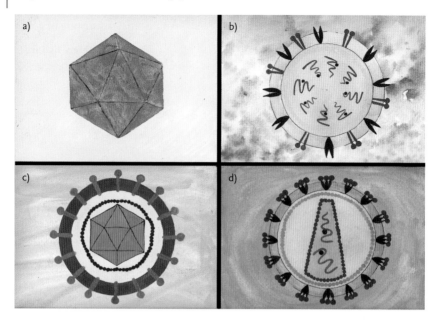

Figure 66 Four viruses: (a) polio, (b) influenza, (c) herpes, and (d) HI (human immunodeficiency) virus. The relative sizes are actually quite different: (a) 28 nm in diameter, (b) 120 nm, (c) 200 nm, and (d) 100 nm. Composition: (a) single-stranded RNA in an icosahedron; (b) eight single-stranded RNA segments surrounded by proteins and a membrane with hemagglutinin and neuraminidase molecules; (c) double-stranded DNA in an icosahedron surrounded by a membrane with glycoproteins; and (d) two single-stranded RNA molecules, reverse transcriptase, and two other enzymes in a capsid surrounded by a membrane with proteins. (Watercolor and gouache: Anne Kemmling, Goettingen, Germany.)

thousands and millions of them cause certain diseases: colds, influenza, measles, oral herpes, gastroenteritis, polio, hepatitis, or AIDS, just to mention a few. Four of these diseases will be discussed. The viruses responsible for polio, influenza, herpes and AIDS are shown in Figure 66.

The poliovirus consists of a ~7500-base-long, single-stranded RNA molecule encapsidated in an icosahedron around 30 nm in diameter. As soon as a virus particle enters a host cell, the RNA is uncoated. It then performs two functions, serving first as messenger RNA for synthesis of around 10 proteins, followed by serving as a matrix for the synthesis of RNA strands for the next viral generation. The proteins include RNA polymerase, the four proteins that make up the viral capsid, and helper proteins such as proteases. Because the poliovirus genome can function as mRNA, it has, by convention, (+) strand polarity. Humans are infected via the oral route. Polioviruses proliferate in gastrointestinal tissues, predominantly downstream of the stomach, and they sometimes enter the bloodstream. In less than 1 percent of those infected do viruses enter the central nervous system and cause paralytic poliomyelitis by inhibiting protein synthesis. Millions of people were infected and many suffered until two potent vaccines were introduced in the middle of the twentieth century by two American medical scientists: the inactivated virus vaccine (IPV) of Jonas Salk and the live, attenuated vaccine (OPV, administered orally) of Albert Sabin. In 1988 the World Health Organization embarked upon a campaign to globally eradicate poliovirus. Now in its twenty-second year, the campaign has been very successful as it reduced the cases of poliomyelitis to yearly approximately 2000 cases, primarily in Africa and in Southeast Asia. But poliovirus is very resilient and its eradication campaign has already taken twice as many years as the eradication of small pox virus with no end in sight.

In 2002, the poliovirus again made headlines because its genetic material had been synthesized chemically and was shown to give rise to the formation of polioviruses when introduced into a cell-free extract. In fact, not the true viral RNA genome was chemically synthesized its cDNA. What cDNA is and how the exciting experiment was performed will be described by Eckard Wimmer (Stony Brook, NY, USA):

"I am an organic chemist by training, but I was attracted to working with viruses because they seemed to exist at the threshold of dead chemicals and living organisms. Thus, in 1968, I began to work with poliovirus, the cause of poliomyelitis. Poliovirus is a rather simple human virus that was then, and still is today, one of the best known infectious agents. As a chemist I decided to concentrate on solving structural properties of the virus, particularly the nucleotide sequence of its RNA genome. This was accomplished in 1981: a mind-boggling array of ~7450 nucleotides emerged (by then the longest chain produced by sequencing a naturally occurring polynucleotide) that encodes the secrets of "proliferation", the most important property of life. Our structural work even allowed us to determine an empirical formula of the organic matter of poliovirus:

$C_{332,652}H_{492,388}N_{98,245}O_{131,196}P_{7501}S_{2340}$

which we published 1991 in *Science*.

In 1991, we discovered that naked RNA genomes of poliovirus, when mixed with an extract of smashed up, uninfected human cells, woke up (were "booted", a term borrowed from computer programming) to suddenly proliferate by going through all steps required to generate authentic progeny: viral protein synthesis, genome replication, morphogenesis. This was the first "in vitro", *cell-free* synthesis of a virus. Importantly, it depended upon the presence of intact genomes that instructed the biochemical reactions so that authentic virus progeny was formed.

Organic chemists who study natural products have a binding tradition: if they solve the chemical structure of a natural product, say that of a blue dye, they must subsequently generate this structure by chemical synthesis. If their original claim of the blue dye's structure was accurate, the synthetic product better be blue. If it turned out to be red – back to the bench! In the 1990s we entertained the possibility of, and then succeeded in, synthesizing the poliovirus genome chemically from scratch. We then incubated the synthetic RNA in the cell-free extract described above, hoping that the extract would "boot" the RNA to life and produce authentic poliovirus as prove of the accuracy of the original sequence.

Long RNA polymers, like genomes of poliovirus, are very unstable and they cannot be synthesized chemically as yet. But one can chemically synthesize DNA of the size of the poliovirus genome. We needed to synthesize two strands of DNA of which one would be complementary to the genomic RNA in the Watson-Crick sense. The double-stranded DNA, called complementary DNA or cDNA, could then be converted ("transcribed") into the poliovirus genome with an enzyme that can read DNA but transcribes it into RNA. We then incubated the synthetic RNA in the cell-free extract and after ~7 hours–poliovirus emerged! It is important to realize that our chemical synthesis of viral cDNA required merely the information (nucleotide sequence) stored in our lab books or, more precisely, stored in computers. Thus, the cDNA and the corresponding virus were generated in the absence of a natural template: its parent(s) is the computer. Remarkably, the virus was generated in the absence of living cells.

Publication of this work in 2002 caught global attention, high praise, ridicule, and fierce condemnation. It was ridiculed because the synthesis per se was not seen as advancing scientific knowledge. For some deeply religious people it suggested that we challenged God by creating life in the test tube. The work was condemned because the world was suddenly less safe as terrorists could synthesize viruses with malicious intent. Moreover, the noble campaign by the World Health Organization to globally eradicate poliovirus was now rendered futile if evil scientists were to synthesize and release the agent. Disturbingly, these last two concerns are, in fact, correct. However, the poliovirus synthesis served to notify the public that medical research and the techniques in modern molecular biology had reached a level where essentially *all* viruses whose sequences are known, could be regenerated chemically in the absence of a natural template (currently, 2361 viral genome sequences have been deposited in the internet). Indeed, only 17 months after the 2002 poliovirus paper the synthesis *in two weeks* of bacteriophage PhiX174 (5386 base pair-long DNA genome) was reported. In 2005 the "Spanish" influenza virus that in 1918 killed some 30 to 50 million people was recreated from sequence information by chemical synthesis. The experiment led to the discovery of the killer instinct of the 1918 flu which is similar to that of the dreaded bird flu and this will help to protect us against a possible devastating bird pandemic. However, the generation of viruses by whole genome synthesis falls into the complicated realm of "dual use research", that is, research that can be at the same time to the benefit and to the detriment of humankind. The scientific community has begun to define this important issue with the intent to possibly limit its possibly disastrous consequences.

Apart from proclaiming a "proof of principle", the 2002 poliovirus synthesis signaled a new era in biology, namely the chemical synthesis of organisms like viruses or bacteria. This offers a novel strategy to investigate gene function and/or pathogenicity of microorganisms. We have recently shown that recoding poliovirus (2008) or influenza virus (2010) by whole genome synthesis has yielded a new class of excellent vaccine candidates, another step towards fulfilling a promise that *de novo* generation of viral genomes by chemical synthesis in the absence of natural template can be beneficial to us."

What a breathtaking report on a virus consisting of 332 652 carbon atoms and five other bioelements! Various types of viruses can cause colds and influenza, commonly called the flu. Influenza viruses also contain single-stranded RNA. This RNA is not present as a single long strand. Instead, the influenza genome consists of seven or eight RNA segments with a total length of more or less 14 000 bases. Each of these segments is surrounded by proteins in a helical form and all of the segments are embedded in a protein solution and held together by a phospholipid envelope. We will discuss in somewhat more detail how these viruses acquire their envelopes.

Let's look at an intracellular influenza virus particle (Figures 66b and 67). It has an inner capsid harboring the segments and surrounded by a membrane that it stole from its host cell. This viral membrane is studded towards the outside with two very important proteins, the hemagglutinin (HA, also abbreviated just H) and the neuraminidase (NA, or just N). The HA and NA proteins serve important functions during the next infection cycle. The HA binds to the next cell to be infected. The NA, an enzyme, cleaves the glycoproteins in the membrane of the infected cell. Both of these proteins are essential for viral entry.

The two viruses discussed so far are RNA viruses. Polio has been under control ever since the vaccines already mentioned became available. In the case of influenza viruses, a more or less effective vaccine has to be developed at least once a year, but why?

When these RNA viruses proliferate in their host cells, many mutants are produced because DNA polymerase is not involved as in DNA viruses. Because of its proof reading and editing functions in DNA replication, mutants are eliminated to a large extent. The RNA viruses like the errors made by their RNA polymerase because it allows them to be very adaptable to unforeseen conditions, so they rapidly develop resistant viruses when a patient is treated with a drug. Many of the mutants are "less fit" or noninfectious and they are eliminated during proliferation. But RNA viruses make huge amounts of progeny viruses and so they can afford mutated progeny. In the case of polio viruses the surface of the particles is very crisp and chemically stable because they must pass through the hostile stomach with acids and digesting enzymes. Poliovirus must avoid changing the particle surface for fear of losing stability and, hence, their antigenic characteristics remain the same. The influenza viruses are different. They avoid the stomach at all cost and infect the respiratory tract where the conditions are very friendly

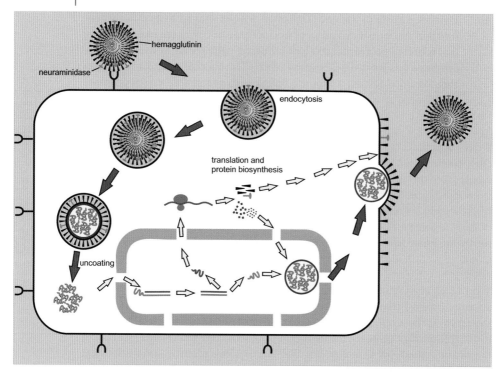

Figure 67 Infection cycle of influenza viruses. Following uncoating, the (−)RNA segments (blue) travel into the nucleus where they are replicated. The (+) strands leave the nucleus and, by translation, the viral proteins are synthesized. Neuraminidase and hemagglutinin molecules are incorporated into the membrane. Capsid proteins travel into the nucleus, where the virus particles (without the membrane) are then completed. (Drawing: Anne Kemmling, Goettingen, Germany.)

compared to the stomach. In this environment, they can afford to change every year the structure of their HA and NA proteins that normally are attacked by our immune system that has memory but in this case the memory is of little use. Hence, we need every year new vaccines. Mutations in HA and NA are called antigen drift. It gets really bad if two different influenza viruses (say one human, the other from swine) that infected the same cell exchange gene segments. This is called reassortment leading to antigen shift. Antibodies that inactivated the progenitor viruses no longer bind to such viruses, which thus are able to escape the immune defense. Influenza viruses are classified on the basis of the antigenic nature of their hemagglutinin (H) and neuraminidase (N) proteins. The 1918 pandemic known as the Spanish flu was caused by the H1N1 type. As mentioned by Eckard Wimmer, the Spanish flu was deadly and killed young adults within 48 hours and more than 50 million people died worldwide. Symptoms were high fever, cough, sore throat, and pneumonia.

Influenza is frequently succeeded by bacterial infection of the respiratory tract, so victims died of a severe viral-bacterial pneumonia. Poultry often serves as the viral reservoir, and this is where the H5N1 influenza pandemic of 2005 had its origin although this pandemic did not cause a large human toll. But we are worried that it may arise again in the future. The 2010 pandemic in humans was caused by a swine H1N1 variant. Fortunately the swine H1N1 of 2010 turned out not to be very pathogenic. Vaccines against the flu usually contain material from three different influenza strains that cocirculate in a human population.

As in the case of the poliovirus, influenza viruses first uncoat their gene segments upon infection. There is, however, an important difference between polio and influenza viruses (Figure 67). Whereas the (+) strand poliovirus genome is immediately translated after entry into the cell, the eight influenza RNA gene segments, once uncoated, cannot function directly as mRNA because they are of opposite polarity than mRNA. By convention, the influenza virus genes have (–) strand polarity. Actually, the (–) stranded gene segments that come in with infecting virus travel right away to the nucleus for replication. First, they must produce mRNA that they perform with an RNA polymerase that they bring along in their virions (because the cell does not have such an enzyme that it could borrow). Eight mRNAs of (+) strand polarity are synthesized that go back to the cytoplasm for translation yielding several proteins, including more polymerase, HA, and NA.

Most of the viral proteins go back to the nucleus except HA and NA that are inserted into the cell membrane. Inside the nucleus, all newly formed viral genes are wrapped up in a protein called NP and they then travel to the cell membrane where HA and NA are waiting for the budding process. It really is frightening but also amazing that simple RNA molecules are able to transform a highly organized animal or human cell into a viral production plant.

We now move to *Herpes simplex* viruses that infect cells of oral or genital mucosa as well as neurons of the ganglia. The infection of epithelial cells may result in sores. Outbreaks of these viruses in the ganglia are triggered by sunlight, certain food components, or stress. Shingles is caused by *Varicella zoster*, which is closely related to *H. simplex*. These outbreaks can be very effectively treated with acyclovir (Zovirax).

Herpes viruses contain double-stranded DNA embedded in an icosahedral capsid and a phospholipid envelope carrying various types of glycoproteins. The size of the genome is approximately 150 000 base pairs, which is nearly 20 times larger than the polio-virus RNA. Proliferation of the herpes viruses is complex and resembles that of influenza viruses. The glycoproteins in the viral envelope bind to certain receptors on the target cell. After entering the cell, the capsid is directed to a nuclear pore and the DNA present in its interior is released into the nucleus. There it serves as matrix for viral messenger RNA as well as for viral DNA synthesis. For the translation process, the m-RNAs have to leave the nucleus because the machinery for protein synthesis is only present in the cytoplasm. The proteins required for viral assembly return to the nucleus. After assembly, the virus particles are escorted to the membrane. They then leave the cell, taking along an envelope loaded with their glycoproteins.

A relative of *Herpes simplex* is the Epstein-Barr virus (EBV). It persists asymptomatically as an extra-chromosomal element in the form of circular DNA, like a plasmid, in human B lymphocytes, and may cause infectious mononucleosis. This virus is of importance for gene therapy, which will be discussed at the end of this chapter.

You have mentioned that RNA polymerase makes more mistakes than DNA polymerase. What is the reason?

The DNA-dependent DNA replication complex, containing the DNA polymerase, like the one in our cells, is sometimes called a factory; it contains several proteins for replication and a proofreading device. The error rate is extremely low, approximately $1:10^{-8}$, meaning an average rate of one error per 100 million base replications. The DNA polymerases of thermophilic bacteria, such as the enzyme from *Thermus aquaticus* used in PCR (see Chapter 25), exhibit a lower fidelity, their accuracy being 100 times lower. However, RNA-dependent RNA polymerases make one error per 10 000 base replications because they lack a proofreading device. Least accurate is the so-called reverse transcriptase of the retroviruses (see below), an RNA-dependent DNA polymerase. It generates a lot of mutants, which causes problems in the development of vaccines or drugs against HIV. Mutations occur so frequently that most HIV strains derived from individual AIDS patients are genetically distinct. In other words, an HIV strain from patient A is different from an HIV strain of patient B.

Influenza pandemics have received considerable public attention, as has the successful fight against smallpox and poliomyelitis viruses by employing effective vaccines. But this can't match the attention that HIV, the human immunodeficiency virus, has received. It was discovered in 1983, a few years after AIDS (acquired immunodeficiency syndrome) was recognized as a new infectious disease. This disease has killed more than 25 million people since 1981. Until five years ago, the perspectives for mankind, especially in Africa, were rather depressing. Campaigns to prevent HIV infections and to develop improved antiviral drugs give reason to be optimistic. The discovery of HIV is associated with the names of Robert Gallo (Baltimore, Maryland, USA), and Luc Montagnier (Paris, France). Both scientists had claimed to have discovered HIV. This became a top issue in the United States and France. Finally, as the result of a meeting in 1985 between the two Presidents, Ronald Reagan and Jacques Chirac, it was accepted that Luc Montagnier had first discovered HIV as the causative agent of AIDS. By no means does this question the epochal contributions made by Robert Gallo on retroviruses, human leukemia viruses, and HIV. In 2008, Luc Montagnier and his colleague Francoise Barré-Sinoussi were awarded the Nobel Prize for their discoveries. They shared the prize with Harald zur Hausen (Heidelberg, Germany), who was awarded for his discovery that human papilloma viruses cause cervical cancer.

What are retroviruses?

Retroviruses are one of the most successful branches of viral evolution. Humans have constantly been in conflict with them, as indicated by the fact that around 8 percent of our genome is genetic material of retroviral origin. This is almost unbelievable. Humans have developed strategies to fight retroviral infections. For example, there are factors that promote hypermutations of viral proteins, rendering the virus particles inactive. Recently, the protein tetherin was discovered that glues the viral particles to the cell surface just after they have budded out, thereby preventing further infections. The HIV-1 protein Vpu helps to release tetherin and increases spread of the virus in humans. The virus from chimpanzees lacks this protein. When the virus jumped over to humans, the Vpu in HIV increased the epidemic. Furthermore, the human immune system had never seen this specific type of infectious agent and, thus, it was unable to antagonize it.

When we take a closer look at retroviruses, it is easy to understand why there is so much retroviral DNA in our genome. Let's use HIV as an example (Figures 66d and 68). Its capsid contains two molecules of (+) single-stranded RNA and a few molecules of so-called reverse transcriptase, a protease, and an integrase. The

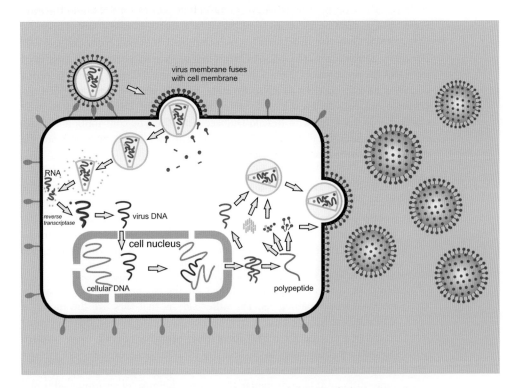

Figure 68 Infection cycle of HIV. Virus particles bind to CD$_4$ receptors of immune T cells. Following fusion with the cell membrane, particles are released and the two RNA molecules are uncoated. As described in the text, each RNA strand is transcribed into a DNA strand and further into a piece of double-stranded DNA, which integrates into a chromosome. Transcription yields (+)RNA molecules, which serve as matrix for the viral proteins as well as genetic material for progeny. (Drawing: Anne Kemmling, Goettingen, Germany.)

capsid is surrounded by a phospholipid membrane into which viral spike proteins are inserted. At first sight, HIV hardly differs from influenza and polio viruses. The most striking difference is the occurrence of a reverse transcriptase. We will see in a moment what this enzyme can do. HIV binds to the CD4–and a second receptor of immune T cells, fuses with the cell membrane, and releases its core particles into the cytoplasm. The cores dissolve, then something unusual happens: the reverse transcriptase uses the (+) RNA as a matrix for synthesis of a complementary DNA strand, essentially a reversal of the "normal" transcription process (when DNA is read but RNA is synthesized). Therefore, the name reverse transcriptase is quite appropriate. Not only is this reaction unusual but so also is the next step, in which the original RNA strand is hydrolyzed and replaced by a DNA strand. As a result, the original genetic information of this retrovirus disappears, having been converted into a piece of double-stranded DNA (a kind of cDNA). This is a fantastic feat because this DNA then enters the nucleus and, with the help of the viral integrase, is incorporated into a chromosome. What follows is rapid transcription by cellular enzymes and the export of (+) RNA into the cytoplasm. There, viral proteins are synthesized by translation of the information on the (+) RNA. Once the virus particles are assembled, they bud from the host cell. *Essentially,* retroviruses are (+) RNA-viruses that require transcription of their RNA into DNA for propagation. This is the reverse of the usual flow of genetic information in nature, from DNA to RNA. Essential for this process is the reverse transcriptase, which was discovered by Howard Temin and David Baltimore, who shared the Nobel Prize with Renato Dulbecco in 1975. The contribution of retroviruses to evolution has probably been underestimated as is apparent from the statement of Karin Moelling, Zurich (Switzerland).

> "Retroviruses are known today as a pathogen, because one of them, the Human Immunodeficiency Virus HIV, causes AIDS. They are also known as oncogenic viruses, because they can cause cancer. They can take up cellular genes and by horizontal gene transfer carry a cellular gene into the recipient cell. If they transport a gene, which gives a growth advantage to the cell, the cell can turn into a cancer cell.
>
> However, this is only one side of the coin, the bad one. Retroviruses are by no means only pathogens. The other side is that they are the motor of evolution – and shaped our genome. They are so ubiquitous, that they must have been around for a very long time – probably almost forever, which means they contributed to the origin of life. Our genome looks like a graveyard of former retrovirus infections. The longer ago this happened, the more difficult it is to recognize the viruses, because of accumulated mutations and deletions. They degenerated to shorter virus-like elements all the way down to solitary viral promoters – making up a considerable percentage of our genome. In the case of the fruitfly Drosophila, these retroelements make up 80%. These elements play a role in evolution and sometimes may contribute to cancer.

Endogenization of a virus into the genome is even detectable today and can be much faster than everybody thought. In Australia, the Koala bears got unintentionally infected by a retrovirus about a 100 years ago – and now the virus has become an integral part of their genome and the Koala bears are no longer sick.

What the viruses are good for, is not yet known, but they are no longer harmful.

This gives a perspective for HIV. One day HIV may become endogenized, integrated into our germline, and will no longer be harmful. This may have happened to monkeys already, because they do not get sick of a relative of HIV. But how long do we have to wait for this process? In the case of the Koala bears it took 100 years. It may take much longer in the case of HIV. Therefore, we cannot wait and have to develop effective therapies."

HIV infection is not yet the disease AIDS. There is something mysterious about the transition. The HIV infection with symptoms like those of a "serious flu" is followed by a latency period, which may last for years. This period is characterized by a slow but continuous decrease of the number of CD4 cells (in which the virus is proliferating) of the immune system. When the CD4 cells reach a certain low level the immune system is severely compromised and the number of virus particles increases dramatically. At this stage the immune deficiency syndrome becomes apparent, making patients vulnerable to opportunistic infections by other viruses, including *Herpes simplex* and the Epstein-Barr virus, or other infectious agents such as *Candida* yeasts or bacteria (e.g., *Mycobacterium tuberculosis* – bacteria causing pneumonia). Furthermore, AIDS patients often develop Kaposi sarcoma, which are tumors caused by human *Herpes virus 8* (HHV-8).

There has been considerable progress in the treatment of AIDS patients. A potent vaccine is not yet available because of the continual appearance of new mutants of the virus. But there are targets for development of drugs against HIV. We recall that three enzymes are associated with the RNA molecules within the capsid, reverse transcriptase, integrase, and a protease. The function of the latter is to cleave the viral precursor protein gp160 into two proteins (pg41 and gp120), which then become major viral envelope proteins. Virus production is not possible without these proteins. HAART, <u>h</u>ighly <u>a</u>ctive <u>a</u>ntiretroviral <u>t</u>herapy, is the treatment of patients with a combination of three distinct inhibitors targeted at the reverse transcriptase, protease, and/or other viral proteins. The combination therapy is based on the following consideration: If mutations in a patient's viral strain occur at the rate of 10^{-4} in one protein, then the probability that the patient's virus population has already three resistance mutations in the three target proteins at the onset of treatment is 10^{-12}, making it highly likely that the virus load can be lowered. HAART treatment of AIDS patients has been very successful. However, successful does not mean complete cure because drug treatment does not remove the virus from the patient's genome. Moreover, drugs must be taken throughout the life span of the patient in order to alleviate symptoms and extend the life expectancy with an improved quality of life. It should be stressed that avoidance of contact with this life threatening virus is the best preventive measure.

Viruses and gene therapy, this sounds like a contradiction.

Stephen Gottschalk (Houston, Texas, USA) reports on systems that are truly promising:

> "Gene therapy refers to a therapeutic approach that aims to treat diseases by replacing, removing or introducing genetic material into cells. Therapeutic applications include for example, gene repair to treat inherited diseases like congenital immunodeficiencies or the delivery of genes that are cytotoxic or redirect the immune system to treat cancers. As described above, viruses are ideally suited to deliver genes into cells. For gene therapy, several viruses have been already used in the clinic, including adenoviruses, adeno-associated viruses and retroviruses. For clinical use of viruses, the therapeutic gene is inserted into the genome of the virus and for most applications the virus is also 'disarmed' so that it still can deliver genes into cells, but cannot replicate (so-called 'replication-incompetent recombinant virus') and, thus, cannot cause viral disease..
>
> Gene therapy with replication-incompetent recombinant viruses has been used successfully to correct severe combined immunodeficiency. In addition, patients with a rare inherited eye disease, Leber's congenital amaurosis, had dramatically improved vision after gene transfer into the retina. One promising strategy of cancer gene therapy includes the genetic modification of the patient's own lymphocytes to recognize cancerous cells. For example, infusion of patients' lymphocytes, genetically engineered to recognize melanoma cells, has resulted in meaningful clinical responses including complete cures.
>
> While these successes have highlighted the great potential of gene therapy with replication-incompetent recombinant viruses, studies have also raised safety concerns. On two out of more than forty clinical studies, in which replication-incompetent recombinant viruses were used, patients developed cancers. These cancers were linked to the retroviruses used in the respective gene therapy since they inserted in or near to genes in the genome of the cell, which are able to induce cancers (so-called 'oncogenes'). Further analysis revealed that not only the integration site, but also the uncontrolled expression of the delivered therapeutic gene played an important contributing factor. Thus continued basic research into the normal life cycle of viruses is clearly needed to develop better viral vectors for gene therapy, which are less toxic, have better targeting capabilities and allow for transcriptional control of the delivered gene."

Where did all the viruses come from?

Viruses evolve in continuous interaction with their hosts. Virologists speak of a viral cloud in which the members of the three domains of life are embedded. The number of viruses is astronomical. Oceans contain 10 times as many viruses as bacteria. Billions of infections occur every second, and viral DNA contributed and contributes to evolution. So there is "much life" in viruses, and we come back to the title of this chapter. If viruses are not alive, are they then just chemicals causing epidemics? – as was asked by Louis Villareal (Irvine, California, USA).

We quote Eckard Wimmer:

> "When I am asked whether poliovirus is a nonliving or a living entity, my answer is yes. I regard viruses as entities that alternate between nonliving and living phases. Outside the host cell, poliovirus is as dead as a ping-pong ball. It is a chemical that has been purified to homogeneity and crystallized. Once poliovirus, the chemical, has entered the cell, however, it has a plan for survival. Its proliferation is then subject to evolutionary laws: heredity, genetic variation, selection towards fitness, evolution into different species, and so forth."

We conclude, also in the light of Chapter 2: yes, viruses are not living organisms, but they have the unique ability to express a viral life by recruiting essentials of life from their host organisms.

There is another important issue that came up in the recent years: did viruses bring DNA into our world? We asked Patrick Forterre (Paris, France) for his opinion:

> "I was travelling by train to visit a colleague working on ribonucleotide reductase, an enzyme essential to form DNA precursors from RNA precursors (removing the oxygen on carbon atom 2 of the ribose) when I got an idea. Could it be that the first ribonucleotide reductase and other enzymes essential to produce DNA (such as reverse transcriptase) were first selected in an RNA virus to transform its genome into a DNA genome? Indeed, DNA can be considered as a modified form of RNA. This modification was a wise strategy for a virus to escape the arms elaborated by its victims. These victims often aimed at destroying RNA genomes.
>
> RNA contains ribose, a "normal" sugar, like glucose, whereas DNA contains an "abnormal" ribose, deoxyribose, lacking the famous oxygen removed by ribonucleotide reductase. Evolution led to a second change in DNA: the base uridine (U) in RNA was changed to the chemically closely related base thymidine (T) in DNA. There are parallels in today's world. The DNA of viruses infecting bacteria is modified in such a way that it cannot be recognized by the restriction enzymes of their bacterial preys. Thus, I suggest that a similar process occurred billions of years ago leading to DNA as the most extensively modified form of RNA."

So, the hypothesis is that DNA evolved first in viruses. How did it get into microbial cells? Patrick Forterre continues:

> "The viral hypothesis means that at some point in the history of life, DNA should have been transferred from viruses to cells. This might have occurred in an RNA cell, infected by a persistent DNA virus whose genome was stably maintained in the RNA cell. It has been recently discovered that some large DNA viruses harbor in their genome the genes of retroviruses, including those encoding reverse transcriptase (the enzyme required to transcribe RNA into DNA). If the viral DNA present in an ancestral RNA cell harbored such retroviruses, cellular genes present in the RNA genome might have been progressively transferred to the viral DNA genome. At the end of the process, the latter would have completely replaced the former chromosome of the RNA cell, a new entity would have been born, a DNA cell, possibly a microbial cell.
>
> I am not sure that we will be able to prove one day that DNA indeed originated in the viral world and why, but for me the interest of this hypothesis is to encourage us to explore in more depth the viral world, not only for medical, but also for fundamental purposes. Viruses have played a crucial role in the history of life, which has been underestimated for a long time. Life could not have evolved to its present state without viruses and their incredible ability to invent new proteins and new functions."

Viruses are dangerous, but they are also of fundamental importance for life on our planet. The hypotheses on viral life presented here are fascinating, and they put into perspective some of the propositions made in Chapters 2 and 4.

> Nothing has such power to broaden the mind
> as the ability to investigate systematically and truly
> all that comes under thy observation in life
>
> Marcus Aurelius

Chapter 31
The "omics" era

Even the cleverest geneticist, microbiologist, or molecular biologist would have had problems 20 years ago to explain what an "omics" era is. This era actually began with the complete sequencing of the genome of the pathogenic bacterium *Haemophilus influenzae*, the pioneering work done by a 38-member team of TIGR, The Institute for Genomic Research (Rockville, Maryland, USA). This team was led by J. Craig Venter, Hamilton O. Smith, and Claire M. Fraser. The genome of *H. influenzae* consists of a circular chromosome with 1 830 140 base pairs, the information for 1740 genes. The completion of this sequence was a milestone in biosciences. Before 1995, it was only possible to determine the sequences of genes containing 3000 base pairs or of plasmids with 10 000 base pairs. It was a real breakthrough when sequencing methods for handling two million base pairs were developed. Thus, a new era was ushered in by the year 1995 with the publication of the genome sequence of *H. influenzae*. In that era, insights into biological processes were gained in a depth that was unforeseeable 20 years earlier.

The complete genetic information of an organism is called its genome. In humans, the genome comprises 22 pairs of chromosomes plus the sex chromosomes XX in women and XY in men, altogether 23 pairs. In bacteria, the genome may consist of one circular chromosome (as in *E. coli*), of one chromosome and one plasmid (*Clostridium tetani*), or of one chromosome and several plasmids (*Ralstonia eutropha* or *Borrelia burgdorferi*). *Streptomyces* species, described in Chapter 22, contain linear chromosomes because the composition of the two ends does not allow them to unite to form a ring.

Genomics is the technology to unravel the complete genetic information of a particular organism. This includes sequencing of all the chromosomes and plasmids that make up the genome, analysis of the data, determination of the genes present, and assignment of biological functions to these genes. Especially the identification of genes and their functions required the development of appropriate bioinformatic methods. Just imagine: the results produced are an endless sequence of the four bases A, T, G, and C. Recognizing where a gene actually begins and where it ends is no trivial matter, and this has to be done for thousands of genes. After 1995, genomics was a big challenge in terms of automation of sequencing, bioinformatics, and further "omics" technologies to be discussed later.

Discover the World of Microbes: Bacteria, Archaea, and Viruses, First Edition. Gerhard Gottschalk.
© 2012 Wiley-VCH Verlag GmbH & Co. KGaA. Published 2012 by Wiley-VCH Verlag GmbH & Co. KGaA.

How is it possible to determine exactly the sequence of millions of DNA base pairs?

The technologies employed are in the process of rapid development. At this point we will describe the classical approach worked out by Frederick Sanger, who in 1980 was awarded the Nobel Prize for his ingenious method – his second one, by the way. The first step is preparation of a genomic library; then the DNA is physically broken at random ("shotgun" sequencing) or cut with restriction endonucleases. The resulting population of fragments of various sizes, in the order of 3000 to 5000 base pairs, is collected by gel electrophoresis (as in Figure 51). The conditions are selected so that these fragments are representative of the whole genome of the organism under study. These fragments are inserted into plasmids and transferred into host cells by transformation. After bacterial growth on Petri dishes, thousands of the resulting colonies are picked. These represent what is called a "shotgun library". In such a library, a number of fragments of the chromosome is overrepresented, so it must be of a size to cover all the sequences of a particular genome.

The next step is that thousands of plasmids from the library are isolated and subjected to PCR with primers so that only the inserts are amplified. But this is not a PCR as discussed in Chapter 23, but a PCR according to the method of Frederick Sanger. In addition to the four deoxyribonucleotides, the reaction mixture contains a small percentage of di-deoxyribonucleotides. When these compounds are incorporated into the growing DNA-strand, this particular PCR-reaction will stop. Now imagine the millions of PCR reactions proceeding in the test tube, and now and then the chain elongation stopping because of incorporation of di-deoxyribonucleotide. The result is a population of chains whose lengths vary from 2 to 2000, assuming that 2000 was the size of the insert. The last nucleotide will always be the di-deoxynucleotide, which, when labeled with a base-specific dye, makes it possible to determine the base at the end of the chain, shown in Figure 69. When the reaction mixture is subjected to an agarose gel electro-

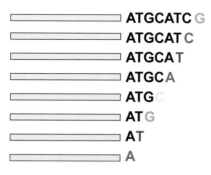 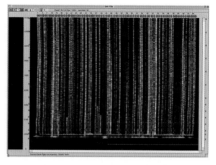

Figure 69 Sequencing DNA fragments. Left, principle of the chain-termination method. Right, sequencing DNA fragments on a gel with 96 lanes. The four colors, blue, green, yellow, and red, represent the four bases C, G, T, and A of DNA. (Diagram and photograph: Anne Kemmling, Goettingen.)

phoresis in which the chains are separated according to their lengths, the sequence of a stretch of 500 bases or so can be determined with the help of a dye-specific laser. A resulting gel as produced in the early days of sequencing is depicted in Figure 69. In modern automated sequencers, this separation of the chains of varying lengths is no longer done in gels, but in capillary tubes. Subsequent steps include assembly of the small fragments on the basis of identical sequences on certain fragments. These represent overlaps by which sequenced fragments can be joined. There are of course many technical problems to overcome, but after a few months or a year, a particular genome is complete and work can begin to identify genes and to reconstruct the metabolism of a certain microbial species. There are always surprises, with genes of unknown function perhaps representing more than 30 percent of the genetic information on a particular microbial genome. Genomics and the research arising from it are really exciting. Here we have asked Claire Fraser-Liggett (Baltimore, Maryland, USA) to explain how genomics began with the *H. influenzae* project at TIGR:

> "From the time that it was established in 1992, TIGR began working on developing experimental and computational tools for the large-scale sequencing and analysis of human cDNAs. After two years of effort on this project, a more comprehensive view of human gene expression and diversity emerged, and discussions began to take place about whether this technology could be applied to other areas of science. Hamilton O. Smith suggested that the same approaches that had successfully been applied to the sequencing and assembly of hundreds of thousands of human cDNAs could also be used in a whole genome shotgun strategy with a bacterial chromosome, and suggested that the *Haemophilus influenzae* genome, at ~2 Mbp (megabase pairs = one million base pairs), would be an ideal project to test this strategy. Between 1994 and 1995, most of the work at TIGR was directed towards completion of the *H. influenzae* genome sequence. The sequencing effort, which represented ~26 000 reads, required several months of work for a large team of laboratory staff. Today, an equivalent amount of data could be generated in a few minutes on one of the next-generation sequencing platforms. After a first attempt at genome assembly, Granger Sutton and his colleagues, who had developed the TIGR Assembler, made a number of substantial revisions to the algorithm to improve the output, and the rest of the team anxiously awaited its delivery. These modifications significantly improved the output and facilitated genome closure, which required another three months. As the genome annotation team began its work to identify genes and assign putative functions, the rest of the team again waited impatiently to see what would be revealed about the biology of *H. influenzae*. Much to everyone's surprise, approximately one-third of the predicted open reading frames encoded novel proteins of unknown function. We had predicted that we would be able to place essentially all of the genes in the *H. influenzae* genome in well-characterized biochemical pathways, and that the genome sequence would reveal all of

the biochemical and metabolic complexities of this organism. The results of this study were published in *Science* in July 1995 and received much attention from the microbiology community. There was a sense that this accomplishment was going to have a substantial impact on the field; however, I don't think that anybody would have predicted that 15 years later we would have thousands of genome sequences available. Amid all of the excitement that the completion of the *H. influenzae* genome generated, the TIGR team was already planning its second microbial genome project focused on *Mycoplasma genitalium*, a self-replicating bacterium with the smallest known genome, estimated to contain between 400 and 600 genes. We had selected *M. genitalium* based on the assumption that we would be able to assign putative function to all of the genes in a minimal genome. But, when this genome project was completed, it was again revealed that nearly one-third of the predicted open reading frames encoded proteins of unknown function. From that point on, we realized how much there is still to be discovered about the diversity of all life on Earth."

Exciting, this view into the lab of the pioneers of genomics and the description of one of the milestones in biosciences: the sequencing of the whole genome of *H. influenzae*. Claire Fraser-Liggett first describes how TIGR developed methods to sequence thousands of human cDNAs. The "c" stands for complementary. We saw in Chapter 2 that the messenger RNA of eukaryotic organisms is not simply a transcript of the corresponding gene. The transcript contains introns, or noncoding segments, which have to be removed by splicing in order to generate messenger RNA. If the sequence of such a messenger-RNA is to be determined, then it is practical to transcribe it first into DNA. This kind of DNA is then designated as cDNA. For this kind of transcription process you need the reverse transcriptase, described in the previous chapter. Such cDNA sequencing was the first goal of TIGR, then the team concentrated on microbial genomes described by Claire Fraser-Liggett.

A completely sequenced microbial genome of *Clostridium tetani* will serve as an example. This organism causes the tetanus disease outlined in Chapter 29. The genome consists of a circular chromosome with 2 799 250 base pairs and a plasmid with 74 082 base pairs, both shown in Figure 70. The two outermost rings represent the two strands of the DNA double helix. The color intensity of these strands is a measure of the number of genes on both strands. A strong bias of the distribution of genes on the two strands is apparent. The outermost strand contains the information of numerous genes clockwise from 12 o'clock until about 5:30, and the inner strand counterclockwise starting at 12 o'clock. This is a very nice demonstration of a bidirectional replication of a chromosome. The origin of replication is at 12 o'clock, from which the DNA replication factories run in both directions.

The prominent genetic element of *C. tetani* is the plasmid, because the tetanus toxin is encoded there. The corresponding gene is located between 11 and 12 o'clock. Other genes could also be identified on the plasmid as well as on the chromosome, genes for replication, transport, and other functions.

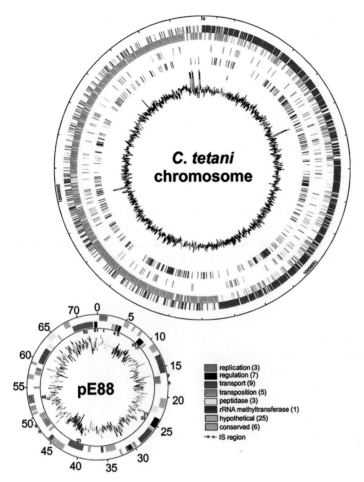

Figure 70 The genome of *Clostridium tetani* in a circular presentation on the basis of the sequence. The chromosome is 2 788 250 base pairs long. The genes on both strands are colored in blue and green. The origin of replication is at twelve o'clock. The plasmid consists of 74 082 base pairs. The gene for the tetanus toxin (red) is located between 11 and 12 o'clock. The other "red" gene between 6 and 7 o'clock codes for a protein-degrading collagenase. (Holger Brueggemann et al., Proceedings of the National Academy of Sciences 100, 1316–1321, 2003.)

What percentage of genes of a bacterial genome can be assigned to a certain function?

Roughly between 50 and 70 percent of the genes present can be identified, but scientists still have a long way to go to completely understand genomic data. This ambitious goal has had a stimulating effect, leading to the development of other omics such as transcriptomics, in which the profile of messenger RNAs is followed in relation to the physiological conditions of a growing bacterial cell. Proteomics aims to analyze all of the proteins an organism produces. The global approaches of metabolomics and fluxomics have similar goals that concentrate on the intermediary metabolism of organisms.

Whereas genomics generates the blueprint of life of a particular organism, proteomics allows insights into life itself. The players, the active enzymes in a cell are recognized and many of them have been identified. The technical effort to separate thousands of proteins and identify them is immense. A typical procedure is a two-dimensional electrophoresis involving all the proteins of a microbial cell extract, which are separated by size and by charge. A characteristic pattern of proteins on such a gel can be seen in Figure 71. This pattern is characteristic of a particular microorganism and for the physiological conditions under which the organism is growing. Even more exciting, of course, is the identification of individual proteins, which is made possible by using modern techniques. Camera-directed robots pick the proteins one by one, then the individual proteins are partially digested and their amino acid sequences are determined. It is fascinating

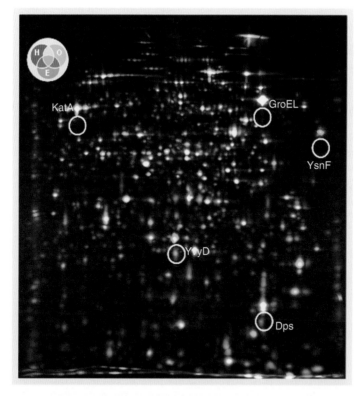

Figure 71 The proteome of *Bacillus subtilis*. Proteins are separated by two-dimensional gel electrophoresis. Colors represent patterns of the synthesis of proteins under certain stress conditions: heat shock, red; ethanol shock, yellow; oxidative stress, blue; all three stress factors, green. (Photograph: Michael Hecker and Decodon GmbH, Greifswald, Germany.)

to go from the amino acid sequence and its corresponding DNA sequence into the genome and to detect the corresponding gene there.

We have asked Michael Hecker (Greifswald, Germany) why proteomics is so exciting for him. This is his response:

> "No doubt, in the omics era our knowledge of life was raised to a completely new level by the tremendous progress made in genomics. Nevertheless, the genome sequence is just the blueprint of life, not life itself, so functional genomics is required to be able to describe real life. Here, proteomics plays a special role because it deals with the true players of life, the proteins. With the improved techniques available, high-resolution two-dimensional gel electrophoresis and mass spectroscopy, it is now possible to make the majority of bacterial proteins visible directly. A clever combination of gel-based and gel-free procedures allows us to identify and quantify 90 percent of the proteins synthesized in a bacterial cell. One result is that bacterial life apparently is not so complex, only requiring 1000 to 2500 different proteins. In some bacteria, for example, members of the genus *Mycoplasma*, even fewer are needed for microbial life. Protein expression profiling is the basis for a better understanding of life in a microbial cell. A highly sophisticated network of gene regulation guarantees that each protein in the cell is present in the right place at the amount required for the fulfillment of its function. Life is more than just a mesh of different proteins. Following their synthesis at the ribosomes, the "dance of proteins" begins. They find each other and form protein aggregates; they are integrated into structures such as the cytoplasmic membrane or sent outside; they are activated or inhibited. The dynamics of the synthesis of proteins as well as their degradation, the story of their life from birth until death can be followed by proteomic analysis, which allows us to make a picture of a living microbial cell that would have been unimaginable 20 years ago. In view of the progress made in understanding microbial life, we should not become too euphoric when studying the lives of higher organisms, especially of man. It is more than questionable whether we will ever be able to describe such complex processes as joy, sorrow, or grief as a dance of proteins."

The "dance" of omics technologies will end with a discussion of metagenomics. Why should microbiologists isolate microorganism after microorganism from soil or other habitats, grow and characterize them, isolate their DNA, all for genomics? Another approach is appealing: take some soil, for instance, lyse all the microbes, isolate all the DNA present, and analyze it according to state-of-the-art genomic technologies. Huge amounts of sequence data of course are obtained, which then can be examined with the help of rather sophisticated bioinformatic methods. This kind of approach is called metagenomics, a term introduced by Jo Handelsman (Madison, Wisconsin, USA) in 1998. One of the findings of a metagenomic soil analysis revealed that 99 percent of the microflora present could neither be cultivated nor (of course) characterized. It's not that microbiologists are untalented when it comes to isolating organisms. These microbes are rather sedentary and so perfectly well adapted to their environment that they simply refuse to grow in flasks of nutrient broth or on nutrient agar plates.

How useful are such metagenomic libraries?

These libraries comprise all the microbial genes of a particular habitat. In fact, not just thousands but sometimes millions of genes are detectable on a microbial genome. The wealth of microbial genes and an almost unbelievable microbial diversity becomes increasingly apparent as more and more metagenomes are studied, but there are other advantages. Genes coding for enzymes of biotechnological interest become available. Novel fat-hydrolyzing lipases and protein-hydrolyzing proteases have been discovered. If appropriate assays are available, the genes for enzymes required to convert substrate X into product Y may be found in metagenomes. A new discipline known as gene mining has emerged, and the "miners" are in the process of discovering one treasure after the other.

Metagenomics has become widely known due to a number of spectacular studies by Edward F. DeLong and colleagues (Monterey, California, and later, Cambridge, Massachusetts, USA), published in 2000. A gene encoding a protein named proteorhodopsin, which is related to bacteriorhodopsin (see Chapter 8), was recovered on a DNA fragment from an oceanic bacterium. It was shown that this protein is found in a large number of bacteria and archaea living in our oceans. Another famous metagenomics experiment known as the Sargasso Sea experiment was done by J. Craig Venter and Hamilton O. Smith and published in 2004. Water samples were taken south of the Bermudas, total DNA was recovered, and more than one billion base pairs were sequenced. A million previously unknown genes were detected in the samples. The human microbiome project is also full of surprises relating to the microbial diversity of our intestinal tract (see Chapter 10). We have asked Edward DeLong to write about his experiments:

> "The field of metagenomics – more specifically, microbial community genome sequencing – is exciting because it offers us a new window into the natural microbial world. The approach can reveal the types of microbes that are found in diverse habitats, and reveals "community blueprints" – the list of combined genes that make up any given community, that provides important clues about how they function in the environment. Some new genes we're now finding, and their presence in certain habitats, were unknown to Science until metagenomics came along.
>
> The approach really stems from early ideas from Norman Pace in the mid-1980s, about how one can identify microbes in an environment, without having to cultivate them one by one. Given the large diversity and numerous habitats of microbes, this was an important idea. What Pace figured out was that it is possible to collect total microbial biomass from any given habitat, extract the total microbial community DNA, and sequence all the taxonomic marker genes (like ribosomal RNA, rRNA) found in each differ-

ent microbe, in one fell swoop. Each rRNA is like a name tag – it tells you who the microbe is – so by reaching into a sample, and grabbing and identifying (by DNA sequencing) all the microbial rRNAs, you can in essence take a census of the entire microbial community. This allows one to answer the question of "Who's out there' more quickly and comprehensively than trying to grow and identify individual microbes one by one. The "cultivation independent" approach has dramatically changed our views of microbial diversity.

Now imagine sequencing not only one marker gene, but *all* the genes from all the microbes within a given community! Now you have not only a census of "Who's out there", but a complete gene "parts list" that can tell you, in any given habitat, "What are they doing?" For a microbiologist, that's pretty exciting, because this sort of information in the past has been really hard to obtain. When I was a postdoc in Norman Pace's lab in the early 1990s, we realized this potential, when we constructed large genome fragment "libraries" that archived big chunks of microbial genomes from ocean waters. Back then, we were more interested in looking at rRNA genes to figure out "Who's out there?" But we knew with these libraries we could "walk" along the chromosome fragments from marker genes like rRNA, into the functional gene components that make up microbes – and so learn "What are they doing?" It wasn't really until later, in the late 1990s, that improvements in DNA sequencing technologies made this approach practically possible.

I hadn't fully realized the power of this approach until my then postdoc, Oded Beja, and I found ion-pumping rhodopsins in oceanic bacteria. Until then, ion-pumping rhodopsins were not know to occur in bacteria at all. The only microbes known to contain ion-pumping rhodopsins were a few unusual salt-loving *Archaea* that can grow in saturated brines. (It had also been shown that these archaeal light-driven, ion-pumping rhodopsins make energy for the cell from light.) Then came our discovery in 2000, enabled by metagenomics. When "walking" (by DNA sequencing) along large chromosome fragments of oceanic bacteria collected directly from seawater, we were surprised to find the genes for rhodopsins where they shouldn't be – in common marine bacteria! This was exciting to us, because it suggested that marine bacteria, previously unknown to interact with light, might actually be using the sun to gain energy. By expressing the rhodopsin gene in the common "lab rat," *E. coli*, we were able to demonstrate that it biochemically functions as a light-driven ion pump, allowing the cell to make energy from light. We now know, through later metagenomic studies, that many, perhaps half or more of all surface water bacteria, have these rhodopsins. So vast numbers of microbes previously unknown to interact with light, now appear to be growing using energy from the sun. In this way, metagenomics has changed our view of ocean ecology in major ways. Future metagenomic gene surveys are likely to change our world view of microbial function in the environment, in similarly profound ways."

Light-driven ion-pumping rhodopsins in marine bacteria, what a discovery! It documents the power of metagenomics for an understanding of microbial diversity on our planet.

Scientists are generating metagenomic libraries from environments all over our planet. These include oceans, sediments, fresh-water sites, various soils, deserts, and even glaciers, studied by Rolf Daniel (Goettingen, Germany). He writes:

> "Glacier ice is a unique habitat for bacteria as this habitat is melting away. It is assumed that most of the European glaciers will disappear in the next 15 to 30 years. The discovery of ice on the planet Mars and Jupiter moon Europa led to an increased research interest in glaciers as some sort of an equivalent of extraterrestrial habitats. Investigations on ice samples from the Northern Schneeferner glacier (Germany) showed that it is the habitat of around 150 different psychrophilic (cold-loving) bacterial species. Most of the glacier bacteria exhibited optimum growth in refrigerators. Sequencing of about 250 million base pairs from a glacier ice metagenome allowed a deep insight into the adaptation of microbes to survive at low temperatures in an extremely nutrient-deficient environment. A whole array of genes for the synthesis of antifreezing agents was detected, which prevent these microorganisms from freezing. In addition, these glacier microbes have special transport systems for the uptake of nutrients present in extremely low concentrations."

Isn't it intriguing that glaciers contain so much life? There, microbial life goes on at an extremely slow pace, but enzyme systems of interest are also present in these microorganisms. Perhaps proteases discovered by such gene mining will soon find their way into areas of application.

These examples show that omics technologies have revolutionized our knowledge not only of the functions of the bacterial cell, including its response to changes in the environment, but also of microbial diversity, where all the new findings have been overwhelming.

Needless to say that there has been a breathtaking progress in the further development of the technologies of sequencing, of proteome analysis, and data processing. For a better understanding, we have introduced here the classical sequencing procedure and gel-based proteomics. Today's sequencing equipment is more effective by a factor of at least 1000, and gel-free proteomics is on its way. It will allow the detection of low-abundance proteins.

> It is impossible not to be
> thrilled by incredible microbes
>
> *modified slogan for Edgar Wallace, Goldmann Publishers*

Chapter 32
Incredible microbes

Microbiologists are probably the only ones to admire microbes.

Well, after having studied the thirty-one chapters of this book, you have nearly become a microbiologist yourself. Just think back over the large numbers of microbes on our planet whose role in evolution cannot be overestimated. They were the driving force in the transition from an anaerobic microbial world to an aerobic world, allowing higher organisms to evolve in the first place. Bacteria and archaea have conquered some extreme living sites that are too hostile for higher organisms: temperatures above 100 °C (212 °F), alkaline or acidic environments, bodies of water with extremely high salt content, to mention just a few. Bacteria not only produce gases in amounts exceeding the capacity of the world chemical industries: methane, molecular hydrogen, and hydrogen sulfide, but also consume these same gases as well as carbon monoxide and carbon dioxide. When it comes to biotechnology, they are real workhorses, producing alcohols and organic acids in huge amounts. The elucidation of basic pathways and reactions in biochemistry as well as laying the foundations for molecular biology and genetic engineering are inconceivable without the study and development of microbial model systems.

Certain microbes are capable of degrading trichloroethylene, even dioxin. They use selenate as an electron acceptor, tolerate extremely high nickel concentrations, and are highly radiation resistant. Radiation resistance is a good cue for us to examine a few of these most incredible microbes.

Deinococcus radiodurans was isolated from cans, or tins, that had been sterilized by radiation. Its resistance is unbelievable. *D. radiodurans* can even withstand extremely high doses of UV radiation or of the gamma radiation originating from radioactive decay of chemical elements, which is used to sterilize foods such as spices and also in cancer treatment. *D. radiodurans* survives radiation up to 10 000 Gy (Gy stands for Gray, the unit dose of radioactive radiation). This is twenty-fold above the radiation tolerated by other bacteria, and 2000 times the dose that could kill a human. The question arises as to how *D. radiodurans* is able to cope with such a radiation exposure because its DNA strands are as subject to breakage as the DNA of other organisms. The secret of *D. radiodurans* is the presence of multiple copies of its DNA, so the probability is high that essential genetic information is still present, even if the individual DNA strands are no longer intact.

Discover the World of Microbes: Bacteria, Archaea, and Viruses, First Edition. Gerhard Gottschalk.
© 2012 Wiley-VCH Verlag GmbH & Co. KGaA. Published 2012 by Wiley-VCH Verlag GmbH & Co. KGaA.

Moreover, *D. radiodurans* has one of the most efficient DNA repair system available. In addition, its pigments, primarily red carotenoids, also help withstand damage by absorbing radiation energy.

Actually, DNA repair is important for all organisms, for bacteria as well as for humans. DNA damage is frequently induced by UV light, by which two adjacent Ts on the DNA are linked to give what is called a thymine (T) dimer. These dimers, of course, interfere with replication and transcription processes. There is a special DNA repair mechanism that recognizes the TT dimers, removes and replaces them with two Ts. This is only one example for a DNA repair system found in many organisms, apparently also in the archaeon *Sulfolobus solfataricus*. Exposure to UV light for 45 seconds is a disaster for its chromosome (Figure 72). But repair systems are at work for hours, and as an incredible result, DNA with the size of the chromosome rises from a mash of DNA fragments like a phoenix.

"Life in hot carbon monoxide" is the title of a publication on *Carboxydothermus hydrogenoformans*. Carbon monoxide (CO) is very toxic to us and to many bacteria

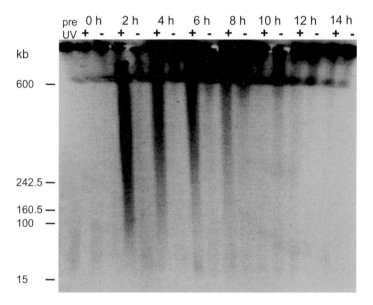

Figure 72 Repair of UV-damaged DNA in the hyperthermophilic archaeon *Sulfolobus solfataricus*. A culture of *S. solfataricus* grown at 78 °C was treated with ultraviolet (UV) light with a wavelength of 245 nm for 45 s. The treated and the control cultures were stored in the dark at room temperature for 15 minutes before they were reincubated at 78 °C. To determine the extent of DNA double strand breaks (DSBs), total genomic DNA from UV-treated (+) and control (−) cells were analyzed by gel electrophoresis. The DNA in both samples was intact before (pre-UV) and shortly after UV treatment (0h). But after 2h, DNA breakdown was pronounced, as a result of DNA DSBs, visible as a smear on the gel. Then, a remarkable DNA repair proceeded, and after 8 hours small DNA fragments below the compression zone at 600 kilo base pairs (kb) had disappeared in the UV-treated samples. (Sabrina Froels et al., J. Bacteriol., 189, 8708–8718, 2007.)

because it binds irreversibly to hemoglobin, thus interfering with respiration. CO is produced in nature during incomplete oxidation of organic material in smoldering fires, and it is also present in the gases released by volcanoes or hot springs. There are many carbon-monoxide-oxidizing bacteria around, especially in the soil, where they prevent its accumulation. The species *Carboxydothermus* was isolated by Russian microbiologists from hot springs on the island Kunashir, so it more or less lives above hot coals and grows at 78°C (172°F) in a CO atmosphere. It is almost unbelievable that all cellular constituents and cofactors of this bacterium tolerate the hot and thus very reactive CO.

The next habitat is quite comfortable, a small lake in a park, for instance. Some sediment from the lake is put into a glass cylinder with some lake water. The south pole of a magnet is attached to the glass wall, and after some time, a brownish biofilm develops on the inside glass wall adjacent to the magnet. This biofilm is formed by magnetotactic bacteria of the genus *Magnetospirillum*, for example. A cell of the species *Magnetospirillum gryphiswaldense* is shown in Figure 73. These organisms prefer the sediment/water interface where the oxygen concentration is low. *Magnetospirillum* are excellent swimmers, and for them it is no question in which direction to swim. They contain magnetosomes consisting of the iron mineral magnetite surrounded by a membrane. These magnetosomes are aligned along a cytoskeletal structure, much like a string of pearls. These bacteria thus have their own navigational system. They follow the magnetic field of the Earth, in the northern hemisphere in the northern direction, and in the southern hemisphere in the southern direction. Due to the Earth curvature, the orientation of their movements is toward the sediments in either case. So it would be a neat experiment to collect magnetotactic bacteria in Lake Rotorua in New Zealand and watch them swim, let's say, in Lake Tahoe in California.

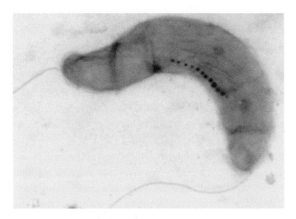

Figure 73   Electron micrograph of *Magnetospirillum gryphiswaldense*. The dark (electron-dense) particles in the center of the cell represent a chain of magnetosomes. Each magnetite crystal is around 50 μm in diameter. On either pole of the cell a flagellum can be seen. (Electron micrograph: Dirk Schueler, Munich, Germany.)

Incredible are also members of the genus *Shewanella* due to their metabolic versatility and because they perform EET. What this means will be explained by Kenneth Nealson (San Francisco, California, USA).

> "In the mid-1980s geochemical data began to appear showing that rates of solid metal oxide reduction (primarily iron and manganese oxides) were far too fast to be accounted for by the chemistry of the environment. I became interested in this because of the challenge that must be presented to bacteria to respire a solid substrate ("breathing rocks" as I call it). At this time, the possibility of direct electron transport to extracellular substrate was regarded as highly unlikely, if not impossible. Using anaerobic sediments from Oneida Lake in New York, we enriched samples with solid manganese oxide as the only electron acceptor. To my delight, many bacteria grew in these enrichments. The one we chose for study was eventually named *Shewanella oneidensis*, and would become one of the model organisms for the study of what is now referred to as extracellular electron transport or EET. More than 20 years later, EET is known to be distributed through many bacterial genera, the process is accepted as evidenced by its discussion in textbooks, and the ramifications of it are beginning to be felt in areas like bioremediation, corrosion, and bioenergy (microbial fuel cells). Recent work has revealed the somewhat surprising fact that microbes also have the ability to take up electrons directly, using the energy for cellular metabolism. This area is a very exciting area of the unknown—a great place for a young microbiologist!"

EET, what an accomplishment by these bacteria. Some sort of a wire is formed between them and the surface of metal oxides, and through this wire, electrons are transferred.

Some bacteria can swim, while others are even capable of gliding. Myxobacteria are interesting gliders, moving across solid surfaces like a loose formation of ships; they degrade any cellulose or lyse any bacteria that are in their way. These organisms form swarms, as shown in Figure 74a, then they move together to form fruiting bodies like the one seen in Figure 74b. This is the most spectacular structure that microbiology has to offer. Under self-abandonment, some individuals form the stem at the tip of which other cells differentiate to form reproduction cells, which are blown away to perhaps give rise to another population of myxobacteria somewhere else.

In Chapter 14 it was pointed out that oceans are almost as productive in the primary production of biomass as the continents. There is one microorganism, *Prochlorococcus marinus*, which is considered to be the most numerous living creature on Earth. Cells of this species are major constituents of the picoplankton. The number of cells in the upper layer of oceans is in the range of 400 000 cells per ml. What an astronomically number of cells per km^2 of the ocean surface. There are two variants of this species: one lives in the water layers close to the surface, and the other one in water depths up to 150 meters (490 feet). Their chlorophyll is

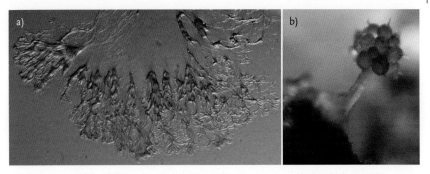

Figure 74 Myxobacteria. (a) a swarm of *Sorangium cellulosum* on an agar plate, showing the typical vein system. The swarm is gliding toward the center of the picture. (b) fruiting body of *Chondromyces robustus*. (Photographs: Hans Reichenbach, Braunschweig, Germany.)

special: it can absorb light of the wavelength 490 nm that reaches such depths. Isn't that a remarkable adaptation? The chlorophyll contains two vinyl residues by which its absorption maximum is shifted to the already-mentioned 490 nm.

Another extremely common marine microorganism received considerable attention recently. It is *Pelagibacter ubique*. Cells of this organism have a diameter of 0.2 µm; they are smaller than *Prochlorococcus* cells, which have a diameter of approximately 0.5 µm. This microorganism is so much adapted to its life in sea water that it is very difficult to grow in the laboratory. Estimations indicate that there are 10^{28} cells of this organism around on our planet, so it may be as common as *Prochlorococcus*. Even if we assume that *P. ubique* has one-tenth the weight of *E. coli* (that would be 10^{-13} g/cell), then the 10^{28} cells would weigh one billion tons, which corresponds to the weight of 13 billion humans. So there is microbial life around us, in incredibly high numbers.

Whereas the microbes mentioned previously represent the safe and sound marine life, other microorganisms are in demand after extensive oil spills occurring in oil-tanker accidents or drilling-rig disasters, as in the Gulf of Mexico. Microorganisms utilizing hydrocarbons are widespread in soil as well as in oceans. Their occurrence was even taken by Claude E. ZoBell (1904–1989) (La Jolla, California, USA) to indicate the presence of underwater oil fields that release small amounts of oil into the ocean water. Oil degradation by microorganisms in oceans is a relatively slow process because it occurs primarily in the interface of oil and water, but not inside of oil drops. In addition, oxygen is required for respiration of these organisms, but also minerals, phosphate and nitrogen sources for growth. Microbial degradation will be extremely slow if oil is present in the form of clumps. Typical hydrocarbon-utilizing microorganisms belong to the genera *Acinetobacter*, *Mycobacterium* and *Alcanivorax*. Yeasts, for example, *Candida* species, are also able to degrade oil. All these organisms must release a biosurfactant at their surfaces so that the water-insoluble hydrocarbons can be taken up and metabolized. There is hope for our oceans after oil-spill accidents due to the occurrence and activity of oil-degrading microorganisms.

For a long time it was assumed that degradation of hydrocarbon requires oxygen and that therefore, only aerobic organisms are able to grow with hydrocarbons. This notion has to be revised in light of studies performed by Friedrich Widdel (Bremen, Germany). He writes:

"Following our studies on the nutritional diversity of sulfate-reducing bacteria, we came across the question what these organisms feed on in oil tanks filled with ballast water. In such systems, sulfide produced from sulfate can cause corrosion and affect oil quality. First, we thought that fatty acids, which are often present in such water, were the nutrients. But then, in growth experiments under the strict exclusion of oxygen, we found that oil-containing particles alone were sufficient to maintain bacterial growth and to catalyze the conversion of sulfate to hydrogen sulfide. The assumption that oil hydrocarbons can serve as substrates was confirmed when growth experiments were carried out with a defined hydrocarbon, with hexadecane instead of the oil-containing particles. To prove that growth with the hydrocarbon was strictly anaerobic, the bacteria were grown in glass vials that had been fused shut. These experiments took several months.

Further types of such anaerobic hydrocarbon degraders were found when in 1991, I participated in an excursion to the Gulf of California where hydrothermal activity converts sedimented biomass to hydrocarbons. Samples taken by means of the submersible ALVIN at 2000-meter (650-foot) depths proved to contain sulfate-reducing bacteria that grew with defined hydrocarbons as well as with oil. They were not closely related to species that we described a decade ago. Also, depths of sediments from other marine sites rich in oil from pollution or natural seepage harbor such bacteria. Maybe they even thrive in oil reservoirs. Their contribution to natural remediation is unknown. They grow much more slowly than aerobic hydrocarbon degraders. Also, they produce a toxic product, hydrogen sulfide. Fortunately, this is easily biodegraded when it comes into contact with air. But it may become a problem upon massive oil spills in marine environments when aerobic bacteria consume the oxygen in seawater and degradation of oil hydrocarbons is taken over by sulfate-reducing bacteria."

Oceans are the habitat of a great variety of microorganisms. In addition to the organisms already mentioned we will discuss bacteria specialized in the oxidation of hydrogen sulfide and of sulfur. A giant among these bacteria is *Thiomargarita namibiensis*, which has a diameter of around 0.3 mm. The cells are spherical, and they shine like crystal balls because of the sulfur drops present (Figure 75). These microorganisms have their habitat just above the sediment layers along the coast of Namibia, where they were discovered by Heide Schulz-Vogt (Bremen, Germany). The mass of *Thiomargarita* cells is in the range of 50 g per m^2 sediment surface, which makes 50 tons per km^2. Cells of this microorganism contain a vacuole that is not filled with gas, but with a nitrate solution. In conditions under which air is

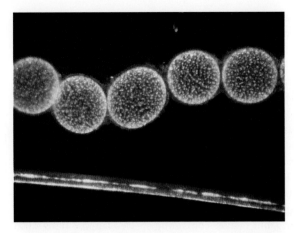

Figure 75 Micrograph of a chain of five *Thiomargarita* cells together with a human hair (0.1 mm thick). The cells have a diameter of about 0.3 mm and they shine because of several sulfur globules in their interior. There is a slime sheath outside the cells that holds the *Thiomargaritas* together. (Micrograph: Heide Schulz-Vogt, Bremen, Germany.)

not available in sufficient amounts, these organisms take nitrate as an oxygen substitute for respiration. This is something that neither we nor fish are able to do.

Other fascinating sulfur oxidizers are also found in the ocean. Douglas Nelson (Davis, California, US) is an expert on them. Concerning these bacteria and *Thiomargarita* he writes:

> "The largest *Thiomargarita namibiensis* cells (0.75 mm in diameter) have a per cell volume that is roughly 200 000 000 times that of a typical rod-shaped bacterium such as *Escherichia coli*. Remarkably, between 98 and 99 percent of this total volume is a membrane-bound central vacuole, which acts as a storage tank for nitrate. As discussed in Chapter 12, nitrate serves many bacteria as an oxygen-substitute when anaerobic life is the only option, and that's its role in *Thiomargarita* and other large vacuole-containing sulfur bacteria. Why nitrate? The ambient concentration of oxygen in seawater is modest, and oxygen cannot be further concentrated in vacuoles because, as a noncharged molecule, it would leak freely across a membrane as rapidly as it could be pumped in. By contrast, negatively charged nitrate molecules, although present at a lower ambient concentration, can be accumulated in these vacuoles to reach a concentration (0.5 moles per liter) that is 2500-fold higher than maximum available oxygen. The large size of the *Thiomargarita* "scuba tank" and high concentration of nitrate allow it to sustain a lifestyle as an anaerobic oxidizer of sulfur compounds, while only needing to refill its "tank" about 4 times per year! Contrast this with a typical one-hour dive time for a person using an air tank and you will begin to appreciate this marvelous adaptation! Except inside *Thiomargarita* cells, nitrate is missing from their typical sediment environment and this presents a real problem

for these cells because they cannot move on their own to access the more or less constant nitrate supply in the overlying water. *Thiomargarita* cells seem to rely on gas bubbles arising occasionally from their sediment niches to re-suspend them into the overlying water, where they can rapidly refill the vacuole with nitrate before descending again to the sediment. An almost bewildering array of other types of vacuolate, nitrate-accumulating sulfur bacteria can move through sediments by a process known as gliding. Species in the genus *Beggiatoa* glide as single filaments (= chain of cells). Those in the genus *Thioploca* glide as many filaments within a common sheath, which can form a sort of freeway that guides their vertical migration while they alternately fill vacuoles with nitrate at the sediment surface and glide as deep as 10 to 20 cm where hydrogen sulfide is available. The diameter of cells forming these filaments is constant within a particular type of gliding sulfur bacteria, but vacuole-containing species ranging from 0.005 to 0.150 mm diameter exist. Different sediment grain sizes, different types of sediment material and the need to commute to different depths in various sediments undoubtedly drove this diversification. Because of the high concentration of sulfate in seawater, sulfate reduction (which produces hydrogen sulfide as discussed earlier in this chapter) is extremely vigorous in marine sediments. This, in turn, puts a high value on the ability to preemptively consume hydrogen sulfide at depth (using nitrate) before it is more widely available to aerobic sulfur oxidizers when it contacts oxygen in the overlying water."

The microorganisms described by Friedrich Widdel and Douglas Nelson are players in the global sulfur cycle (▶Study Guide) that also comprises "acid lovers" such as *Sulfolobus* and *Thiobacillus* species mentioned below.

Many bacteria thrive under somewhat luxurious conditions. Think of the microflora in a decaying sugar beet or of the happy bacteria in milk. We also have admired the microbes preferring extremes such as temperatures above 100 °C (212 °F) or hot carbon monoxide. There is a group of microbes, notably anaerobes, which have a lifestyle at the thermodynamic limit. What this means will become clear by looking at two examples. The first one is *Propionigenium modestum*, which was isolated by Bernhard Schink (Konstanz, Germany) from the sediments of the Canal Grande in Venice, Italy. This microorganism is able to grow at the expense of the conversion of the four-carbon compound succinate into the three-carbon compound propionate. The energy yield the microbe is getting out of this conversion is extremely low, only 0.33 ATP per succinate. Just for comparison, we and also *Escherichia coli* get a yield of around 30 ATP per glucose. Yeast performing the alcohol fermentation will get 2 ATP per glucose. *P. modestum* lives on the most meager diet thermodynamically possible. This is almost like a car running on water.

The second microbe is *Syntrophus aciditrophus*, which represents the syntrophic bacteria. These are microorganisms that cannot grow without the help of other microbes. How this works is explained by Michael J. McInerney, (Norman, Oklahoma, USA).

"To isolate *Syntrophus aciditrophicus*, we had to add a second organism, a molecular hydrogen-using sulfate reducer called G11, to the agar medium. This is because, in nature, microorganisms do not act as individuals but as tightly coupled multispecies units to perform certain functions such as breakdown of fatty and aromatic acids in the absence of oxygen. A good analogy is the human body, which has many different organs like the heart, stomach, etc. Each organ has a different function and the function of all of the organs is needed to have a healthy person. In a syntrophic culture, each microbial species has a specific job and each species must do its job for the syntrophic culture to grow. This is the meaning of the term syntrophy; each species must feed together not individually to grow. The challenge was how to get *S. aciditrophicus* away from its natural partner so we could study it in the laboratory. To do this, we added large amounts of G11 to the agar medium to take the place of *S. aciditrophicus*' natural partner. Individual cells of *S. aciditrophicus* would be next to many G11 cells and start growing as a coculture. We used a technique called roll tubes to get *S. aciditrophicus* to grow with G11. Molten agar in a sealed glass tube was inoculated with the above organisms. The tube is then rapidly spun on its side to solidify the agar onto the inside surface of the tube. The medium in the roll tube stays reduced for a long time and any colonies are easily seen without opening the tube. After about two months, colonies with both *S. aciditrophicus* and G11 cells developed in roll tubes with benzoate but not in those without benzoate. Once we got *S. aciditrophicus* to grow with G11, we repeated the procedure to get *S. aciditrophicus* to grow with its natural partner, the hydrogen-using methanogen, *Methanospirillum hungatei*.

S. aciditrophicus eats fatty and aromatic acids and excretes hydrogen, which is the food used by G11 and *M. hungatei*. Very little energy is released during fatty and aromatic acid degradation so even small amounts of molecular hydrogen make fatty and aromatic acid degradation unfavorable (positive change in free energy). The hydrogen user must be present to keep H_2 levels low so *S. aciditrophicus* can continue to eat fatty and aromatic acids and feed hydrogen to G11 or *M. hungatei*. Initially, syntrophy was thought to involve only H_2 transfer between the two partners, but we now know that transfer of formate between the two partners also occurs. Even under optimal conditions, the free energy changes involved in syntrophy are very small, only enough to make one-fourth or one-third of an ATP per fatty or aromatic acid. Syntrophic cultures grow very slowly with doubling times measured in days to weeks rather than in minutes to hours as found for most microbial species. Syntrophy is an essential intermediary step in the conversion of sewage, industrial wastes, agricultural residues, and decaying plant and animal material to CO_2 and CH_4. Thus, it is an indispensable component of the global carbon cycle and an extreme lifestyle that exists on a marginal energy economy."

Here is one example: think of a biogas plant (see Chapter 13). Oxygen is rapidly consumed, the system becomes anaerobic, and various fermentations take place. Only a few of the fermentation products are direct precursors of methane (acetate, $H_2 + CO_2$). All other products as butyrate, propionate and ethanol have to be converted to acetate and $H_2 + CO_2$ by syntrophic interactions. The existence of syntrophs is only known to experts, but without them, the carbon cycle on our planet would be incomplete.

The microbial diversity under strongly alkaline conditions is also remarkable. A few "soda lovers" such as the archaeon *Natronomonas pharaonis* were mentioned in Chapter 6. But there are many more around, having been searched for by Koki Horikoshi (Tokyo, Japan), who writes:

> "In 1968, I was looking at the autumnal scenery in Florence, Italy. Centuries earlier, no Japanese could have imagined this Renaissance culture. Suddenly, I heard a voice whispering in my ear, "There could be a whole new world of microorganisms in different, unexplored cultures." The acidic environment was being studied, probably because most food is acidic. However, very little work had been done in the alkaline region. Science, as much as arts, relies strongly upon a sense of romance and intuition. Upon my return to Japan, I introduced a bouillon medium containing sodium carbonate (1 percent weight/volume) to increase the pH to 10 and placed a small amount of soil in it, and incubated it overnight at 37 °C (99 °F). To my surprise, various microorganisms flourished in all 30 test tubes. I isolated a great number of alkaline-loving bacteria and purified many alkaline enzymes. Here was an undiscovered world, an alkaline world, which was utterly unlike the neutral world discovered by L. Pasteur. The first paper on alkaline protease was published in 1971.
>
> Using such a simply modified nutrient broth, I found thousands of new microorganisms (alkaliphiles) that grow optimally well at pH values of 10, but cannot at the neutral pH of 6.5.
>
> My studies indicated that these alkaliphiles are living all over the Earth, including deep-sea deposits at the frequency of 10^2 to 10^6 cells per gram of soil. Many different kinds of alkaliphilic microorganisms have been isolated including bacteria belonging to the genera *Bacillus*, *Micrococcus*, *Pseudomonas* and *Streptomyces*, and eukaryotes, such as yeasts and filamentous fungi.
>
> Then my interests became focused on the enzymology, physiology, ecology, taxonomy, molecular biology, and genetics. It is noteworthy to write in this book that alkaliphiles had a great impact on industrial applications. Biological laundry detergents contain alkaline enzymes including alkaline cellulases and/or alkaline proteases from alkaliphilic *Bacillus* strains. Another important application is the industrial production of cyclodextrin with alka-

line cyclomaltodextrin glucanotransferase. This enzyme reduced the production costs and opened new markets for cyclodextrin use in large quantities in foodstuffs, chemicals, and pharmaceuticals.

A series of my work established a new microbiology of alkaliphilic microorganisms, and the studies performed spread out all over the world."

This is an impressive report, also touching upon application aspects: enzymes working at alkaline pH values and production of cyclodextrins. These are cyclic structures consisting of six to eight glucose molecules. Cyclodextrins find applications in food industries as gelling agents.

Acid lovers live at the other end of the pH scale. An incredible archaeon, in this respect, was already mentioned: *Picrophilus torridus*. But there are also bacteria that are fond of sulfuric acid, producing it by oxidation of sulfur-containing ores. In this context, we should mention organisms such as *Thiobacillus thiooxidans* and *Th. ferrooxidans*. Imagine a copper mine. You see nothing but ore and sprinklers providing water, also oxygen and carbon dioxide. There is no vegetation, but myriads of thiobacilli are at work oxidizing sulfides to sulfuric acid, which dissolves the copper. A dilute solution of copper sulfate is drained off, a microbial product in an inorganic world.

Finally, we return to Karl Stetter's hot microbe kitchen. We have already met *Thermoproteus tenax* and *Pyrodictium occultum*, but not yet a remarkable couple that was isolated in this kitchen, *Ignicoccus hospitalis* and *Nanoarchaeum equitans*, an archaeal horse and rider team (Figure 76). *Ignicoccus hospitalis* lives in the deep sea near hot volcanic vents at temperatures around 90 °C (195 °F) and uses molecular hydrogen, sulfur, and carbon dioxide as substrates. The role the archaeal dwarf rider plays in this partnership is still unknown.

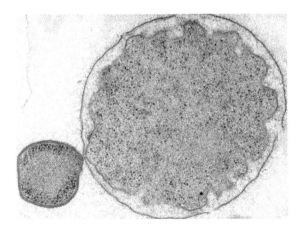

Figure 76 Electron micrograph of an *Ignicoccus hospitalis* cell together with its satellite *Nanoarchaeum equitans*. (Electron micrograph: Harald Huber and Reinhard Rachel, Regensburg, Germany.)

In this parade of incredible microbes, we shouldn't fail to mention the strain of *Mycoplasma mycoides* constructed by the team of The J. Craig Venter Institute. The 1.08-Mbp genome of this organism was synthesized and amplified in the yeast *Saccharomyces cerevisiae* as a plasmid. It was carefully isolated and transplanted into *Mycoplasma capricolum*. The introduced genome took over and, as a result, a population of *M. mycoides* developed out of the *M. capricolum* cells.

With this breathtaking experiment we will end our parade of incredible microbes. However, somewhere out there in nature, there are many more yet to be discovered.

Epilog

To quote the British economist and writer Barbara Ward (1914–1981, In: Maurice F. Strong (ed.) Who Speaks For Earth. W.W. Norton & Company (1973)):

> "We cannot cheat on DNA. We cannot get round photosynthesis. We cannot say, "I am not going to give a damn about phytoplankton." All these tiny mechanisms provide the pre-conditions of our planetary life. To say we do not care is to say in the most literal sense that "we choose death"."

It is the aim of this book to substantiate the visionary insights of Barbara Ward and to fill them with life. These "tiny mechanisms" rule our lives, they may even kill some of us – but they also provide the conditions required for life on our planet.

We had better learn as much as possible about them. You may do so with the help of the study guide (Part two).

Part Two
Study Guide

Overview to the Study Guide

The Study Guide facilitates a comprehensive understanding of the various subjects in microbiology. Each section in the Study Guide refers to one or more chapters in the Reading Section.

Section 1   Microbial growth (read Chapter 1)
                    1.1) Batch and continuous culture
                    1.2) Growth conditions
                    1.3) Microbial shapes and sizes

Section 2   Molecules that make up microbes (read Chapter 2)
                    2.1) The informational macromolecules DNA, RNA, and proteins
                    2.2) The cell membrane and cell wall
                    2.3) The role of ATP

Section 3   Evolution – from the RNA world to the tree of life (read Chapters 3 to 5)
                    3.1) The RNA world
                    3.2) Possible properties of LUCA
                    3.3) Prokaryotes represent two of the three domains of the tree of life

Section 4   Archaea (read Chapters 6, 7 and 13)
                    4.1) The four archaeal phyla
                    4.2) Habitats
                    4.3) Archaeal metabolism
                    4.4) Methanogenesis
                    4.5) Anabolic metabolism

Section 5   Bacterial diversity (read Chapters 8, 12 and 14)
                    5.1) The phylogenetic tree of bacteria
                    5.2) Cycles and food chains
                    5.3) Survival strategies

Discover the World of Microbes: Bacteria, Archaea, and Viruses, First Edition. Gerhard Gottschalk.
© 2012 Wiley-VCH Verlag GmbH & Co. KGaA. Published 2012 by Wiley-VCH Verlag GmbH & Co. KGaA.

Section 6 Membranes and energy (read Chapters 2, 8, 9 and 26)
    6.1) Transport
    6.2) Principles of ATP synthesis in microbes
    6.3) MCPs and motility
    6.4) Two-component systems
    6.5) Quorum sensing

Section 7 Carbon metabolism (read Chapters 8, 16 and 18)
    7.1) Aerobic heterotrophic growth
    7.2) Incomplete oxidation
    7.3) Autotrophic growth
    7.4) Bacterial fermentations

Section 8 Regulation of microbial metabolism
    8.1) Regulation at the DNA level
    8.2) Transcription and provision of messenger RNA for the translationary machinery
    8.3) Translational regulation
    8.4) The regulation of enzyme activity

Section 9 Genomes, genes, and gene transfer (read Chapters 20 and 31)
    9.1) Genomes
    9.2) Gene transfer

Section 10 In-depth study of special subjects
    10.1) Antibiotics (read Chapters 22 and 23)
    10.2) Biotechnology (read Chapters 15, 16, 24 and 28)
    10.3) Pathogenic microorganisms (read Chapters 10 and 29)
    10.4) Viruses (read Chapter 30)

Section 1
Microbial growth

1.1 Batch and continuous culture

In an adequate environment in nature or in the laboratory, microbes will grow by synthesizing their cellular constituents. Eventually they will undergo cell division, most commonly by binary fission: two daughter cells arise from one mother cell.

In the growth experiment described in Chapter 1, it was assumed that *Escherichia coli* would grow exponentially (also called logarithmically) for 48 hours. As explained, this is impossible in reality because logarithmic growth can only proceed for a few hours. A growth curve of bacteria in batch culture performed in an Erlenmeyer flask or a fermentation vessel looks more or less like the curve shown in Figure S1.

The following growth phases can be seen:

- Lag phase (growth of individual cells, few cell divisions).
- Log phase (optimal division of cells, logarithmic growth).
- Stationary phase (further growth limited by shortage of nutrients or accumulation of inhibitors).
- Cell death.

Key terms defined

- The generation time g (*hours*) is the time between two cell divisions.
- The division rate n ($hours^{-1}$) is the number of divisions per hour.

A typical g value for *E. coli* is about 20 min. In some species g is almost 10 minutes (*Vibrio alginolyticus*); in others, it is 60 to 90 minutes (*lactic acid bacteria*) or even a few days (*Mycobacterium tuberculosis*).

Another way to describe microbial growth is to follow the increase of microbial cell mass. Here, the equations applied have to take into account that increase of cell mass is an autocatalytic process: the cell mass just synthesized will immediately participate in the synthesis of even more cell mass. Therefore, the rate of increase of cell mass is proportional to the amount of cell mass present, and the growth rate, µ, is defined as follows:

$$\frac{dx}{dt} = \mu \cdot x; \mu = \frac{1}{x}\frac{dx}{dt}; \mu = \text{growth rate } (hours^{-1}); x = \text{cell mass } (grams)$$

Discover the World of Microbes: Bacteria, Archaea, and Viruses, First Edition. Gerhard Gottschalk.
© 2012 Wiley-VCH Verlag GmbH & Co. KGaA. Published 2012 by Wiley-VCH Verlag GmbH & Co. KGaA.

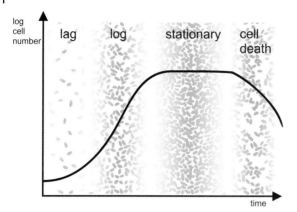

Figure S1 Growth curve of bacteria in batch culture. The growth phases are explained in the text.

The doubling time of the cell mass, termed t_d, can be determined by withdrawing samples from a growing culture at given time intervals. The dry weight of the bacterial cells in these samples is then measured. The growth rate can be calculated from the t_d value.

$$\mu = \frac{\ln x_1 / x_0}{t - t_0}; \mu = \frac{\ln 2}{t_d}$$

If the mass of the individual cells remains constant during growth, then t_d is equal to g. If the cells become smaller (and lighter), a common feature of logarithmic growth, then t_d is larger than g.

Important for applications is the cell yield x, the amount of cells harvested once growth is completed. Depending on the microorganism and the culture conditions, x may be a gram of cells per liter or only a tenth of a gram or as much as 100 g per liter (dry weight) in cultures reaching a high cell density.

So far we have discussed growth in batch culture. Irrespective of the size or the form of the culture vessel, growth will eventually come to an end. This is not the case when we consider a continuous culture. One way to maintain such a culture is the chemostat principle. Culture medium is employed in which one of the nutrients is growth limiting, for example, the energy source (e.g., glucose) or the nitrogen source (e.g., ammonia). The growth chamber (Figure S2) is connected to a medium reservoir. Following inoculation, the medium initially supports growth of the microorganisms until limiting conditions are reached. A continuous culture is then started by allowing fresh medium to drip from the reservoir into the growth chamber. The excess volume of culture medium leaves the growth vessel by way of an overflow. Under these conditions and within a certain range of medium feed, the growth rate of the organisms adjusts itself to the dilution rate. As a result, the amount of bacteria washed out equals the amount replaced by growth. We say the culture is in a steady state.

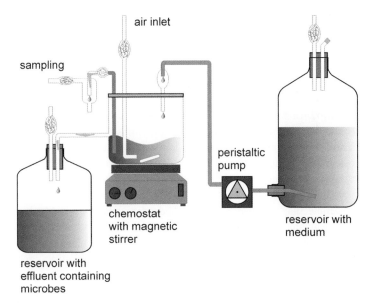

Figure S2 Apparatus for continuous culture of bacteria in a chemostat. Fresh medium is pumped into the culture vessel, and the culture volume in the vessel is kept constant by an overflow. Sterile air may be passed through the culture to support growth of respiring bacteria. Likewise, oxygen-free gas mixtures or molecular nitrogen may be used when culturing bacteria that carry out fermentations.

Continuous cultures have various applications. Microbial cells can be kept in a defined physiological state such as the logarithmic growth phase. The protein profiles of these cells can be investigated (see Chapter 31) as well as the effect of altering parameters such as temperature, pH, and availability of certain nutrients. Furthermore, not only maximal growth rate and cell yield can be determined but also the physiology of microbes growing at the low substrate concentrations found in lakes or oceans. One example for the employment of a continuous culture for enzyme evolution is shown in Figure 37.

1.2 Growth conditions

Conditions vary widely under which microbes are able to survive and even grow. The incredible extremes under which microbial life is possible will be highlighted later. The majority of microbes grow at ambient temperatures and at pH values close to pH 7. They are grouped into two basic nutritional classes:

Phototrophic microorganisms use light as an energy source. Its utilization can be connected to the evolution of oxygen. This kind of photosynthesis, oxygenic photosynthesis, is performed by the cyanobacteria. Photosynthesis without oxygen evolution is carried out by the anoxygenic phototrophic bacteria, of which the purple sulfur bacteria mentioned are an example, (see Figure 8a).

Chemotrophic microorganisms obtain metabolic energy from various conversions of organic or inorganic compounds. These microbes can be aerobic and respire like we do, or they can be anaerobic and carry out fermentations, for example, lactic acid fermentation. Some of them even oxidize compounds such as hydrogen sulfide or ammonia. As a whole, the thousands of species of microorganisms perform an enormous variety of biological processes. Like the green plants, several species are able to produce biomass from CO_2. Most species derive their metabolic energy from degradation of practically all organic compounds occurring on Earth. Their habitats comprise not only oceans, rivers, lakes, fertile soils, leaf surfaces, and intestinal tracts but also hot, acidic, or alkaline environments. Such examples are mainly presented in Chapters 6 and 7. Another characteristic of some microbial species is the ability to oxidize or reduce inorganic compounds, including ammonia, nitrate, molecular nitrogen, molecular hydrogen, sulfate, sulfur, hydrogen sulfide, and ferrous ions.

Furthermore, the ability or inability of microbes to utilize molecular oxygen is reflected by dramatic differences in their physiology, energy metabolism, and biochemistry. We distinguish between the following groups:

Aerobic microorganisms respire like we do. Substrates are oxidized and the reducing equivalents are transferred to molecular oxygen via the respiratory chain located in the cytoplasmic membrane.

Anaerobic microorganisms such as the lactic acid bacteria carry out fermentations. Several species are strict anaerobes and may even die in the presence of oxygen (some *Clostridium* species). Others are more or less aerotolerant, such as some of the lactic acid bacteria.

Facultative anaerobic microorganisms are aerobes but switch to fermentations in the absence of oxygen. A typical organism of this type is *Escherichia coli*.

1.3 Microbial shapes and sizes

Most microbes are spherical or rod-shaped. For example, *Staphylococcus aureus* and the lactic acid bacterium *Streptococcus lactis* are spherical. Organisms like *Streptococcus pyogenes* form chains of coccoid cells; sarcinas are little parcels consisting of four coccoid cells, an example is *Methanosarcina barkeri*. *Escherichia coli* and the anthrax-causing *Bacillus anthracis* are rod-shaped. Important representatives of spiraled rods called spirilla are *Azospirillum lipoferum*, which fixes molecular nitrogen, and the phototroph *Rhodospirillum rubrum*. Comma-shaped microbes are called vibrios; examples are *Vibrio cholerae* and *Desulfovibrio vulgaris*, a sulfate-reducing anaerobe. There are stalked bacteria such as *Hyphomicrobium vulgaris*, which often grows on the walls of water baths in laboratories. Several microbial species form filaments. Such microbes are found among the cyanobacteria, including the beautiful *Spirulina* or among the sulfur-oxidizers, for example, *Beggiatoa alba*. Microbial cells may stay together in a common sheath that is typical of *Sphaerotilus natans*, often abundant in waste water plants. Rarely, disc-shaped or even square-shaped cells occur, such as *Pyrodictium occultum* and *Haloquadratum walsbyi*. An obvious question, of course, is what determines the shape of a microbial cell. Apparently it is the cell wall into which the cell is stuffed, much like a sausage into its casing or skin. There are also microbes that are not surrounded at all by a cell wall, like the *Mycoplasma* species, including the recently created *Mycoplasma capriolum*. These cells are irregular in shape and stabilized solely by skeleton-like structures.

Regarding cell size, 1 µm in length is a good benchmark, although there is considerable variation. Smaller ones include the pathogenic *Chlamydia* and *Rickettsia* species and, notably, *Pelagobacter ubique* (see Chapter 32), with a diameter of 0.2 µm. *Bacillus* species are up to 10 µm long. Giants among the microbes are *Epulopiscium fishelsoni*, which is 600 µm long, and *Thiomargarita namibiense*, a spherical microbe with a diameter of 500 µm (see Figure 75).

Questions

1) What is the difference between generation time g and doubling time t_d?
2) Describe the characteristic features of the logarithmic growth phase.
3) What are the features of chemostats?
4) What is the difference between oxygenic and anoxygenic photosynthesis?
5) How is chemotrophy defined?
6) You see rods while viewing a sample through the microscope. Name two bacterial species that could be present in the sample.

Section 2
Molecules that make up microbes

2.1 Informational macromolecules

Microbes consist of macromolecules, primarily informational macromolecules whose biological function is determined by the sequence of building blocks present.

- Deoxyribonucleic acid (DNA)
  DNA consists of chains of deoxyribose molecules linked by phosphate bridges. The deoxyribose molecules carry one of four different bases, adenine (A), thymine (T), guanine (G), and cytosine (C). Their formulas are given in Figure S3. One base, one deoxyribose molecule, and one phosphate molecule together comprise one building block of DNA. The four building blocks are also given in Figure S3. Their precursors in DNA synthesis are the corresponding triphosphates. DNA synthesis proceeds according to the general formula:

$$DNA\,(n) + dATP \xrightarrow{DNA-dependent\ DNA\ polymerase} DNA\,(n+1) + PP_i$$

The equation describes the lengthening of DNA by one building block, in this case, deoxyadenosine 5′-monophosphate, PP_i stands for pyrophosphate.

- Ribonucleic acid (RNA)
  RNA differs from DNA because RNA contains ribose instead of deoxyribose. Furthermore, T is replaced by uracil (U), as seen in Figure S4. All organisms have three types of RNA: In microbes, these are

Ribosomal RNAs form, together with more than 50 different proteins, the ribosomes. These are the factories producing proteins. Microbes contain three types of ribosomal RNA that vary in size, namely, 5S, 16S, and 23S. S means Svedberg units. These sizes correspond to molecules containing around 120, 1500, and 2900 bases, respectively. Examples of 16S rRNA sequences are shown in Figure 5.

Messenger RNAs are the transcripts of the genes on the DNA. RNA polymerase is responsible for mRNA formation and release. They represent the information that is used by the ribosomes for protein synthesis (see Figure 3).

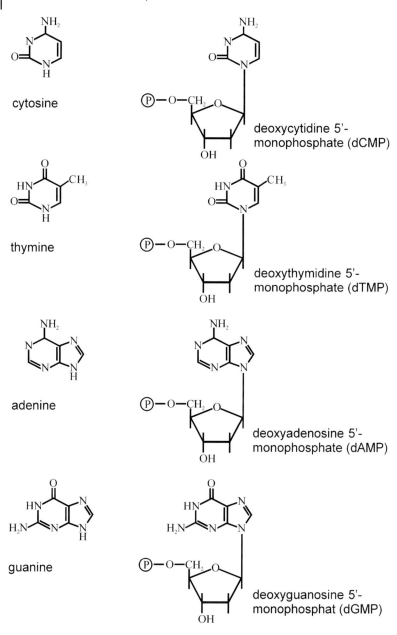

Figure S3 The building blocks of DNA. Note that when cytosine is linked with deoxyribose it is called deoxycytidine; thymine then becomes deoxythymidine, adenine, deoxyadenosine, and guanine, deoxyguanosine.

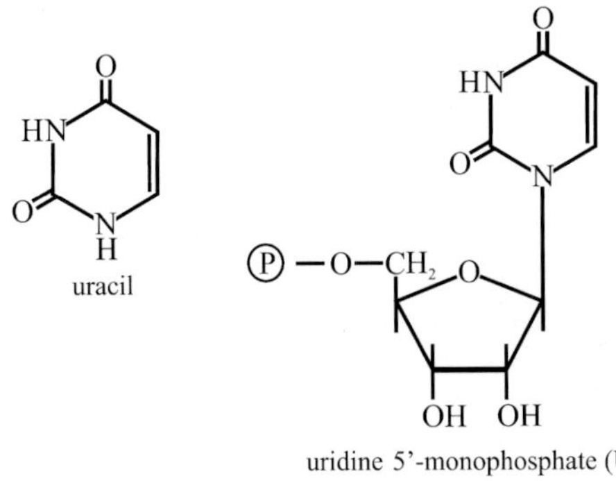

Figure S4 The characteristic building blocks of RNA. Compare the formulas of T and U and those of deoxyribose and ribose.

Transfer RNAs. There are specific tRNAs for each of the 20 amino acids present in proteins. Enzymes known as aminoacyl tRNA synthetases specifically link an amino acid with "its" tRNA (see equation b in Figure S10). In a cell actively synthesizing proteins, as in Figure 3, at least 20 different aminoacyl tRNAs are available to deliver the "right" amino acids in the course of protein synthesis. Further information on the processes of replication, transcription and translation is given in Figure 3.

Actually, a fourth group of RNA is present in all microorganisms, the so-called small RNAs that fulfill regulatory functions; see Section 8.

The principle underlying all these processes is base pairing, whose principle is depicted in Figure S5. We should take notice that the strands carrying the bases A and G, and T and C run in an antiparallel or opposite direction. One strand has a phosphate residue at the top, designated as the five prime end (5'-end), whereas the other strand has an OH-group at the top, known as the 3'-end. It is inherent that DNA replication can only proceed from the 5'-end to the 3'-end. This has important consequences, as depicted in Figure S6. When single strands become available for the replication process, the left strand in Figure S6 can be replicated directly. This strand is called the leading strand. The strand on the right is called the lagging strand because the DNA-polymerase has to "wait" until a piece of single-stranded DNA is available. Only then can replication proceed. An additional important feature is that DNA-polymerase requires a primer, an oligonucleotide with a 3' OH-group, to start replication. This is important for PCR, as discussed in Chapter 25.

- Proteins, mostly enzymes
  The proteins consist of 20 different amino acids. Their chemical formulas are depicted in Figure S7. The amino acids are linked to each other by peptide bonds formed between the carboxyl group (-COOH) of one amino acid and the amino group ($-NH_2$) of a second amino acid. Chains of linked amino acids

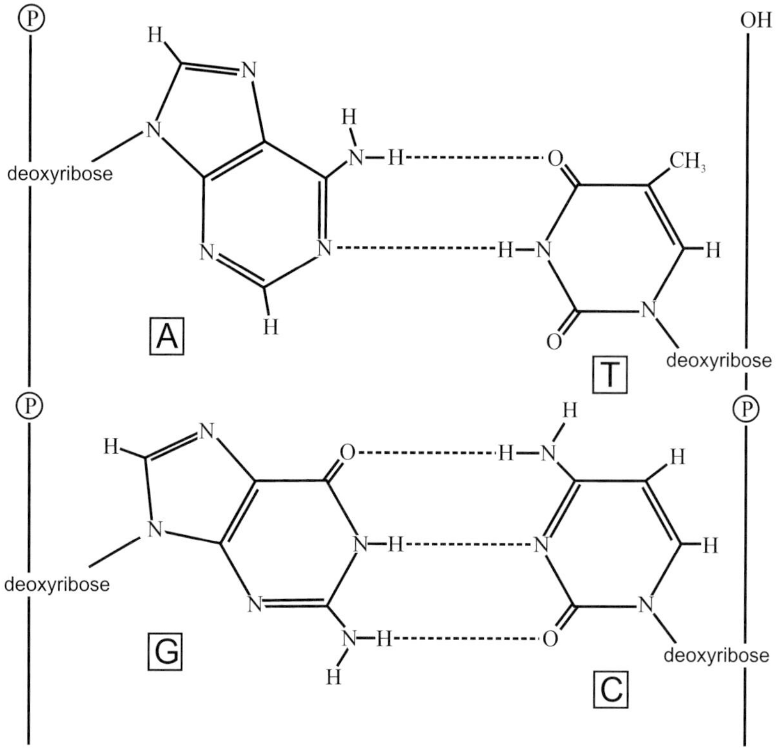

Figure S5 Principle of base pairing. Pairing is achieved by hydrogen bonds. P (phosphate group) is at the 5'-end and OH is at the 3'-end of the DNA strands.

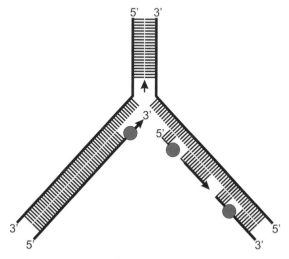

Figure S6 Basic scheme of DNA replication. Several proteins are involved, but only the DNA polymerase is shown here. It can only function in the direction from the 5'-end to the 3'-end. The DNA fragments formed along the lagging strand are called Okazaki fragments. The gaps between them are closed by the enzyme ligase.

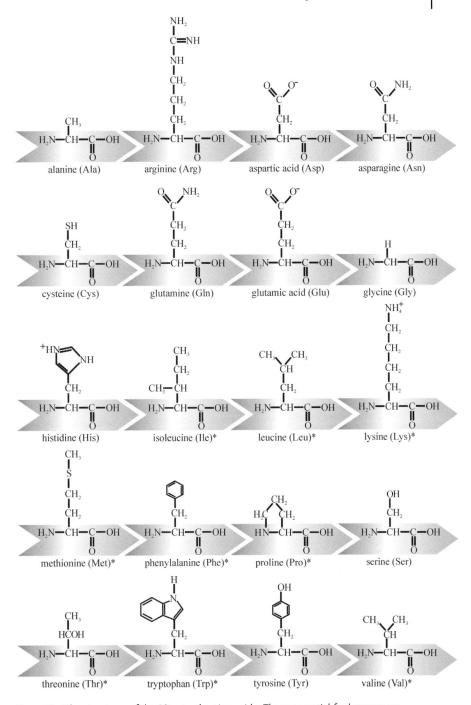

Figure S7 The structures of the 20 natural amino acids. Those essential for humans are indicated with an asterisk.

					AGA
					AGG
	GCA				CGA
	GCG				CGG
start	GCC	GAU		AAU	CGU
AUG	GCU	GAC		AAC	CGC
Met	Ala	Asp		Asn	Arg
				GGA	
				GGG	
UGU	GAA	CAA		GGU	CAU
UGC	GAG	CAG		GGC	CAC
Cys	Glu	Gln		Gly	His
	UUA				AGU
	UUG				AGC
	CUA				UCA
AUA	CUG				UCG
AUU	CUU	AAA		UUU	UCU
ATC	CUC	AAG		UUC	UCC
Ile	Leu	Lys		Phe	Ser
CCA	ACA			GUA	
CCG	ACG			GUG	UAA
CCU	ACU		UAU	GUU	UAG
CCC	ACC	UGG	UAC	GUC	UGA
Pro	Thr	Trp	Tyr	Val	stop

Figure S8 The genetic code. The triplets given are at the messenger RNA level. At the DNA level, U is replaced by T.

are called polypeptides. It is apparent from Figure S7 that all of the amino acids except glycine contain an asymmetric carbon atom with four different substituents. Such compounds occur as two optical isomers called the L- or D-forms. It should be noted that the amino acids found in proteins only contain the L-form.

Translation proceeds as shown in Figure 3. As mentioned, the genetic code comprises 64 triplets. In microbes, there is one start codon, three stop codons, and 60 codons for the amino acids. These are listed in Figure S8. There is only one codon for the amino acid tryptophan but six each for the amino acids arginine, leucine, and threonine. A certain codon preference has been observed in microorganisms; for example, *Escherichia coli* has a preference for two of the arginine-coding triplets but rarely uses the other ones. The use of more than 20 codons by a particular organism for protein synthesis means, of course, that there have to be more than 20 different tRNAs. Microbes often contain 50 different tRNAs. The first code word discovered was UUU, for phenylalanine.

2.2 The cell membrane and cell wall

The cytoplasmic membrane of the cell is the osmotic barrier between the inside and the outside. As already mentioned, it is electrically charged. The basic structure of the membrane is a bilayer (see Figure 2c) consisting of phospholipids. The properties of phospholipid molecules are responsible for formation of this double layer. As depicted in Figure S9a, phospholipids have hydrophilic heads: phosphate esters linked to a glycerol molecule on one side and to a methylated ethanolamine on the other. The head carries a positive and a negative charge and is hydrated. The

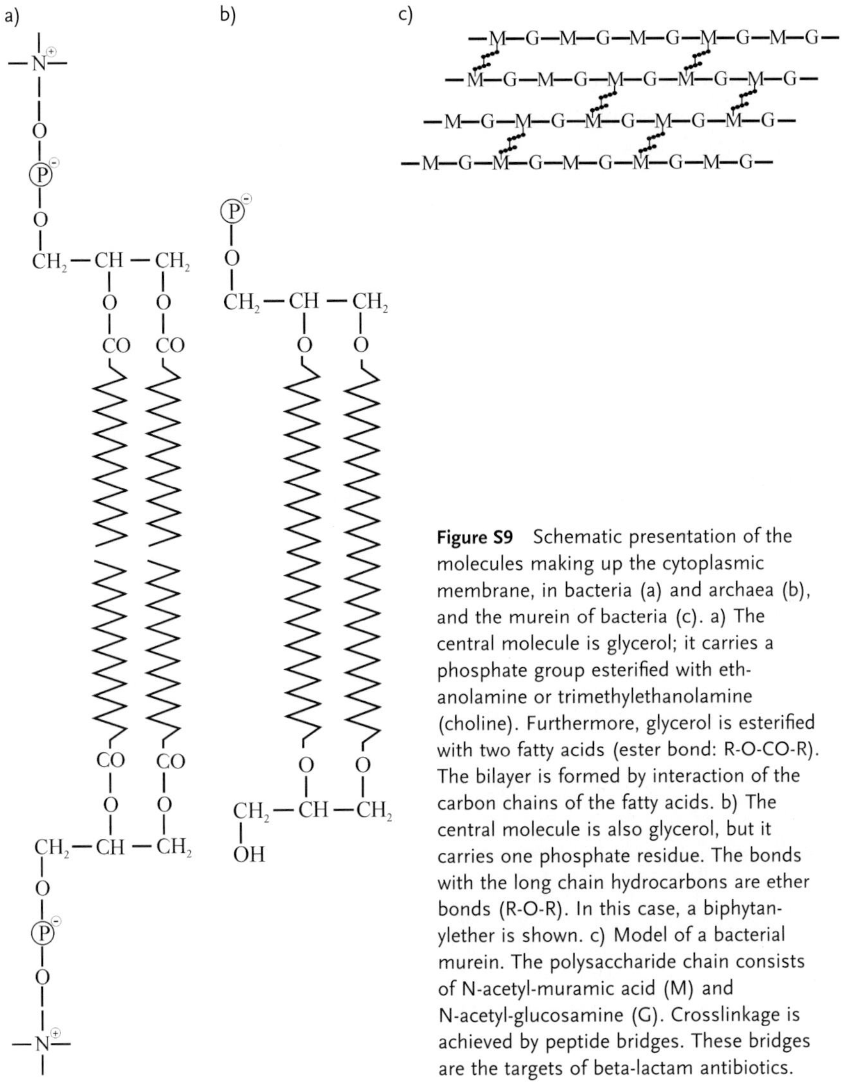

Figure S9 Schematic presentation of the molecules making up the cytoplasmic membrane, in bacteria (a) and archaea (b), and the murein of bacteria (c). a) The central molecule is glycerol; it carries a phosphate group esterified with ethanolamine or trimethylethanolamine (choline). Furthermore, glycerol is esterified with two fatty acids (ester bond: R-O-CO-R). The bilayer is formed by interaction of the carbon chains of the fatty acids. b) The central molecule is also glycerol, but it carries one phosphate residue. The bonds with the long chain hydrocarbons are ether bonds (R-O-R). In this case, a biphytanylether is shown. c) Model of a bacterial murein. The polysaccharide chain consists of N-acetyl-muramic acid (M) and N-acetyl-glucosamine (G). Crosslinkage is achieved by peptide bridges. These bridges are the targets of beta-lactam antibiotics.

hydrophobic tails of the phospholipids are long-chain fatty acids (16 carbon atoms or so) esterified with the other two hydroxyl groups of glycerol. A biologically active membrane not only consists of phospholipid molecules but also of proteins that are embedded within the membrane and fulfill transport functions. Otherwise, a bacterial cell would be unable to import or export molecules.

The cell wall determines the shape of a bacterial cell and functions like chain mail. Its rigidity withstands the osmotic pressure of the cell interior exerted against the cytoplasmic membrane. We must keep in mind that the microbial cytoplasm is a highly concentrated solution of metabolites and salts, so the concentration gradient of solutes decreases from the inside to the outside. Bacterial cell walls are of two basic types known as Gram-positive and Gram-negative cell walls. Why these walls were named after Christian Gram is explained in Chapter 22. Gram-positives like *Staphylococcus aureus* or *Bacillus subtilis* have walls consisting of a thick murein sacculus (sac or pouch). Murein or peptidoglycan is a macromolecule typical of bacteria and is not found in higher organisms (Figure S9c). Gram-negatives such as *Escherichia coli* have walls consisting of a rather thin murein sacculus. This is surrounded by a second membrane, the outer membrane, and the space in between, called the periplasmic space.

The composition of the cytoplasmic membrane and the cell walls described here only apply to bacteria but not to archaea, which will be discussed in Section 4.

2.3 The role of ATP

Cellular processes are primarily driven by ATP hydrolysis, an exergonic reaction by which energy is released (structure of ATP, Figure S10). The products of ATP hydrolysis are ADP and inorganic phosphate. All organisms require processes by which ATP is resynthesized from ADP and inorganic phosphate. These processes are discussed in Chapters 8 and 9 and in Section 5, where the reactions driven by ATP hydrolysis are introduced (Table S1). One type is the glutamine synthetase reaction by which glutamine is formed from glutamate, ammonia, and ATP.

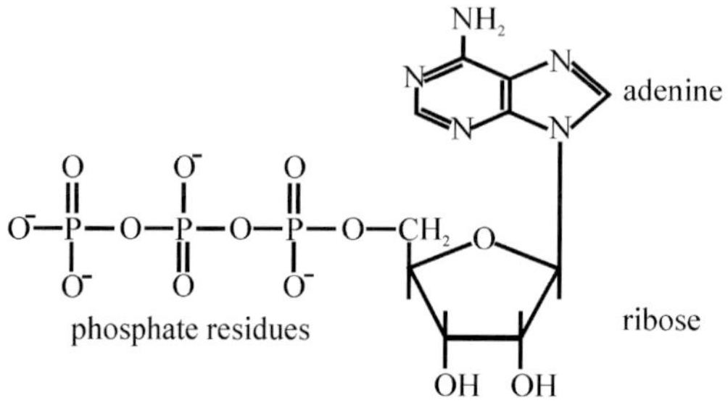

Figure S10 Adenosine 5′-triphosphate (ATP). The bonds between the phosphate residues are energy-rich bonds.

Table S1 ATP as energy currency.

General role
$$ATP + H_2O \xrightarrow{cellular\ processes} ADP + P_i$$
ATP, adenosine 5′-triphosphate
ADP, adenosine 5′-diphosphate
P_i, inorganic phosphate

Typical ATP-consuming reactions
   a. Glutamine synthetase
      Glutamate + NH_3 + ATP → glutamine + ADP + P_i
   b. Hexokinase
      Fructose + ATP → fructose 6-phosphate + ADP
   c. Alanyl-tRNA synthetase
      Alanine + ATP → alanyl − AMP + PP_i
      Alanyl − AMP + tRNA → alanyl − tRNA + AMP
      PP_i, inorganic pyrophosphate
      AMP, adenosine 5′-monophosphate

This is the type of reaction by which amino acids are activated for the translation process
Further conversion of AMP and PP_i:
$$AMP + ATP \xrightarrow{adenylate\ kinase} 2\ ADP$$
$$PP_i + H_2O \xrightarrow{pyrophosphatase} 2\ P_i$$

Sugars are generally activated through phosphorylation by ATP. Glucose is thus converted to glucose 6-phosphate, or fructose into fructose 6-phosphate. Another important type of reaction driven by ATP is the reaction linking each tRNA with its specific amino acid. As apparent from Table S1, ATP is converted in a two-step-reaction to AMP (adenosine 5′-monophosphate) and pyrophosphate. Subsequent enzyme-catalyzed reactions then convert AMP to ADP, and pyrophosphate to inorganic phosphate.

Questions

1) Define an informational macromolecule.
2) Why does A pair with T and G with C?
3) What is a hydrogen bond?
4) What are the functions of the three types of RNA?
5) Why does the genetic code comprise 64 triplets?
6) What is the function of an aminoacyl-tRNA synthetase?
7) What is the function of the cytoplasmic membrane and what are the constituents of a phospholipid molecule?
8) What is the difference between Gram-positive and Gram-negative cell walls?
9) Describe ATP-requiring reactions.

Section 3
Evolution, from the RNA world to the tree of life

3.1 The RNA world

We recall that RNA molecules were most likely the first self-replicating systems on Earth, and that the DNA world evolved from the RNA world. The important features of RNA are:

- RNA molecules exhibit RNA-replicase activity.
- RNA molecules may also have an RNA-ligase activity.

Ligases are enzymes that join fragments, such as RNA fragments. A hypothesis on the transition of the RNA world into the DNA world is presented at the end of Chapter 30.

3.2 Possible properties of LUCA

- An anaerobe because, if at all, there were only traces of molecular oxygen in the atmosphere and in bodies of water.
- A moderate thermophile growing between 50 °C and 80 °C.
- A microorganism deriving metabolic energy in the form of ATP from the reaction of H_2 with elemental sulfur to yield hydrogen sulfide. This microorganism may be of a type similar to *Thermoproteus tenax*, which still exists.
- Or alternatively, a microorganism growing on reactive gases such as carbon monoxide and H_2. Such organisms still exist today; examples are *Carboxydothermus hydrogenoformans* and *Clostridium ljungdahlii*.

3.3 Prokaryotes represent two of the three domains of the tree of life

The breakthrough came from 16S ribosomal RNA sequencing, which had several advantages:

- This molecule is present in all prokaryotes.
- 16S rRNA has a constant function.
- The size (around 1500 nucleotides) allows rapid sequencing, especially with automated equipment.
- The molecule has enough "room" for mutative changes that may have occurred during evolution.

The breathtaking result of 16S-rRNA sequencing: several prokaryotes belong to a separate domain, for which the name "archaea" was coined. Why are archaea not identical with bacteria? Members of both domains have several features in common. They lack a membrane-enclosed nucleus and have a large metabolic diversity, but archaea are more specialized. They grow in extreme habitats, including saturated salt brines as well as under conditions with the highest temperatures and the lowest and the highest pH values at which life has been observed. Members of the archaea are the only organisms capable of producing large amounts of methane, and archaeal ammonia oxidizers are apparently widespread in oceans (see Section 4).

The principal differences between bacteria and archaea, however, lie in their molecular biology and the composition of their cell membranes and cell walls. This will be discussed in Section 4.

Developmental milestones of microbial processes following LUCA are

- Fermentations.
- Anoxygenic photosynthesis.
- Sulfate reduction and related processes.
- Oxygenic photosynthesis.
- Respiration with oxygen as terminal electron acceptor, alternatively with other terminal electron acceptors, such as nitrate or ferric ions (Fe^{3+}).

Problems arose in the oxygen world due to formation of ROS (reactive oxygen species). These species are formed when electrons are transferred via flavoproteins to oxygen. The three species mentioned in the text are detoxified in the cells as follows:

*O_2^- is detoxified by superoxide dismutase

$$2\,{}^*O_2^- \rightarrow O_2 + O_2^{2-}$$

O_2^{2-} is the anion of H_2O_2 (hydrogen peroxide)

OH* is not detoxified enzymatically and reacts rapidly with organic molecules in the cell

H_2O_2 is detoxified by catalase

$$H_2O_2 \rightarrow H_2O + \tfrac{1}{2}O_2$$

ROS may cause damage to proteins and nucleic acids, so it is especially important that the above-mentioned enzymes are active in microorganisms that respire with oxygen or are exposed to oxygen.

Questions

1) What are the features of the RNA world?
2) Give several reasons why *Thermoproteus tenax* or *Clostridium ljungdahlii* could be microorganisms of the LUCA type.
3) What are the advantages of using 16S rRNA for the construction of phylogenetic trees?
4) Name the microbial milestones associated with the following microorganisms: Lactobacillus bulgaricus (Chapter 18), Clostridium acetobutylicum (Chapter 15), Chromatium okenii, Chlorobium limicola (Chapter 4), Desulfovibrio vulgaris, Archaeoglobus fulgidus (Section 4), Streptomyces griseus (Chapter 22), Mycobacterium tuberculosis (Chapter 29), Prochlorococcus marinus (Chapter 32), Anabaena azollae (Chapter 11).
5) How long did it take for oxygen to accumulate in the atmosphere after the appearance of oxygenic photosynthesis?
6) Summarize the reactions by which ROS are detoxified in organisms.
7) What is the principle of the Ames test?

Section 4
Archaea

It has been outlined that the archaea represent one of the three domains of life. Archaea are prokaryotes but are quite distinct from bacteria. Important differences include:

- Archaeal membranes contain ether-linked isoprenoids but no phospholipids (see Figure S9b).
- Murein (peptidoglycan) is not a constituent of archaeal cell walls, which contain S-layer proteins, heteropolysaccharides or pseudopeptidoglycan.
- Molecular biological processes more closely resemble the eukaryotic type than the bacterial type (see below).

4.1 The archaeal species

A phylogenetic tree is depicted in Figure S11. Four phyla have been identified so far:

- The Korachaeota comprise hyperthermophilic microorganisms present in hot springs. They have not yet been characterized.
- The Nanoarchaeota, represented by *Nanoarchaeum equitans*, the "rider" on *Ignicoccus hospitalis* (see Figure 76).
- The Crenarchaeota, including thermophiles and hyperthermophiles such as the Thermoproteales, Desulfurococcales and Sulfolobales, but also the psychrophilic (cold-loving) *Cenarchaeum symbiosum* living in marine sponges.
- The Euryarchaeota, to which the large and important group of the methanogens belongs.

The Crenarchaeota include *Thermoproteus tenax*, mentioned several times, and *Sulfolobus acidocaldarius*, an archaeon living at low pH values and high temperatures that was isolated by Thomas Brock and colleagues in 1972. Both organisms differ in the way they metabolize sulfur compounds. *T. tenax* is an anaerobe that lives on sulfur plus molecular hydrogen and carbon dioxide. In the absence of H_2, sulfur can also be reduced with organic compounds such as sugars or organic

Discover the World of Microbes: Bacteria, Archaea, and Viruses, First Edition. Gerhard Gottschalk.
© 2012 Wiley-VCH Verlag GmbH & Co. KGaA. Published 2012 by Wiley-VCH Verlag GmbH & Co. KGaA.

Euryarchaeota

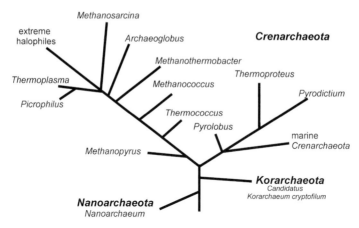

Figure S11 The phylogenetic tree of archaea. Only one representative of the Nanoarchaeota is known, *Nanoarchaeum equitans*, see Figure 76. The phylum Korarchaeota was postulated on the basis of unique DNA sequences. Members have been observed in mixed cultures, but pure cultures are not yet available.

acids. *S. acidocaldarius* is an aerobe that oxidizes H_2S or sulfur to sulfate. In this respect, it resembles the thiobacilli of the bacterial domain (see Chapter 32). All of these organisms can live on CO_2 as carbon source.

An interesting species is *Acidianus ambivalens*, which reduces sulfur under anaerobic conditions but also oxidizes sulfur to sulfate. It is named after the two-headed Roman god, Janus, because it combines the sulfur metabolism of *Thermoproteus* (one head) with that of *Sulfolobus* (the second head).

The most spectacular members of the archaea are the Desulfurococcales. They comprise the most hyperthermophilic species to date, the already-mentioned *Pyrodictium occultum* (Figure 11) and *Pyrolobus fumarii* that has a maximum growth temperature of 113 °C.

Euryarchaeota include the methanogens, the largest and ecologically most important group of the archaea. They currently comprise the major classes Methanobacteriales, Methanomicrobiales, and Methanosarcinales. Most species are able to produce methane from $CO_2 + H_2$. Only members of the Methanosarcinales grow on acetate, methanol, or methylamines, and they contain cytochromes. Well-known species are *Methanosarcina barkeri*, *Ms. mazei*, *Ms. acetivorans*, *Methanosaeta concilli* (grows on acetate only), and *Methanohalobium evestigatum*, which is halophilic and alcaliphilic (preferred substrate: trimethylamine). Extensively studied species of the other classes are *Methanothermobacter thermoautotrophicus* and *Mt. marburgensis*, moderate thermophiles and preferred species for the elucidation of the biochemistry of methanogenesis. Furthermore, there are the hyperthermophiles *Methanopyrus kandleri* and *Methanocaldococcus jannaschii*, and the inhabitants of the digestive tract, *Methanobrevibacter smithii* and *Methanosphera stadtmanae*. The latter only grows on a mixture of methanol and H_2. Other euryarchaeotes are the *Halobacteriales*, in which bacteriorhodopsin was discovered (see Chapter 8, Figure 18). Organisms such as *Haloarcula marismortui* and *Natronomonas pharaonis* are adapted to high salt concentrations and the latter to alkaline pH values up to pH 10. The Pyrococci such as *Pyrococcus furiosus* grow above 100 °C and their metabolism resembles that of the Thermoproteales. Extreme acidophiles belong to the Thermoplasmata, including *Picrophilus torridus*, previously mentioned in the text (Figure 12), and a species that grows at 55 °C, *Thermoplasma acidophilum*.

4.2 Habitats

Most sites where archaea thrive are rather exotic in comparison with our living conditions. The sites include salt lagoons, soda lakes, volcanic areas such as Yellowstone National Park (USA), the Solfatara volcanic crater (Italy), various geysers (Iceland), hydrothermal vents, and the black smokers on the ocean floor. The methanogens, which represent the major archaeal cell mass on our planet, are an exception because they are metabolically active wherever there are anaerobic conditions. We recall that around one billion tons of methane is produced annually.

4.3 Archaeal metabolism

Prokaryotic metabolism is usually considered under two aspects:

1) The type of compounds and reactions leading to the synthesis of ATP, in other words the energy metabolism of a particular microorganism.
2) The type of compounds and reactions leading to the building blocks of cellular material, the anabolic metabolism of a microorganism.

Because of their outstanding importance for nature, processes of the sulfur cycle will be discussed with reference to both archaeal and bacterial energy metabolism

- Dissimilatory reduction or oxidation of sulfur. These processes are widespread among the archaea, especially the Crenarchaeota. One process already introduced is the reduction of sulfur by molecular hydrogen as carried out by *Thermoproteus tenax*. The equation for this reaction is presented once more, but "ATP, metabolic energy" is replaced by a ΔG^{0l} value, the free-energy change. The more negative this value, the more exergonic the reaction.

$$H_2 + S^0 \rightarrow HS^- + H^+ \quad \Delta G^{0l} - 33.6 \text{ kJ/mol (kJ = kilojoule)}$$

The hydrolysis of ATP to ADP and inorganic phosphate has a ΔG^{0l} value of −32.6 kJ/mol, so the synthesis of ATP would have a value of +32.6 kJ/mol. Under the conditions of cell growth and proliferation with ATP as the driving force, this value is much higher, around +70 kJ/mol, because it is concentration dependent. This means that per mol of H_2 and sulfur approximately half a mol

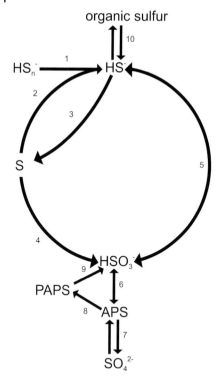

Figure S12 The sulfur cycle. 1, polysulfide reductase. Polysulfide contains a chain of sulfur atoms between HS and SH; 2, sulfur reductase; 3, sulfide: cytochrome c oxidoreductase; 4, sulfur oxygenase; 5, sulfite reductase; 6, adenylylsulfate (APS) reductase; 7, ATP sulfurylase; 8, APS kinase; 9, PAPS reductase; 10, cysteine desulfurase.

of ATP becomes available. This process requires a hydrogenase and a sulfur reductase. How the electron flow from H_2 to sulfur is coupled to ATP synthesis will be explained in Section 6.

The reaction shown above is part of the sulfur cycle depicted in Figure S12. Several archaea are able to perform parts of this cycle:

- Oxidation of sulfur to sulfate, an aerobic process, is carried out by *Sulfolobus acidocaldarius* and related species, but also by *Acidianus ambivalens*.

- Reduction of sulfate to sulfide, with sulfite (HSO_3^-) as an intermediate, is performed by *Archaeoglobus fulgidus*, an euryarchaeote phylogenetically related to the methanogens.

The sulfur cycle as a whole is dominated by bacterial rather than archaeal processes. The "bacterial sulfur cycle" is more likely of greater ecological importance because it predominates under ambient conditions in oceans, freshwater bodies, and soil. Major bacterial groups involved in the sulfur cycle are:

- *Thiobacillus thiooxidans* and relatives, rod-shaped or coccoidal microorganisms occurring in soil and ore (Chapter 32).
- *Beggiatoa alba* and relatives, which thrive in aquatic habitats and grow as filaments consisting of many cells enclosed in sheaths (Chapter 32).
- Sulfide- and sulfur-oxidizing bacteria that perform an anoxygenic photosynthesis (Chapter 9); they comprise the purple sulfur bacteria (e.g., *Chromatium okenii*; Chapter 4) and the green sulfur bacteria (Chapter 9).
- Sulfur-reducing bacteria such as *Desulfuromonas acetoxidans*.
- Sulfate-reducing bacteria, a large group of bacteria, of which *Desulfovibrio vulgaris* and *Desulfobacterium autotrophicum* have been described.

The global sulfur cycle, one of the three major cycles of matter on Earth, operates on a billion-ton scale per year. The carbon and the nitrogen cycles will be discussed in the next section.

4.4 Methanogenesis

Whereas the sulfur cycle is operated by members of the archaea and the bacteria, methanogenesis is dominated by the archaea. Minimethane producers have been described as well as a yet undefined methane production in forests, but methane production by nonarchaea is negligible. Without exaggeration it can be stated that the biochemistry of methanogenesis and its coupling to ATP synthesis is one of the most impressive examples of microbial energy metabolism. Two processes will be discussed: methanogenesis from H_2 and CO_2 and methanogenesis from compounds containing preformed methyl groups, such as methanol, methylamines, and acetate. These compounds are only utilized by members of the *Methanosarcinales*. Methanol serves as an example; it is metabolized according to the following equation:

$$4\,CH_3OH \rightarrow 3\,CH_4 + CO_2 + 2\,H_2O \quad \Delta G^{0\prime} = 319.5\,kJ/mol$$

This equation is actually the sum of two equations. One molecule of methanol is oxidized to CO_2 to generate the reducing equivalents for reduction of three methanol molecules to methane. First we should look at the latter process, which involves two novel coenzymes, coenzyme M and coenzyme B (Figure 13a). In a reaction catalyzed by a methyl transferase, methanol and coenzyme M are combined to yield methyl-coenzyme M. This is a central intermediate of methanogenesis in general. The reaction leading to methane is the methyl-coenzyme M reductase reaction. This is not a simple reduction process: coenzyme B provides the reducing power and forms together with coenzyme M what is called a heterodisulfide, while methane is liberated at the nickel-containing factor 430 located within the active center of the methyl-coenzyme M reductase. Reducing equivalents are required for reduction of the heterodisulfide. These are produced by

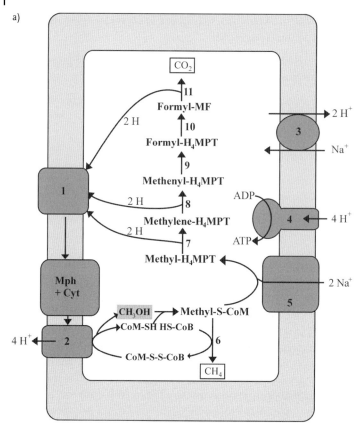

Figure S13 The two methanogenic pathways. (a) Formation of methane and CO_2 from methanol. Membrane-integrated complexes: 1, coenzyme $F_{420}H_2$ dehydrogenase; complex containing cytochromes and methanophenazin; 2, heterodisulfide reductase; 3, sodium ion/proton antiporter; 4, A_1A_0 ATPase, which translocates protons as well as sodium ions; 5, methyl-H_4MPT: CoM-SH methyl transferase. Soluble enzyme systems: 6, methyl-CoM reductase; 7, methylene-H_4MPT reductase; 8, methylene-H_4MPT dehydrogenase; 9, methenyl-H_4MPT cyclohydrolase; 10, formyl-MF: H_4MPT formyltransferase; 11, formyl-MF dehydrogenase. MF, methanofuran; H_4MPT, tetrahydromethanopterin. Note that the oxidation of one molecule of methanol generates $3 \times 2H$ that are used to reduce three molecules of methanol via methyl-CoM to methane. Acetate molecules are converted by the carbon monoxide/acetyl-CoA synthase to methyl groups and CO. CO is oxidized to CO_2 and the methyl groups enter the pathway at the level of methyl-H_4MPT. *(continued)*

methanol oxidation and are stored in the reduced form of coenzyme F_{420}, which is functionally in methanogens what NADH is in bacteria. From there they travel via an electron-transport chain to the heterodisulfide. This electron-transport chain involves cytochromes and a novel electron carrier, methanophenazine. The process yields ATP, as will be discussed in Section 6.

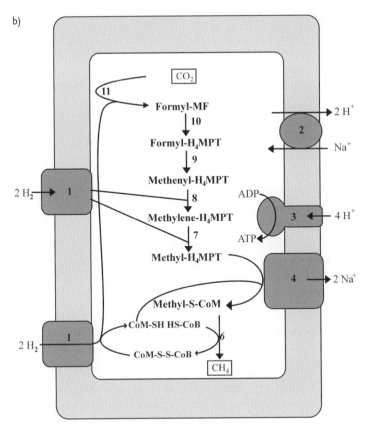

Figure S13 (*continued*) (b) Methane formation from $CO_2 + H_2$. The biochemistry differs from Figure S13a as follows: reaction 11 is driven by the heterodisulfide reductase (5); cytochromes and methanophenazin are not involved; reaction 4 is the only energy-conserving reaction.

As noted before, most methanogens reduce CO_2 to methane according to the following equation:

$$CO_2 + 4\,H_2 \rightarrow CH_4 + 2\,H_2O \quad \Delta G^{0\prime} = -136 \text{ kJ/mol}$$

The final reactions correspond to those discussed except that the heterodisulfide is not the acceptor for electrons from an electron-transport chain as above. In this case, the heterodisulfide is involved in coupling reactions to achieve the energetically very unfavorable first step in the reduction of CO_2 (Figure S13b). As a consequence, ATP has to be synthesized at a different site of the pathway, a methyl transferase reaction that functions as a pump for sodium ions. Thus, there is a fundamental difference in energy metabolism between the Methanosarcinales, the most advanced methanogens, and the other groups living on H_2 and CO_2. Only the former employ an electron-transport chain, whereas the latter employ a sodium ion pump for ATP synthesis.

4.5 Anabolic metabolism

There is close correspondence between the pathways employed by the archaea and the bacteria for synthesis of building blocks of the cells, but there are dramatic differences in the machinery used for synthesis of informational molecules. Examples:

- The RNA polymerase. The bacterial type of this enzyme has the subunit composition $\alpha_2\beta\beta'$. This core enzyme, together with a so-called σ-factor, represents the enzyme that transcribes DNA sequences into RNA sequences. The archaeal type consists of at least eight subunits and resembles the eukaryotic enzyme. All of these enzymes do the same job, but the bacterial enzyme and that of archaea and eukarya have little in common with regard to composition and structure.

- Initiation of protein synthesis. In bacteria, this is carried out by a tRNA linked to N-formylmethionine, so the first amino acid in a synthesized protein is always N-formylmethionine. The formyl group is usually removed in subsequent reactions. In archaea and eukarya, unsubstituted methionine is at the start position of the growing polypeptide chain.

- Composition of ribosomes. The 70S ribosomes in bacteria and archaea are composed of 30S and 50S subunits but differ in protein composition. Classic antibiotics of the streptomycin type do not inhibit archaea.

- Elongation factor 2 (EF-2). This factor is important in translation. In archaea and eukarya, EF-2 is inactivated by the diphtheria toxin via ADP-ribosylation. However, this toxin has no effect on bacteria, so the modes of action of diphtheria toxin on bacteria and on archaea are completely different.

The archaea are really amazing. It is remarkable that their molecular biology corresponds in part with eukaryotic systems, and it is even more remarkable that they differ dramatically from bacteria.

Questions

1) Summarize the differences in composition of archaeal and bacterial cells.
2) Name representatives of the Crenarchaeota and the Euryarchaeota.
3) Which archaea hold the record with respect to maximal growth temperature?
4) Where would you take samples to isolate *Haloarcula marismortui*?
5) Which methanogens reside in our intestinal tract?
6) Describe the archaea involved in the sulfur cycle.
7) Formulate the equations mentioned in the text for conversion of methanol to methane and CO_2.
8) What are the differences in energy metabolism between the Methanosarcinales and the methanogens growing exclusively on H_2 and CO_2?

Section 5
Bacterial diversity

First of all, we should recall the enormous amount of microbial biomass present on our planet, nearly 500 billion tons of carbon, and that 40% of this amount is decomposed annually. However, at the same time, an equivalent amount present in 2×10^{30} cells is generated by an astronomical number of cell divisions. The microbial biomass is present in the form of about one million species. Approximately 8000 of these species have been isolated and characterized, including assumedly the major players of microbial activity on Earth.

5.1 The phylogenetic tree of bacteria (Figure S14)

- Under extreme conditions, archaeal life outweighs bacterial life. Nevertheless, one branch of the bacterial tree includes organisms that grow at temperatures of up to 95 °C. We met their most important representatives previously: *Thermus aquaticus*, the anaerobe *Thermotoga maritima*, and the obligate aerobe *Aquifex aeolicus*. Together with related species, they represent the most hyperthermophilic bacteria known so far. Interestingly, the radiation-resistant *Deinococcus radiodurans* presented in Chapter 32 also belongs to this branch, although it is not a thermophile but grows at ambient temperatures. Its closest relatives are *Thermus thermophilus* and *Thermus aquaticus*.

- The phylogenetic tree has grown with the analysis of more and more 16S-rRNA sequences, giving some surprising insights into evolution. One of the surprises was that phototrophic bacteria do not cluster on one branch or phylum but are found on various branches. The *Chloroflexus* species are located within the thermophilic cluster just discussed. They are thermophilic green nonsulfur bacteria; their representative is *Chloroflexus aurantiacus*, which forms microbial mats in hot springs.

- The cyanobacteria form their own branch. They have already been discussed in connection with the appearance of oxygen on our planet and their significant role in the biomass production in oceans. This branch comprises the most abundant organism, *Prochlorococcus marinus* (Chapter 32), the filamentous species of the genera *Oscillatoria* and *Anabaena*, and *Trichodesmium*, which is

Discover the World of Microbes: Bacteria, Archaea, and Viruses, First Edition. Gerhard Gottschalk.
© 2012 Wiley-VCH Verlag GmbH & Co. KGaA. Published 2012 by Wiley-VCH Verlag GmbH & Co. KGaA.

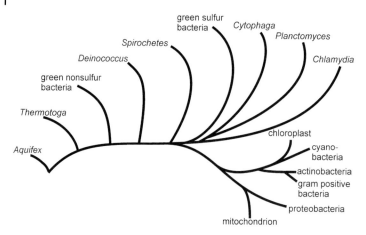

Figure S14 The phylogenetic tree of bacteria.

often responsible for cyanobacterial blooms in oceans. There are also single-celled cyanobacteria, whose well-known species is *Gloeocapsa*. All these organisms contain a photosynthetic apparatus to perform oxygenic photosynthesis with CO_2 as carbon source. The cyanobacteria – sometimes still called blue-green algae – are typical pioneer organisms since many of them are also able to fix N_2, converting molecular nitrogen in the atmosphere into bound nitrogen (Section 5B). In addition to minerals, they only require light, CO_2, and N_2.

- The next branches comprise the Gram-positive bacteria. Their cell walls primarily consist of a thick murein sacculus, a feature that is important in connection with the effect of penicillin and other antibiotics on bacteria (Chapter 22). Here we find extremely dangerous microorganisms, pathogens such as *Clostridium botulinum*, *C. tetani*, *Mycobacterium tuberculosis*, and *Bacillus anthracis* (Chapter 29) as well as organisms of biotechnological significance, such as *Bacillus licheniformis*, *Clostridium acetobutylicum*, and most importantly *Streptomyces*, the major antiobiotics producers. *Bacilli* and *Clostridia* form endospores, dormant forms with impressive properties. They are able to withstand high temperatures (well above 100 °C), dessication, and irradiation (Section 5C). Another group of organisms of this branch are the lactic acid bacteria, major players in the dairy industry (Chapter 18), but also organisms such as *Staphylococcus aureus* (Chapters 23 and 29) or *Propionibacterium acnes* (Chapter 10). It is interesting that phototrophic bacteria have been found within the group of endospore formers, species of *Heliobacterium*, which have been isolated from soil. Of course, this raises the question how they manage to get enough light.

- The "purple group" of phototrophic bacteria belongs to the next rather large branch, the Proteobacteria. All these organisms are Gram-negatives, so their murein sacculus is rather thin but surrounded by an additional protective shield called the outer membrane. So far, they are subdivided into alpha, beta, gamma, delta, and epsilon (α, β, γ, δ, and ε) proteobacteria. Some of the "alphas" are the *Rickettsia* species from which the mitochondria originated (Chapter 27). Here we find typical representatives of the anoxygenic phototrophic bacteria such as *Rhodospirillum rubrum*, a nonsulfur purple bacterium, the ethanol producer *Zymomonas mobilis*, the ethanol utilizers belonging to the genus *Acetobacter*, and *Nitrobacter* species performing the terminal reaction in the process of nitrification. Among the beta proteobacteria are the thiobacilli, *Nitrosomonas* species that catalyze the first step in nitrification, and the methylotrophs that oxidize methane or the methanol liberated in leaves (Figure 65). The gamma proteobacteria comprise the enterobacteria with *Escherichia coli* and *Salmonella typhi* as important members, the genus *Pseudomonas*, of which *Pseudomonas aeruginosa* (Chapter 26) and *Pseudomonas syringae* (Chapter 29) were mentioned in the text, furthermore *Vibrio cholerae*. Prominent members of the deltas are the bdellovibrios that invade bacteria and feed on them, the desulfovibrios and the myxobacteria. *Helicobacter pylori* (Chapter 29) belongs to the epsilon proteobacteria as well as *Agrobacterium tumefaciens* (Chapter 24).

The final branches of the phylogenetic tree comprise the green sulfur bacteria, including *Chlorobium limicola*, spirochaetas, and important freshwater microorganisms belonging to the *Planctomyces*.

5.2 Cycles and food chains

Nature has examined every niche on Earth for sustainability of microbial life. Useful niches must have the following prerequisites:

- Physical conditions such as pH, temperature, and salinity within a tolerable range (Chapters 6, 7 and 32).
- Substrates capable of supporting a given microbial energy metabolism.
- Precursors available for the synthesis of biomass.

On Earth, there are thousands of different niches that provide conditions for microbial life. The overwhelming microbial diversity that has evolved comprises thousands of species and metabolic types. They may be described as follows:

- Primary biomass producers, which are able to synthesize all their cell carbon from CO_2, for example:

    - Cyanobacteria, which perform oxygenic photosynthesis.

    - Bacteria that perform anoxygenic photosynthesis (*Chromatium okenii, Chlorobium limicola, Thiospirillum jenense, Rhodospirillum rubrum*).

    - Chemolithotrophic bacteria, which derive their energy from the oxidation of H_2, CO, NH_3, or H_2S, including *Ralstonia eutropha* (Chapter 28), *Oligotropha carboxydovorans, Nitrosomonas europaea*, and *Nitrobacter winogradskyi* (Chapter 12), *Sulfolobus solfataricus*, and *Thiobacillus thiooxidans* (Section 4).

- Homoacetogenic anaerobic bacteria such as *Clostridium aceticum, C. ljungdahlii*, and *Acetobacterium woodii*, which convert H_2 and CO_2 into acetate.

- Methanogenic archaea such as *Methanosarcina mazei* or *Methanocaldococcus jannaschii*, which produce methane and cell mass from CO_2 and H_2.

• Microorganisms that are involved in major cycles on Earth, the carbon cycle (Chapters 8 and 14), the nitrogen cycle (Chapter 12), and the sulfur cycle (Section 4). Practically all organisms are members of the carbon cycle. They respire and thereby produce CO_2. Especially the anaerobes tend to form food chains. Here is one example: sugars are fermented by certain lactic acid bacteria to lactate, which is then fermented by propionic acid bacteria to propionate, acetate, and CO_2. The propionate is subsequently converted by a syntrophic culture (Chapter 32) to acetate, CO_2 and H_2. These compounds are finally converted into methane. All reactions proceed under anaerobic conditions. The methane diffuses into aerobic zones, where it is oxidized to CO_2 by methanotrophic bacteria (Section 7) or oxidized anaerobically (Chapter 7). In biofilms (Chapter 26), the product of one member of the film may serve as substrate for another member. Ultimately, CO_2 will also be the final product.

The cycles have some leaks; the more important ones lead to the evolution of methane and dinitrous oxide (Chapter 14).

Microbial diversity is fantastic, it is the result of evolution. Just as nature abhors a vacuum, evolution abhors leaving behind any natural products. Nature has created ways and organisms to feed all these compounds into the appropriate cycles.

5.3 Survival strategies

Microbes like to grow and proliferate, and they do so in their habitats as long as conditions permit. Lactic acid bacteria living in milk have plenty of food available, so they grow rapidly until the substrates or the decreased pH become limiting. Subsequently, a considerable percentage of the population undergoes lysis. There still are survivors that may attach themselves to casein molecules or make themselves comfortable in biofilms developing on surfaces. These ups and downs of microbial populations can also be observed when we look at *E. coli* and other intestinal microorganisms, which live in some sort of semicontinuous culture. In aquatic environments, cell populations may also be very high (Figure 4) but are subject to considerable fluctuations, depending on substrate availability and physical conditions. Soil is a special habitat where the conditions can vary extremely. Soil may dry out or become muddy; growth of vegetation may provide a food supply, but this food may not be available for several months or even years. Perhaps this is the reason why many soil microbes are able to change into a dormant form. The most spectacular ones are the endospores.

- Endospores have fantastic properties. Because of their heat resistance, spores of many species tolerate temperatures well above 100 °C. They are also resistant to desiccation and radiation, and to a number of chemicals.

  The reasons for their extraordinary properties are:

  — Low water content (10%–25%).

  — High content of dipicolinic acid, which is also responsible for the refractility of spores seen under the microscope.

  — High content of calcium ions.

  — Thick coat and cortex layers around the core of the spore.

  Whenever growth conditions become less favorable, special regulatory factors induce spore formation inside the cell, a process that usually takes several hours. Sporulation is initiated by unequal division of the cytoplasm into two compartments: one is the mother cell and the second, smaller one first develops into a forespore containing one copy of the chromosome and several other factors (Figure S15). The content of the forespore increases in density as the calcium salt of dipicolinic acid accumulates. In addition, thick coats and envelopes are formed around the spore, which is finally released from the cell. Endospores may persist for years. If conditions become favorable for the development of a bacterial population, the spores will germinate. Water is taken up and the dipicolinic acid is released, so the spores lose their refractility. In a process that only takes some minutes, a bacterial cell develops that is capable of growth and proliferation.

  Sporeformers belong to the *Firmicutes*. The most prominent genera are the bacilli and the clostridia, among them several pathogenic organisms such as *Bacillus anthracis*, *Clostridium tetani*, and *Clostridium botulinum*. The spores are usually the infectious agents. They are taken up as aerosols, in contaminated food, or when wounds come in contact with soil. This is the unpleasant side of the spores. Since they remain viable for many years, they are a permanent threat. However, there are also "good" spores, for instance, those of *Clostridium acetobutylicum*, which germinate to vegetative cells that produce acetone and butanol, as we have seen (Chapter 15). *Bacillus subtilis* or *Bacillus licheniformis* are valued as exoenzyme producers and *Bacillus thuringiensis* produces the Bt toxin, a constituent of Bt corn (Chapter 24). Other spore-forming organisms belong to the genus *Sporomusa*, anaerobes that produce acetate from various substrates, or *Heliobacterium*, a rather special group of phototrophic bacteria.

- Soil bacteria such as the nitrogen-fixing *Azotobacter vinelandii* form what are called cysts under conditions unsuitable for growth. These cysts are dormant forms that are not as resistant as spores, but they do allow the organisms to survive under dry or nutrient-poor conditions. *Myxobacteria* such as the

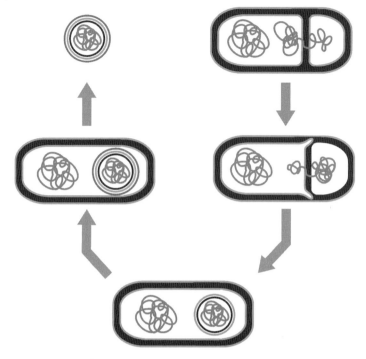

Figure S15 Sporulation. Asymmetric septation is followed by growth of the cytoplasmic membrane around the "spore compartment" which is then surrounded by two membranes. Transfer of the chromosome is completed and the cortex between the two membranes is formed as well as the core layer. Finally, the vegetative cell disintegrates.

Chondromyces robustus shown in Figure 74 form so-called myxospores, which have properties similar to cysts. Myxospores may be released and carried to other sites where they may find conditions suitable for germination and growth.

- Repair systems. On the molecular level, bacteria are also prepared to counteract the damage caused by hostile conditions. We already discussed *Deinococcus radiodurans* (Chapter 32) and the repair of UV-damaged DNA in *Sulfolobus solfataricus* (Figure 72). Additional repair systems are present in bacteria and archaea to excise damaged nucleotides from the DNA and to replace them. There are mismatch repair systems that recognize and remove wrongly paired bases. Known as the SOS system, it derepresses genes for special DNA polymerases that repair single-stranded portions of damaged DNA.

- The synthesis of proteins slows under conditions in which bacteria are unable to grow for one reason or another. There are mechanisms to attenuate transcription and translation processes and a stringent control system capable of largely reducing metabolic activity of microorganisms under nongrowing conditions (Section 8).

This is just a small selection of strategies to help cells survive. Other strategies include phobotactic responses, in which organisms simply swim away from repellents, or biofilm formation, the adhesion of microorganisms to surfaces in a favorable environment.

Questions

1) What is approximately the upper temperature limit for growth of bacteria and archaea?
2) Why are cyanobacteria pioneer organisms?
3) What are fundamental differences between *Firmicutes* and proteobacteria?
4) What are anaerobic primary producers?
5) Why aren't the major cycles completely closed, and where are the "leaks?"
6) Where does dipicolinic acid occur?
7) How are thymine dimers formed?
8) What is a mismatch?

Section 6
Membranes and energy

6.1 Transport

The microbial cell with its ion-tight cytoplasmic membrane is like a fortress. It has to be ion tight in order to provide the proton motive force required for ATP synthesis. Nevertheless, cells have to take up nutrients as well as secrete certain products, so traffic is heavy across the cytoplasmic membrane. The transport proteins involved are highly specific, and they exhibit a high affinity for the compound to be transported. We may recall the example involving a bacterial cell in the ocean. In order for the cell to survive or even grow and proliferate, the cellular concentration of all required metabolites must be least 100 or 1000 times higher than the concentration in sea water, a remarkable achievement. For osmotic reasons, the cell must be capable of taking up water or releasing it. This is carried out by tiny channels in the cytoplasmic membrane called aquaporins, which are filled with water and able to release it inside or outside the cell. Gases may enter or leave the cells by diffusion processes, but nearly everything else has to be actively transported, which means up-hill with respect to concentration. There are two types of transport systems that occur in practically all organisms, not only bacteria.

- Primary transport systems include the ABC transporters, the largest family of transporters. Substrates are specifically bound, and the active uptake of substrates is catalyzed by the energy of ATP hydrolysis. Sugars, amino acids, and especially inorganic ions such as phosphate, molybdate, sulfate, or nickel are taken up by such systems (Figure 16a). It is extremely important for a living cell to maintain an intracellular pool of potassium ions. KdpFABC, a high-affinity uptake system of K^+ represents another important group of primary transport systems, it is a K^+-dependent ATPase. ATPases of this type differ in composition from ABC transporters.
- Secondary transport is performed by symporters (Figure S16b). The principle of these systems is to allow the influx of protons together with the solute to be transported. There are many symporters, for instance, for the transport of amino acids. In addition to the ATPase for uptake of potassium ions, there is also a potassium symporter present in *E. coli*, called the TrkG/TrkH system (Figure S16b). It is of low affinity as compared to the ATPase.

Discover the World of Microbes: Bacteria, Archaea, and Viruses, First Edition. Gerhard Gottschalk.
© 2012 Wiley-VCH Verlag GmbH & Co. KGaA. Published 2012 by Wiley-VCH Verlag GmbH & Co. KGaA.

Antiporters are related to symporters. Best known is the Na^+/H^+ antiporter, which in *E. coli* and many other microorganisms couples the uptake of protons with the export of sodium ions. This is important because the physiology of the microorganisms requires maintenance of a low intracellular Na^+ concentration but a high concentration of K^+ in the cytoplasm.

- Finally, there is a transport system that is only present in prokaryotes, the phosphotransferase system (PTS), primarily for sugars. Such transport systems employ phosphoenolpyruvate (PEP) instead of ATP as the driving force for substrate transport. Via a cascade of reactions, PEP phosphorylates a membrane protein that takes in glucose under phosphorylation, so instead of free glucose, glucose-6-phosphate appears inside the cells. When we look at the metabolic scheme of lactic acid fermentation (Figure S17), we can see how very important this is. Of course, only one PEP molecule remains for ATP synthesis because the second one has been used for the phosphorylation of glucose. But nevertheless, one ATP is saved in the initial phosphorylation reactions and the breakdown of glucose is coupled to its specific uptake. This system is widespread among bacteria and is used for the uptake of many sugars, including glucose, mannose, fructose, and sugar alcohols such as mannitol.

Not only uptake but also secretion is important for microorganisms. This becomes apparent when we look at the growth substrates. Usually, glucose is considered to be an important substrate. After all, sugars, especially fructose

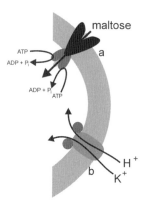

Figure S16 Active transport. (a) Maltose uptake by the high-affinity ABC transporter MalFGK$_2$. (b) K^+ uptake by the TrkG/TrkH symporter. ABC is a commonly used abbreviation for the ATP-binding cassette, the ATP-binding region of the protein. The conformational changes in the protein that are caused by ATP hydrolysis bring about the uptake of maltose.

Figure S17 Fermentation of glucose to two molecules of lactate via the Embden-Meyerhof-Parnas pathway. 1, PEP: glucose phosphotransferase; 2, glucose phosphate isomerase; 3, phosphofructokinase; 4, fructose bisphosphate aldolase; 5, triose phosphate isomerase; 6, glyceraldehyde 3-phosphate dehydrogenase; 7, 3-phosphoglycerate kinase; 8, phosphoglycerate mutase; 9, enolase; 10, pyruvate kinase, 11, lactate dehydrogenase. Note that the dihydroxyacetonephosphate formed in reaction 4 is isomerized in reaction 5 to glyceraldehyde-3-phosphate so that two molecules of the latter are converted by reactions 7 to 9 to two molecules of PEP.

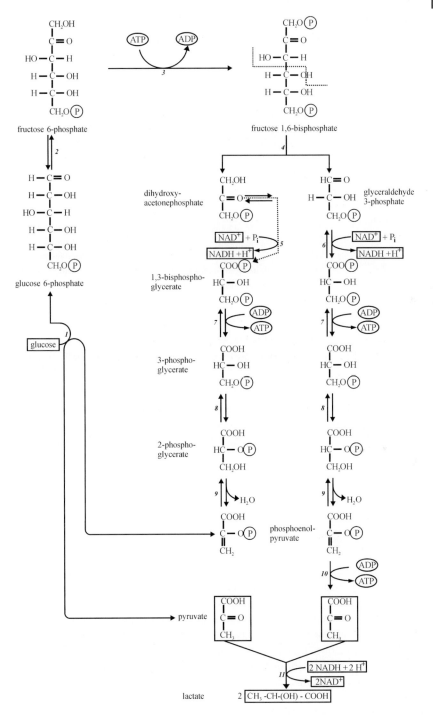

Table S2 Some typical exoenzymes secreted by microorganisms.

Enzyme	Reaction catalyzed and products
α-amylase	Cleaves the α-1,4-glycosidic bonds in starch and glycogen, produces maltose, glucose and branched-chain oligosaccharides
β-amylase	Common in plants, less common in microorganisms, cleaves α-1,4-glycosidic bonds in starch starting from the so-called nonreducing end of starch, maltose and branched-chain oligosaccharides are the products
Pullulanase	Cleaves α-1,6-glycosidic bonds, the branch points in starch, sometimes called "debranching enzyme"
Cellulase	Degrades cellulose; endoglucanases hydrolyze internal bonds of cellulase and produce oligosaccharides, cellobiose and glucose; the cellobiohydrolase produces cellobiose from the nonreducing end of cellulose. Effective cellulase breakdown requires a multienzyme complex containing endoglucanases, cellobiohydrolases and β-glucosidases, and the latter hydrolyze cellobiose to glucose
Xylanase	Hydrolyzes xylan, the polymer of the C5-sugar xylose, xylan is an important constituent of the hemicelluloses in plants
Pectinase	Hydrolysis of pectin, which requires a pectin methylesterase to remove methyl groups in the form of methanol from pectin, and polygalacturonate hydrolases, which cleave the backbone of pectin
Lipase	Hydrolysis of fats to glycerol and fatty acids
Protease	A variety of enzyme types that degrade proteins to amino acids. An important industrial enzyme is subtilisin.
DNAse and RNAse	These enzymes hydrolyze free DNA and RNA and make these compounds available as nutrients
Chitinase	A mixture of enzymes that degrade chitin to glucosamine and acetate. Chitin is a component of the cell walls of fungi and of the exosketeleton (shell) of insects

and lactose, are very abundant in fruits and milk, respectively. However, most of the substrates of microbes are polymers such as proteins, starch, cellulose, and other polysaccharides. Polymers need to be hydrolyzed because they are often larger than the microorganisms themselves. Only then can the products of hydrolysis be taken up, for instance, glucose or maltose. Maltose, a product of starch breakdown, consists of two glucose molecules. For hydrolysis of polymers, the microbes have to secrete exoenzymes such as amylases or proteases (Table S2) into their immediate environment. Moreover, also the respiratory-chain proteins and the ABC transporters somehow have to be escorted to the membrane and integrated.

Proteins to be secreted are tagged with a special signal peptide. The growing polypeptide chain is attached to a protein called Sec B, which prevents folding of the polypeptide chain. The unfolded form of the polypeptide reaches a protein complex in the cytoplasmic membrane, where it is pushed through a channel with the signal peptide headfirst. The signal peptide is cleaved off of the protein, which is released to the outside and folded with the help of a chaperone. The latter represents a class of proteins, also common inside the cell, that assist the proper folding of newly synthesized but inactive proteins to form active enzymes. This general Sec system plays an important role in exoenzyme production in Gram-positive bacteria.

Protein export becomes more complex when we consider Gram-negative bacteria, where a second membrane has to be crossed before proteins can be released (Figure 63). So far, five different protein-secretion systems in Gram-negative bacteria have been described. The type III system was discussed in connection with the injection of proteins of pathogenic bacteria into animal or plant cells. Other export systems are involved in the secretion of hemolysin or toxins, for example. The transformation apparatus for DNA uptake is related to the secretory systems (Figure 38).

6.2 Principles of ATP synthesis in microbes

As outlined in Section 2C, ADP and inorganic phosphate are the products of metabolic energy expenditure. Therefore, the generation of ATP from ADP and P_i is a vital process of all living organisms, including of course the microbes. One type of reaction leading to ATP generation is known under the term "substrate-level phosphorylation."

Substrate-level phosphorylation means that degradation of organic substrates proceeds via intermediates containing high-energy bonds. Intermediates of this type are 1,3-bisphosphoglycerate and phosphoenolpyruvate, for example. We should look again at the diagram of lactic acid fermentation (Figure S17) as an example. The breakdown of glucose proceeds via the above-mentioned molecules, whose further metabolism is coupled to ATP synthesis. The balance is such that one molecule of PEP and one molecule of ATP have to be invested in the activation of the sugar molecule when proceeding from glucose to fructose 1,6-bisphosphate. Four molecules of ATP are synthesized by substrate-level phosphorylation, which means that the net ATP yield of this fermentation is two molecules of ATP per glucose.

Anaerobes carrying out fermentations usually take advantage of this type of ATP synthesis by substrate-level phosphorylation. These processes not only lead to the formation of ethanol or lactic acid as fermentation products. Many organisms, notably the clostridia, produce molecular hydrogen and acids such as butyric acid and acetic acid. The formation of these acids from the corresponding acetyl or butyryl phosphates is coupled to ATP synthesis (Figure S29). These are other important reactions that allow ATP formation by way of substrate-level phosphorylation.

The chemiosmotic mechanism. The second, more complex principle by which ATP is synthesized is universal in nature. We say that ATP is synthesized by a chemiosmotic mechanism. What does this mean? It means that an ion motive force is generated at the cytoplasmic membrane in bacteria, or at the membrane of mitochondria or chloroplasts. This force is in most cases a proton gradient but in some cases a gradient for sodium ions as well. The proton motive force consists of two components, a proton gradient resulting from a higher proton concentration outside the cytoplasmic membrane as well as a membrane potential. Protons are positively charged, so the cytoplasmic membrane is positively charged on the outside and negatively charged on the inside.

As said before, the proton motive force is composed of the membrane potential and a term related to the ΔpH, the proton concentration inside and outside, as follows:

$$\Delta P = \Delta \psi - Z \cdot \Delta pH$$

Z is 2.3 RT/F (R, gas constant; T, absolute temperature; F, Faraday constant). *E. coli* growing in a medium of pH 6.0 has an internal pH of 7.8 and a membrane potential of $\Delta\psi = -95$ mV. The proton motive force then amounts to approximately -200 mV. The highly sophisticated F_0F_1 ATP synthase takes advantage of this proton motive force by coupling the influx of protons back into the cell with the synthesis of ATP (Figure 19).

Since we are already familiar with the ATP synthase, the key question is how to generate the proton motive force. Quite a diversity of reactions are involved:

- The aerobic respiratory chain (Figure S18). It consists of a number of cytochromes and quinones, all integrated or associated with the cytoplasmic membrane. Substrates such as glucose or other sugars are taken up by the organisms and oxidized, eventually, to CO_2. Uptake may involve ABC transporters, symporters, or PEP-phosphotransferase systems, as discussed in Section 6A. The sugar is oxidized to pyruvate, via the Embden-Meyerhof-Parnas (EMP) pathway, and further to CO_2 by the reactions of pyruvate dehydrogenase and the tricarboxylic acid (TCA) cycle (Figure S18a). This oxidation must be coupled to the reduction of a cellular compound, usually via conversion of coenzyme NAD$^+$ to NADH. This compound is a strong reducing agent that feeds electrons into the respiratory chain. There the electrons are passed along, much like traveling down a cascade of waterfalls, finally to oxygen, which is reduced to water. During their journey the electrons pass so-called coupling sites, devices that couple the electron movement through the chain to the export of protons. When the two electrons of NADH travel to oxygen, this is associated with the translocation of 12 protons from the inside to the outside (Figure S18b), whereas the ATP synthase requires about four protons to be able to make one molecule of ATP. Thus, the oxidation of one NADH with oxygen through the respiratory chain is linked to the synthesis of three molecules of ATP. Since ten molecules of NADH and two of FADH$_2$ are formed by oxidation of one glucose to six CO_2, a total yield of 34 ATP per glucose can be expected. This is a fantastic yield in comparison to lactic acid fermentation. It was a big step in evolution when the availability of oxygen allowed fermentative energy metabo-

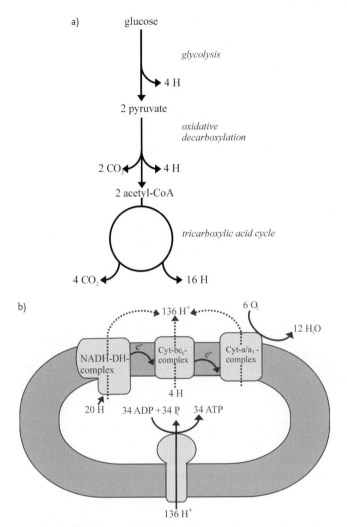

Figure S18 The aerobic respiratory chain. (a) Scheme of the oxidation of glucose to CO_2 and reducing equivalents involving the EMP pathway, the pyruvate dehydrogenase reaction and the TCA cycle. (b) Translocation of protons during transfer of reducing equivalents from NADH and $FADH_2$ to oxygen. The 20 H represent the 10 NADH generated in the course of glucose oxidation to CO_2, as do the 4 H of the 2 $FADH_2$ generated. The oval represents the cytoplasmic membrane.

lism to be replaced by a respiratory type. $FADH_2$ is the reduced form of flavin adenine dinucleotide (FAD), a weaker reducing agent than NADH. So the yield of 30 ATP results from the oxidation of ten NADH, with an additional four ATP from two $FADH_2$ and two ATP from substrate-level phosphorylation. The total is 36 ATP per glucose. This maximal value is reached in mitochondria and in some aerobic microbes, whereas in *E. coli* it is nearly one-third lower.

- Anaerobic electron-transport processes. These processes follow the same principle. We have discussed the growth of anaerobes on H_2 and elemental sulfur. In this case, the electrons of H_2 travel to an enzyme system capable of transferring them to elemental sulfur. One coupling site is present that allows the translocation of probably two protons across the membrane. Sulfate-reducing bacteria such as *Desulfobacterium autotrophicum* have also been mentioned, in which sulfate is first enzymatically activated and reduced to sulfite with the help of two enzyme systems. Here, sulfite plays the role that oxygen plays in aerobic respiration. The sulfite accepts electrons from molecular hydrogen, for example. The sulfite is reduced stepwise to sulfide and, again, a proton motive force is generated.

    There is an important process in nature that microbiologists call denitrification; it is part of the nitrogen cycle presented in the next section. This denitrification closely resembles oxygen respiration, and many bacteria can switch from oxygen to nitrate as electron acceptor during respiration. The principle is the same, but five electrons are required per nitrate. It is first reduced to nitrite (NO_2^-) and further to nitric oxide (NO). Two molecules of NO then proceed to nitrous oxide (N_2O) and, in a final reduction step, to molecular nitrogen (N_2). This process is of great ecological importance not only in soil but also in oceans, as seen in Chapter 32.

    Exciting microorganisms in this context are those capable of reducing metal ions such as manganese (IV), iron (III), or uranium (VI). These microbes grow under anaerobic conditions at the surface of metal oxides such as MnO_2 by coupling the oxidation of organic substrates with the reduction of metal oxides. We have already met *Shewanella oneidensis* in Chapter 32. Another expert in these reductions is *Geobacter metallireducens*.

- Light-driven proton translocation. Bacteriorhodopsin is a relatively simple proton pump that was discussed in Chapter 8. It consists of a membrane-integrated protein to which retinal, as a type of switch, is bound. Proton translocation takes place upon illumination. This light-driven proton pump occurs in halophilic archaea. Recently, related compounds called proteorhodopsins have been detected in marine microorganisms of the *Roseobacter* group and in *Pelagibacter ubique* (Chapter 32). These rhodopsins may also function as proton pumps. However, the photosystems of phototrophic organisms are certainly of overwhelming importance. There are about five types of photosystems in organisms performing an anoxygenic photosynthesis, of which two are presented in detail:

    The system P870 occurs in nonsulfur as well as in sulfur purple bacteria. As depicted in Figure 20, photons are collected by antennae pigments. The excitation energy is transferred to the reaction center P870, a powerful reducing agent in its excited state. The electrons flow back to the reaction center and protons are translocated. This is a cyclic electron transport. The translocated protons are taken up by the ATP synthase to make ATP. The P840 system of the green sulfur bacteria is more complex. The excited P840 feeds electrons into a branched pathway, a cyclic one comparable to the P870 system and a noncyclic one that ends with the reduction of NAD^+ to NADH or of $NADP^+$ to NADPH. These reduced coenzymes are then used in anabolic metabolism to make cellular constituents.

The culmination is the system that first appeared on Earth with the cyanobacteria and allows oxygen evolution. As shown in Figure 21, it consists of two systems. PSI resembles the P840 of the green sulfur bacteria because it allows not only proton translocation but also reduction of $NADP^+$ to NADPH. PSII is an ingenious further development of the photosystem of the purple bacteria. It only accepts light of shorter wavelengths (below 680 nm). The excited P680 releases electrons to the P700 reaction center of photosystem I, where they are again excited and finally used to reduce $NADP^+$ to NADPH. The second advancement of PSII is the presence of a manganese-containing protein. By electron transport, this protein loses its electrons to the reaction center P680. In its oxidized form, it is so reactive that it even reacts with water, whereby it is again reduced and oxygen as well as protons are released.

- Sodium ion pumps. We have seen that an electron-transport chain is involved in proton translocation when *Methanosarcina* species grow on methanol, methylamines, or acetate. The electron donor in this system is reduced coenzyme F_{420} and the acceptor is the heterodisulfide. During methanogenesis from H_2 and CO_2, an intermediate called methyl-tetrahydromethanopterine is formed. The methyl group is subsequently transferred to coenzyme M to yield methyl-coenzyme M, which is then further converted to methane and the heterodisulfide, as discussed above. This methyl transferase is a pump for sodium ions. For each methyl group transferred, two sodium ions are translocated from inside to the outside. As far as we know, this is the only reaction in these organisms by which an ion gradient is generated that can be used for ATP synthesis. A number of other reactions that are coupled to sodium ion translocation have been identified in bacteria. Examples include the decarboxylation reactions that occur in *Propionigenium modestum* (Chapter 32). In this organism, methylmalonyl CoA is decarboxylated to propionyl CoA by a membrane-integrated enzyme system that translocates sodium ions during the reaction. Another example is the NADH dehydrogenase of some pathogenic and alkaliphilic bacteria, *Haemophilus influenzae* to name just one. This enzyme generates a sodium ion motive force that is most likely involved in the expression of virulence factors. This NADH dehydrogenase is quite different in composition from that of complex I of the respiratory chain (Figure S18).

6.3 MCPs and motility

It was pointed out earlier that microbial life, in a way, takes place in little fortresses. Beyond the various transport systems, there is a need for communication with the exterior environment. Certain sensory proteins in some motile bacteria, known as methyl-accepting chemotaxis proteins (MCPs), allow communication between the exterior and the interior of the cell. At least four types of MCPs are present in *Escherichia coli*. They are membrane-integrated and capable of responding to the following substrates:

- Tar, for maltose, aspartate, and nickel or cobalt ions.
- Tsr, for threonine, alanine, and glycine.
- Trg, for ribose, galactose, and glucose.
- Tap, for dipeptides (compounds consisting of two amino acids).

Most of the compounds mentioned are attractants. However, nickel and cobalt ions are repellents, and other repellents are also known. Whenever these compounds bind to the MCPs, a signal is produced and passed along to the intracellular components of the MCPs and ultimately to the flagellar motor. As a result, the cells swim away from the repellent. Attractants have the opposite effect. The binding of aspartate by Tar, depicted in Figure S19, has the following effects:

The conformational change of Tar has an effect on the phosphorylation status of the so-called Che proteins. Upon binding of an attractant by the MCP, the protein Che Y is maintained preferentially in the unphosphorylated form. As a result, the flagellar motor rotates counterclockwise, causing the cells to swim straight ahead. In the absence of an attractant, Che Y becomes increasingly phosphorylated. This form reverses the rotational direction of the flagellar motor. As a result, the flagellae try to rotate clockwise, but this only leads to uncoordinated rotation and causes the cells to tumble.

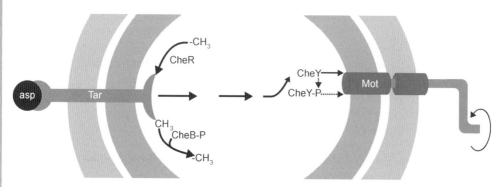

Figure S19 Chemotaxis induced by aspartate (asp) that interacts with the MCP Tar. Intracellularly, the signal is transferred from Tar to CheY, which is kept in the unphosphorylated state. The flagellar motor rotates counterclockwise, so the cells swim smoothly. There is a continuous phosphorylation of CheY. At a certain level of CheY-P, the motors switch to clockwise and the cells tumble. CheR methylates Tar. In its totally methylated form, Tar does not interfere with the CheY/CheY-P ratio, so the cells swim and tumble. Upon demethylation by CheB-P, the tumbling frequency decreases again. Mot, flagellar motor; the methyl groups are provided by S-adenosylmethionine (SAM), they are released as methanol.

There are two further interesting aspects of this sensory system. Cells swimming in the direction of an attractant may also tumble once in a while, but the tumbling frequency is much lower than when they swim away from an attractant. The change in concentration of the attractant with time affects the phosphorylation-dephosphorylation cascade between the MCP and the flagellar motor. The second aspect is the presence of a mechanism that allows the system to relax. If the vessel in Figure 53 contains a solution of aspartate, it would not make sense for the cells to swim rapidly in a given direction. The attractant concentration is the same in all parts of the vessel. The system is able to adapt to such a situation by regulating the methylation status of the MCPs. These MCPs contain four glutamate residues that can be methylated. When fully methylated, the MCPs no longer respond to attractants, so the cells tumble, then swim a little, then tumble again. As soon as conditions change, the methylation status is altered and the system once more becomes alert.

Bacteria are able to move by flagellar rotation (Figure S19). The flagellae are embedded in the outer membrane (in the case of Gram-negative bacteria) and the murein layer, and supported by ring structures. Flagellae are connected with the flagellar motor, which is inserted into the cytoplasmic membrane. Rotation is driven by the proton motive force. This motor resembles the ATP synthase discussed earlier. It rotates at a rate of about 1000 rpm, allowing a bacterial cell to move between 50 and 100 µm per second. The flagellae consist of proteins called flagellin. Flagellar growth proceeds at the tip: flagellin molecules are transported through the hollow core of the flagellum to the tip, where they are inserted by a process called self-assembly. This is comparable to building a chimney by transporting the bricks upward within the chimney instead of using a brick lift or hoist outside the chimney. Synthesis and function of bacterial flagellae are fascinating.

There are further systems to provide microbes with information about the situation outside:

6.4 Two-component systems

These consist of a sensor kinase and a response regulator (Figure S20). Let's take the high-affinity uptake system for potassium ions as an example. It is synthesized and inserted into the cytoplasmic membrane under conditions of K^+ limitation. The conformation of the sensor kinase is altered under these conditions, and this signal activates the kinase function of this protein, it phosphorylates itself with ATP as donor of the phosphoryl residue. This residue is transferred to the response regulator, which in its phosphorylated form functions as a switch for expression of the appropriate kdp genes. Their gene products are inserted into the membrane, where they function as a high-affinity system for K^+ uptake. Two-component systems (a few are listed in Table S3) are widespread in microorganisms. *Bacillus subtilis*, for example, contains the genetic information for the synthesis of 35 such systems. A second system is the ArcBA system, where Arc stands for aerobic respiratory control. This system is important for facultative anaerobic bacteria such

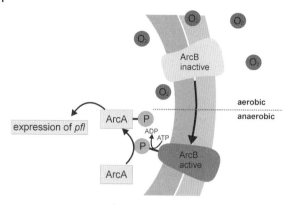

Figure S20 The ArcAB two-component system. ArcB is the sensor kinase, which is inactive in the presence of oxygen. In the absence of oxygen, it is phosphorylated by ATP and the phosphate residue is transferred to the response regulator ArcA that then activates the expression of the genes for pyruvate formate lyase (pfl), an enzyme involved in fermentation (see Figure S30). Furthermore, the expression of genes for the respiratory chain is repressed. Phosphorylation of the sensor kinase occurs at one particular histidine residue. The phosphate residue is transferred to an aspartate constituent of the response regulator.

Table S3 Examples of two-component systems.

Sensor kinase (SK)	Response regulator (RR)	Condition	Function of phosphorylated RR
KdpD	KdpE	Low K^+	Activation of expression of genes for the K^+ transporting ATPase KdpFABC
ArcB	ArcA	Low O_2	Activation of repression of genes for pyruvate formate lyase; inhibition of gene expression of components of the respiratory chain
FixL	FixJ	Low O_2	Activation of gene repression for enzyme systems of rhizobial N_2 fixation
NarX	NarL	Nitrate	Activation of gene expression of nitrate reductase
		Low O_2	Inhibition of gene expression of fumarate reductase
MCP/CheA	CheY	Attractant	Flagellar motor switch adaptation

CheA is cytoplasmic but receives signal from the MCPs.

as the enterobacteria. Under aerobic conditions they respire; under anaerobic conditions they carry out fermentations. The Arc system senses the presence of oxygen, so it remains inactive as long as oxygen is present and respiration is possible. Under anaerobic conditions, the sensor kinase ArcB becomes active and is subsequently converted by ATP into the phosphorylated form, which in turn phosphorylates ArcA. This response regulator represses the synthesis of cytochromes for the respiratory chain and at the same time activates the synthesis of typical enzymes for fermentations, such as the pyruvate formate lyase.

6.5 Quorum sensing

The compounds allowing this kind of communication between microorganisms are:

- Homoserine lactones giving rise to bioluminescence in *Vibrio fischeri* and *V. harveyi* or to exoenzyme production in *Pseudomonas aeruginosa* (Figure S21).
- Small peptides promoting extracellular toxin production in *Staphylococcus aureus*.
- γ-butyrolactone inducing antibiotic production in *Streptomyces*.
- Furanosyl borate esters stimulating cell division in *Escherichia coli*.

The principle of quorum sensing is rather simple. *Vibrio fischeri* serves as another example of this phenomenon. All cells produce homoserine lactone at a given rate, and this compound diffuses to the exterior. Whenever a population of *V. fischeri* cells becomes rather dense, the extracellular as well as the intracellular

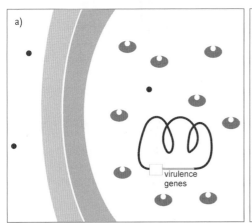 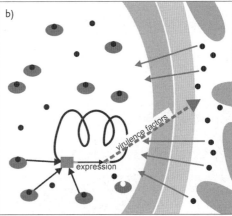

Figure S21 Quorum sensing (QS) and toxin production in *P. aeruginosa*. (a) Low cell density, in which the AHL (acylhomoserine lactone) concentration inside and outside the cells is low, and the QS-regulated genes are not expressed. (b) High cell density under which AHLs are abundant and interact with receptors that serve as activators of gene transcription. The products are toxins, proteases and polysaccharides for biofilm formation.

concentration of this lactone will be high. This is the signal! If the concentration exceeds a quorum value, the lactone binds to the receptor luxR, which in turn activates the so-called lux genes. Consequently, the enzyme luciferase is produced, which then catalyzes oxidation reactions that cause light emission by the cells.

It is apparent that quorum sensing is of special value to organisms in biofilms, where certain activities, polysaccharide production, or exoenzyme or toxin synthesis can be coordinated by quorum sensing (Chapter 26).

Questions

1) What are the characteristic features of ABC transporters?
2) How is PEP involved in transport?
3) What is the basis of substrate-level phosphorylation?
4) What is the driving force for proton translocation in the respiratory chain?
5) What do aerobic respiration and denitrification have in common; and what are the differences?
6) Speculate why different light-driven electron transport systems evolved in purple bacteria and in green bacteria.
7) With which reaction do methanogens pump sodium ions?
8) What are MCPs good for?
9) What is the rotation mode of the flagellae of enterobacteria when they swim smoothly?
10) What is the effect of methylation of the glutamate residues of MCPs?
11) How is a signal transferred from the sensor kinase to the response regulator?
12) Develop your own hypothesis on the structure of the sensor kinase.
13) Why is quorum sensing called quorum sensing?

Section 7
Carbon metabolism

7.1 Aerobic heterotrophic growth

- Carbohydrates are excellent substrates for many microbes. As already discussed, these polymers are hydrolyzed by exoenzymes, then the resulting monomeric or dimeric sugars (glucose, fructose, lactose, or maltose) are taken up and oxidized (Figure S18). Other pathways in addition to the EMP pathway have also evolved for the degradation of glucose to pyruvate. One of these is the Entner-Doudoroff (ED) pathway. It is employed by many *Pseudonomas* species and by the well-known anaerobic alcohol producer, *Zymomonas mobilis*. In this pathway, glucose 6-phosphate is not isomerized to fructose 6-phosphate as in the EMP pathway but oxidized to 6-phosphogluconate. By removal of water, this compound is converted to 2-keto-3-deoxy-6-phosphogluconate, which is then cleaved to two C_3 compounds, to pyruvate and 3-phosphoglyceraldehyde. The latter is subsequently converted to a second molecule of pyruvate. This pathway is of importance because it allows the utilization of gluconic acid and related compounds that are very common in plants. This will be discussed further in connection with alcohol fermentation (Figure S28).

- Growth on organic acids. Many bacterial species are able to grow on just one particular organic compound, for example, acetate. Energy conservation during growth on acetate is relatively simple because acetate is activated by conversion to acetyl-CoA, which is then oxidized via the tricarboxylic acid cycle to give reducing equivalents for ATP production. However, biosynthesis of cellular constituents from acetate is more complicated than when glucose serves as substrate. The breakdown of sugars yields sugar phosphates and, most importantly, phosphoenolpyruvate. It can be converted to oxaloacetate by CO_2 fixation, so that all important precursors of cellular building blocks are available. Growth on acetate requires two additional enzymes that essentially convert the tricarboxylic acid cycle into the glyoxylic acid (glyoxylate) cycle (Figure S22). These two enzymes are isocitrate lyase and malate synthase. Together with reactions of the tricarboxylic acid cycle, these enzymes catalyze the oxidation of acetate to glyoxylate and its conversion, together with a molecule of acetyl-CoA, to malate. Malate can be oxidized within the cycle to oxaloacetate, which can be

Discover the World of Microbes: Bacteria, Archaea, and Viruses, First Edition. Gerhard Gottschalk.
© 2012 Wiley-VCH Verlag GmbH & Co. KGaA. Published 2012 by Wiley-VCH Verlag GmbH & Co. KGaA.

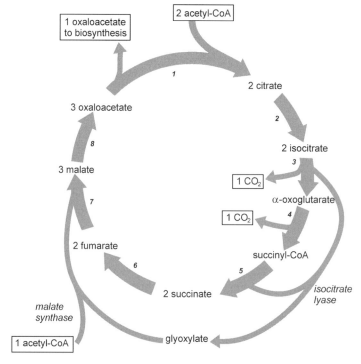

Figure S22 The glyoxylate cycle, a combination of the TCA cycle with the isocitrate lyase and malate synthase reactions. 1, citrate synthase; 2, cis-aconitase; 3, isocitrate dehydrogenase; 4, α-oxoglutarate dehydrogenase; 5, succinate thiokinase; 6, succinate dehydrogenase; 7, fumarase; 8, malate dehydrogenase; 9, isocitrate lyase. Reactions 1–8 represent the TCA cycle.

converted to phosphoenolpyruvate (PEP), a central intermediate for biosynthesis. PEP is involved in the biosynthesis of some amino acids and in gluconeogenesis, the generation of sugar derivatives for cell-wall components.

The importance of the two reactions that convert the TCA cycle into the glyoxylate cycle, and of enzymes such as PEP-carboxykinase, by which PEP can be made from oxaloacetate, should be underlined. During growth on acetate, the oxaloacetate and PEP pools become depleted. These compounds are in demand for anabolic metabolism, so their pools need to be replenished by the above reactions, known as anaplerotic reactions. By the way, during growth on sugars an anaplerotic reaction is also required to replace the oxaloacetate consumed: the PEP carboxylase reaction:

Growth on acetate:

$$\text{Oxaloacetate} + \text{ATP} \xrightarrow{\text{PEP carboxykinase}} \text{PEP} + \text{ADP} + CO_2$$

Growth on glucose:

$$\text{PEP} + \text{CO}_2 \xrightarrow{\text{PEP carboxylase}} \text{oxaloacetate} + \text{Pi}$$

- Growth on methane. We have seen that methane production is the domain of the archaea. The process of methane oxidation to CO_2 with oxygen as electron acceptor, however, is restricted to a rather specialized group of bacteria, the methanotrophic bacteria. They are unable to close the methane cycle completely because the methane produced by ruminants and in rice fields finds its way into the atmosphere, as seen in Chapter 14. We should also recall that there is an anaerobic methane oxidation (Chapter 7) of yet unknown dimensions. Nevertheless, the methanotrophs are capable of oxidizing aerobically at least a major part of the methane produced.

  The methanotrophs are able to perform a reaction that the chemical industry would really like to mimic: the oxidation of methane to methanol, a reaction catalyzed by a membrane-integrated, copper-containing methane monooxygenase (Figure S23). The principle of this reaction is that both methane and NADH are oxidized by molecular oxygen. This is an important enzyme, it is structurally related to ammonia monooxygenase of the nitrifying bacteria such as *Nitrosomonas europaea*, and methane monooxygenase also oxidizes ammonia.

  Methanol is further oxidized to formaldehyde, and this compound is funneled into a pathway that surprisingly employs the C_1 carrier tetrahydromethanopterin, which for a long time was thought to be archaea-specific. The final product is CO_2. Reducing equivalents generated along this pathway are channeled into the respiratory chain for ATP production. Synthesis of cellular constituents is no easy task for the methanotrophs. They operate the so-called serine pathway by which formaldehyde and CO_2 are converted to acetyl-CoA. The acetyl-CoA produced allows the formation of phosphoenolpyruvate by reactions resembling the glyoxylate cycle just discussed, opening the "metabolic door" for synthesis of all required cellular constituents.

- Growth on oil (read the statement by Fritz Widdel in Chapter 32). The spectrum of organisms able to grow on long-chain hydrocarbons or aromatic compounds differs greatly from the methane utilizers. A variety of organisms from different phyla and genera are able to deal with these compounds. Even yeasts, for example, *Candida lipolytica*, or fungi, for example, *Cephalosporium roseum*, are long-chain hydrocarbon utilizers. These also include Gram-positive bacteria such as *Mycobacterium smegmatis*, *Nocardia petroleophila*, or *Arthrobacter simplex*, and the Gram-negative *Pseudomonas fluorescens*, *Acinetobacter calcoaceticus*, and *Alcanivorax borkumensis*. However, the organisms *E. coli* and *B. subtilis* often referred to in this book are unable to utilize hydrocarbons.

  The first step in utilization of hydrocarbons or their mixtures in the form of oil is emulsification. In many organisms, this proceeds in the outer layers of the cell walls, which contain rhamnolipids or trehalolipids, compounds

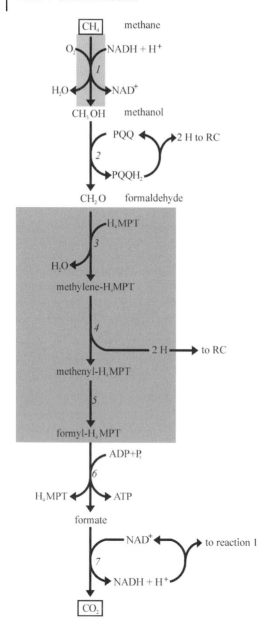

Figure S23 Aerobic oxidation of methane by methanotrophic bacteria. 1, methane mono-oxygenase, membrane-integrated; 2, methanol dehydrogenase. Electrons are transferred to pyrroloquinoline quinone, also called methoxatin. From there, they are transferred to the respiratory chain; 3 to 6, "the archaeal box," see Figure S13; 7, formate dehydrogenase; RC, respiratory chain. Formaldehyde is the precursor for cell mass production in these organisms.

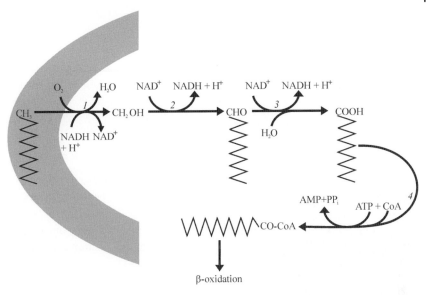

Figure S24 Degradation of long-chain hydrocarbons. The hydrocarbon "dissolves" in the cytoplasmic membrane with the help of detergents produced by the cells. 1, hydrocarbon (n-alkane) mono-oxygenase; 2, n-alkane alcohol dehydrogenase; 3, aldehyde dehydrogenase; 4, acyl-CoA synthetase.

consisting of a hydrophilic sugar portion and a hydrophobic lipid portion. From there, the hydrocarbons are guided to the cytoplasmic membrane, where they are oxidized. What follows is a monooxygenase reaction, as shown in Figure S23. The corresponding alcohol is further oxidized in two steps to a fatty acid (Figure S24). Electrons are transferred to NAD^+. The NADH is used for the initial monooxygenase reaction and for energy conservation in the respiratory chain. The well-known process of beta-oxidation is then employed to degrade the long-chain fatty acid.

With this background on hydrocarbon-utilizing organisms, we are able to draw our own conclusions on ways to improve oil-spill cleanup efforts. I) The bacteria require an aqueous environment and are unable to live in oil itself. Therefore, oil degradation is enhanced by finely dispersing the oil so that there is a large interface between the oil and the water containing the organisms. II) Oxygen is required, so oil degradation is facilitated by conditions under which the water is saturated with air. III) Nutrients are required for growth, especially sources of nitrogen and phosphorus.

7.2 Incomplete oxidations

The advantage of the organisms carrying out incomplete oxidations was discussed in connection with the production of vinegar and vitamin C. These organisms

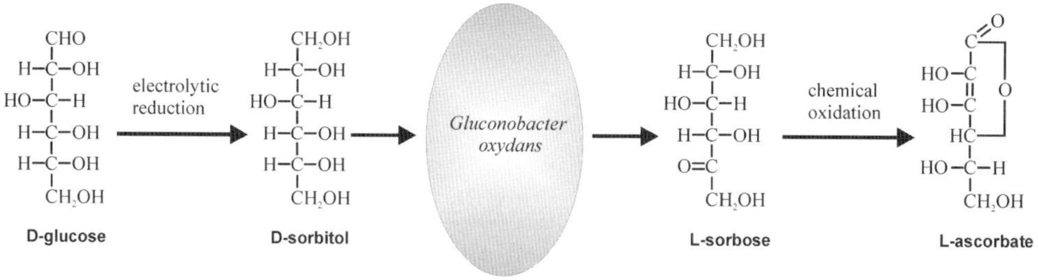

Figure S25 Oxidation of D-glucose to L-ascorbate. Note that sorbitol is designated as having a D configuration because the OH-group located at C-atom 5 stands is on the right. The sorbose molecule has to be turned 180°, then the OH group at C-atom 5 stands is on the left. Therefore, sorbose as well as ascorbate have the L-configuration.

carry out the initial substrate oxidation reactions so rapidly that they outcompete the other groups of microorganisms present. An excellent example is vinegar production. The ethanol oxidizers derive reducing power for operation of the respiratory chain solely from the oxidation of ethanol to acetic acid. As long as ethanol is available, they leave the acetic acid untouched. So oxidation of the substrate is incomplete, which explains the name for this group of organisms. We shall now discuss the synthesis of vitamin C in a little more detail. As shown in Figure S25, D-glucose is first reduced chemically to D-sorbitol, which is the substrate for *Gluconobacter oxidans*. The product, L-sorbose, is ultimately oxidized chemically to L-ascorbate.

7.3 Autotrophic growth

Autotrophic growth means that the cell mass is primarily synthesized from CO_2. There are several autotrophic pathways in microorganisms:

- The Calvin-Benson-Bassham cycle was discussed in Chapter 8. The various reactions of this cycle are depicted in Figure S26. Here we can follow the fate of three molecules of CO_2, which end up as one molecule of glyceraldehyde 3-phosphate (GAP). The reaction of CO_2 with three molecules of ribulose 1,5-bisphosphate (RuP$_2$) gives six molecules of 3-phosphoglycerate, which are reduced to six molecules of glyceraldehyde 3-phosphate. However, five of those molecules (with fifteen carbon atoms) are required to regenerate the three acceptor RuP$_2$ molecules (also 15 carbon atoms). This regeneration involves enzymes known as transketolases and aldolases. We should first look at the lower two GAP molecules. One is isomerized to dihydroxyacetone phosphate (DAP), and a subsequent aldolase reaction results in the formation of fructose 1,6-bisphosphate, which is converted to fructose 6-phosphate by a phosphatase. This fructose 6-phosphate then reacts with the GAP, located in the center of the figure, to give Xu-5P and E-4P. The latter, a C_4 compound, is united with DAP to yield Su-P$_2$. This C_7 compound is subject to a phosphatase reaction to

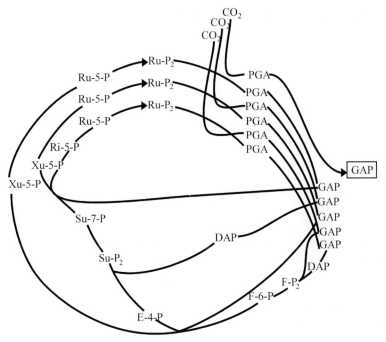

Figure S26 CO_2 fixation via the CBB cycle. Ru-P_2, ribulose 1,5-bisphosphate; PGA, 3-phosphoglycerate; GAP, glyceraldehyde 3-phosphate; DAP, dihydroxyacetone phosphate; F-P_2, fructose 1,6-bisphosphate; F-6-P, fructose 6-phosphate; E-4-P, erythrose 4-phosphate; Su-P_2, sedoheptulose 1,7-bisphosphate; Su-7-P, sedoheptulose 7-phosphate; Xu-5-P, xylulose 5-phosphate; Ri-5-P, ribose 5-phosphate; Ru5-P, ribulose 5-phosphate.

give Su-7P, which reacts with the upper GAP in a transketolase reaction to give a second molecule of Xu-5P and Ri-5P. Therefore, we have generated three C_5 sugars that are converted to Ru-5P, then phosphorylated to give Ru-P_2, the acceptor for another round of the cycle. The overall equation of this cycle is:

$$3\,CO_2 + 9\,ATP + 6\,NADH + 6\,H^+ \rightarrow GAP + 9\,ADP + 8\,P_i + 8\,NAD^+$$

(The electron donor in plants is NADPH)

The number of eight P_i is not a printing error: the ninth phosphate is located in GAP. The high ATP demand of this cycle is apparent, with three molecules of ATP required per molecule of CO_2 fixed. Therefore, the cycle can only be operated by organisms that have plenty of ATP available. These include plants, cyanobacteria, and a number of phototrophic (anoxygenic) and chemolithotrophic bacteria.

- Other CO_2 fixation reactions. Of the five pathways discovered so far, only two will be mentioned here. One is the reductive tricarboxylic acid cycle, which results in the conversion of two molecules of CO_2 to acetyl-CoA. Essentially, the tricarboxylic acid cycle is operated in reverse, whereby several enzymes have to be replaced. One is the citrate synthase, which is replaced by ATP-citrate lyase that cleaves citrate into oxaloacetate and acetyl-CoA. The second enzyme is alpha-ketoglutarate dehydrogenase, which is replaced by alpha-ketoglutarate ferredoxin oxidoreductase. This reversed cycle is in operation in green sulfur bacteria and in some hyperthermophilic chemolithotrophs such as *Hydrogenobacter thermophilus* (bacterium) or *Thermoproteus tenax* (archaeon).

The second pathway is the one used by methanogens growing on CO_2 and H_2. However, a modified form is used by homoacetogenic clostridia, which also grow on H_2 and CO_2 but produce acetate as fermentation product. We shall proceed down the two branches of the pathway depicted in Figure S27. The branch on the right provides CO, which is bound to the key enzyme of this pathway, the CO dehydrogenase/acetyl-CoA synthase. The branch on the left provides the methyl group by reducing CO_2 stepwise via C_1 compounds of the redox level of formate, formaldehyde, and methanol. Then, the key enzyme makes acetyl-CoA from CO, the methyl group, and coenzyme A. This is the Wood-Ljungdahl pathway that is widespread among clostridia, including *Clostridium ljungdahlii* and *Clostridium aceticum*. It is also found in other anaerobes such as *Acetobacterium woodii* and *Moorella thermoacetica* and in the sulfate reducer *Desulfobacterium autotrophicum*.

7.4 Bacterial fermentations

We shall now concentrate on ethanol, lactic acid, acetone-butanol, and mixed acid fermentations, of which the first three were discussed in Chapters 15, 16 and 18.

- Alcohol fermentation as carried out by the yeast *Saccharomyces cerevisiae* is depicted in Figure S28. The key enzyme is pyruvate decarboxylase, by which the pyruvate molecules are converted to acetaldehyde, which in turn is reduced to ethanol by alcohol dehydrogenase. The yield is two molecules of ATP per glucose molecule, as in lactic acid fermentation. *Zymomonas mobilis* employs the Entner-Doudoroff pathway already introduced in Section 7A, in which the ATP yield is only one per molecule of glucose.

- Lactic acid fermentation. In what is called the homofermentative type of lactic acid fermentation, two molecules of lactate are produced per molecule of glucose. This simple process is depicted in Figure S17. It's easy to see that for each molecule of glucose fermented, two molecules of ATP are made available for growth of these organisms. Typical representatives are *Lactobacillus lactis*, *Lactobacillus helveticus*, or *Lactobacillus bulgaricus*, organisms employed for the production of yogurt or cheese. There is another type of lactic acid fermentation, called the heterofermentative type, which is carried out by *Leuconostoc mesenteroides*, for example. It employs a different pathway and yields one molecule each of lactic acid, ethanol, and CO_2 per molecule of glucose fermented.

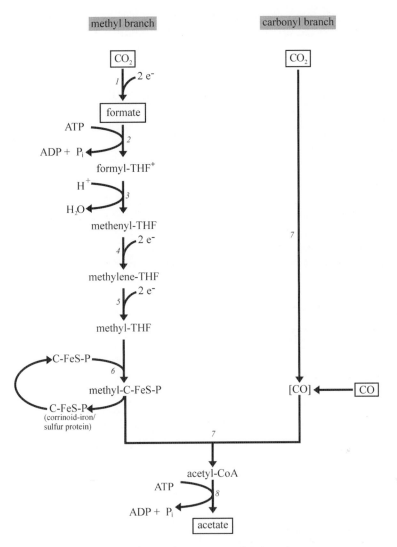

Figure S27 CO$_2$ fixation via the Wood-Ljungdahl pathway. 1, formate dehydrogenase; 2, formyl-THF synthetase; 3, methenyl-THF cyclohydrolase; 4, methylene-THF dehydrogenase; 5, methylene-THF reductase; 6, methyltransferase; 7, CO dehydrogenase/acetyl-CoA synthase, an enzyme system combining CO, the methyl group from reaction 6 and CoA to acetyl-CoA; 8, phosphotransacetylase + acetate kinase. THF, tetrahydrofolate.

320 | Section 7 Carbon metabolism

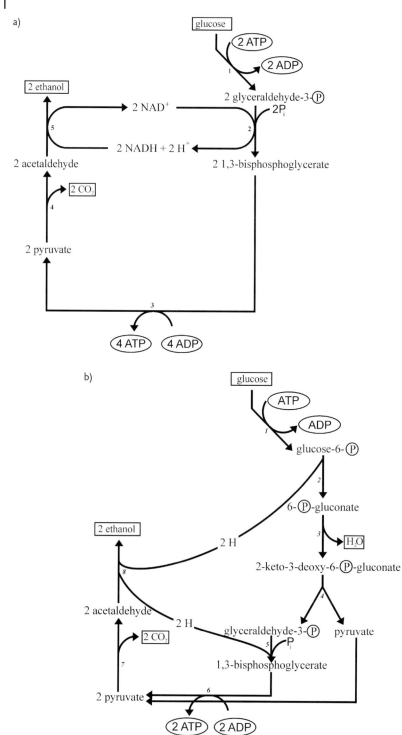

Figure S28 Alcohol fermentation via the EMP and the ED pathways. (a) Fermentation of glucose to ethanol and CO_2 by yeast. 1, initial enzymes of the Embden-Meyerhof-Parnas pathway 2, glyceraldehyde-3-phosphate dehydrogenase, 3, 3-phosphoglycerate kinase; phosphoglycerate mutase, enolase, and pyruvate kinase; 4, pyruvate decarboxylase; 5, alcohol dehydrogenase. (b) Fermentation of glucose to ethanol and CO_2 by *Zymomonas mobilis*. 1, hexokinase; 2, glucose-6-phosphate dehydrogenase; 3, 6-phosphogluconate dehydratase; 4, 2-keto-3-deoxy-6-phosphogluconate (KDPG) aldolase; 5, 3-phosphoglycerate kinase; 6, phosphoglycerate mutase, enolase and pyruvate kinase; 7, pyruvate decarboxylase; 8, alcohol dehydrogenase.

- Acetone-butanol fermentation is carried out by *Clostridium acetobutylicum* and related species. First, a butyrate fermentation takes place by which one molecule of glucose is converted to two molecules each of H_2 and CO_2, and one molecule of butyrate (Figure S29). When the pH value of the culture falls below 5, the fermentation shifts toward butanol and acetone production. The induction of an enzyme, the acetoacetate decarboxylase, leads to formation of the neutral compound acetone. As a consequence, insufficient amounts of acetoacetyl-CoA are available to serve as acceptor for the NADH produced in the upper part of the pathway. To compensate for this, butyryl-CoA is reduced to butanol via butyraldehyde.

- Mixed acid fermentation. Hundreds of thousands of people have grown *Escherichia coli* on agar plates. It has commonly been assumed that the growth of colonies under such conditions is entirely aerobic. However, when fermentable compounds are present in the agar medium, growth will not only be aerobic. Especially those organisms within the colonies can quite easily switch to fermentation. Then the primary products are acids as well as ethanol, CO_2, and H_2. There are several options for carbon flow through the central pathways (Figure S30). Not only lactate but also succinate can be produced; the latter requires a typically "anaerobic enzyme," the fumarate reductase. A key enzyme of mixed acid fermentation is the pyruvate-formate lyase, which catalyzes the cleavage of pyruvate into acetyl-CoA and formate, opening the way for production of acetate, formate, and, by cleavage of the latter, CO_2 and H_2.

- Homoacetate fermentation. This interesting process was also described in Section 7C. The organisms mentioned there convert one molecule of glucose or fructose to three molecules of acetate. Most of them are also able to convert H_2 and CO_2 to acetate or even to H_2 and CO.

- Fermentation of amino acids. These are excellent substrates for anaerobes, although a number of them grow on single amino acids such as glutamate or lysine. Many clostridia, however, prefer the fermentation of mixtures of amino acids by coupling oxidation and reduction of amino acid pairs. This type of fermentation was discovered by L.H. Stickland in 1934 and is called the Stickland reaction. One amino acid is oxidized (alanine in this case) and a second one is reduced (glycine):

$$\text{alanine} + 2 \text{ glycine} \rightarrow 3 \text{ acetate} + CO_2 + 3 \text{ ammonia}$$

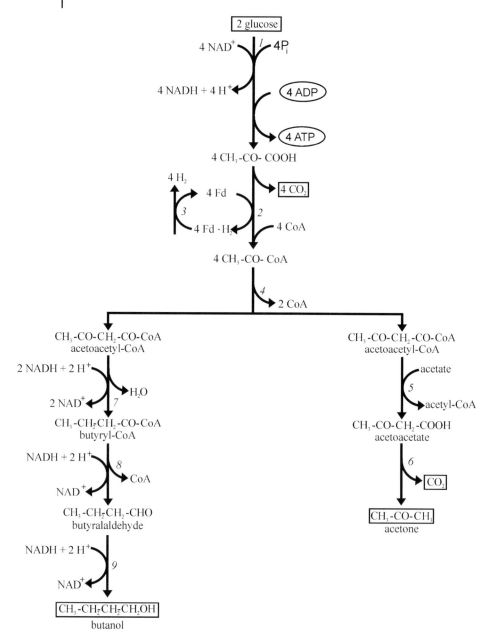

Figure S29 Acetone-butanol fermentation as carried out by *C. acetobutylicum*. 1, reactions as in Figure S17; 2, pyruvate: ferredoxin oxidoreductase, 3, hydrogenase; 4, acetoacetyl-CoA: acetyl transferase (thiolase); 5, acetoacetyl CoA: acetate CoA transferase; 6, acetoacetate decarboxylase; 7, L(+)-β-hydroxybutyryl-CoA dehydrogenase, crotonase, and butyryl-CoA dehydrogenase; 8, butyraldehyde dehydrogenase; 9, butanol dehydrogenase.

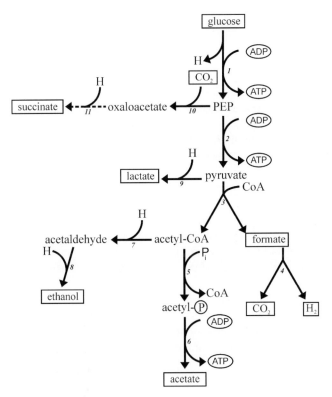

Figure S30 Mixed acid fermentation. 1, enzymes of the Embden-Meyerhof-Parnas pathway as in Figure S17; 2, pyruvate kinase; 3, pyruvate-formate lyase; 4, formate-hydrogen lyase; 5, phosphotransacetylase; 6, acetate kinase; 7, acetaldehyde dehydrogenase; 8, alcohol dehydrogenase; 9, lactate dehydrogenase; 10, PEP carboxylase; 11, malate dehydrogenase, fumarase, and fumarate reductase.

The fermentation according to this equation is coupled to the synthesis of three molecules of ATP from ADP and P_i. *C. sporogenes*, *C. sticklandii*, *C. histolyticum*, and *C. botulinum* are organisms that carry out the Stickland reaction. In addition to alanine, the amino acids leucine, isoleucine, valine, and phenylalanine are usually oxidized, whereas proline and ornithine are reduced.

Questions

1) What is the difference between the EMP pathway and the ED pathway in relation to glucose breakdown?
2) Which two enzymes convert the TCA cycle into the glyoxylic acid cycle?
3) Why is the pathway of methane oxidation of special interest in connection with horizontal gene transfer?
4) What are the prerequisites for effective oil degradation?
5) What do vinegar and vitamin C have in common?
6) Why would it be difficult or even impossible for an anaerobe to operate the Calvin-Benson-Bassham cycle?
7) Why are the findings described in the statement by Andrew Benson considered to be a milestone in life sciences?
8) What determines the direction in which the Wood-Ljungdahl pathway will proceed, toward acetyl-CoA synthesis or CO_2 formation?
9) What connects the oxidative steps of the alcohol and lactic fermentation with the reductive steps?
10) Calculate the ATP yield of the fermentation of glucose to butyrate, CO_2, and H_2.
11) What induces the so-called shift in acetone-butanol fermentation?
12) Speculate on the industrial potential of homoacetogens.

Section 8
Regulation of microbial metabolism

As we have seen in Section 7, microbial metabolism is very diverse. Some specialists are only capable of utilizing a few substrates. Most of the methanotrophic species that have specialized on methane, methanol, and methylamines, serve as examples, as well as *Clostridium acidiurici*, an anaerobe that grows on uric acid and related purine compounds such as xanthine and guanine. Many microorganisms, however, grow on a great variety of substrates. Here are a few examples: *Escherichia coli* is able to grow on various sugars as well as organic acids and amino acids. Sugars are usually utilized under aerobic conditions; they can also be fermented in the absence of oxygen. The aerobe *Ralstonia eutropha* usually grows on mixtures of H_2, O_2, and CO_2 but also utilizes a number of sugars and amino acids. Similarly, an anaerobe such as *Clostridium aceticum* can grow on CO_2 and H_2, thereby producing acetate, but it can also thrive on a variety of sugars. With respect to the substrates utilized, *Paracoccus denitrificans* is comparable with *Ralstonia eutropha*, but it can also grow on a combination of H_2, CO_2, and nitrate. So oxygen-dependent respiration can be replaced by nitrate-dependent respiration. Microbes such as *Klebsiella aerogenes* or *Clostridium pasteurianum* utilize organic nitrogen-containing compounds or ammonia as nitrogen sources. In the absence of such sources, they are able to fix molecular nitrogen by employing the enzyme nitrogenase, as can a considerable number of microbes (see Chapter 11). In all these examples, microbes are confronted with regulatory problems. When *E. coli* is growing under anaerobic conditions, it doesn't make sense for these microbes to synthesize all the enzymes of the tricarboxylic acid cycle and the respiratory chain. Under these conditions, such systems are simply not required by the cells. It also wouldn't make sense to have active nitrogenase present if plenty of ammonia is present. Furthermore, if we look at spore-forming microbes such as *Bacillus subtilis* or *Clostridium acetobutylicum*, the synthesis of compounds linked to sporulation is only required when the cells actually undergo this process.

These examples make it clear that it is not useful for microbes to constantly synthesize all the enzymes encoded in their genomes. As we will see, there are additional reasons as well. Before regulatory mechanisms are discussed, it should be pointed out that the genetic information for the enzymes of a particular pathway is often organized in an operon. In such an operon, several genes are under the control of one regulatory signal. We shall now discuss the different targets for regulatory control.

8.1 Regulation at the DNA level

Methylation of DNA is an extremely important mechanism in eukaryotic organisms. It also occurs in prokaryotes, where for one thing it is responsible for restriction (see Werner Arber's statement in Chapter 25). A further example is methylation of the origin of replication gene oriC, which can trigger the replication of DNA. Differences in the number of copies of a given gene on the chromosome also have an effect. A good example is what is known as ribosomal RNA gene clusters. We may recall that three RNAs are required for the formation of ribosomes: 5S, 16S, and 23S RNA. Their genes are organized in clusters, and the number of clusters varies in certain microorganisms. *E. coli* has seven of these clusters, whereas *Bacillus subtilis* has ten, *Clostridium tetani* six, and *Methanosarcina mazei* three. Although not yet known in detail, the transcription of these clusters may have an effect on the number of ribosomes per cell and the growth rates of these organisms.

8.2 Transcription and provision of messenger RNA for the translationary machinery

A number of mechanisms have evolved at this level. Let's begin with one that is easy to understand. We learned in Section 4E that the bacterial RNA polymerase consists of a core enzyme and a so-called sigma factor. Only when these two components form a complex at the start region of a certain gene can replication take place (Figure S31a). A bacterial species contains a general sigma factor, which is σ^{70} in *E. coli*. These organisms also contain other sigma factors that are required for binding of the core enzyme and initiation of replication of certain operons. For example, σ^{32} is normally inactive, but when *E. coli* is exposed to elevated temperatures such as 41 °C (106 °F), this sigma factor becomes active and promotes synthesis of heat-shock proteins to help the cells survive under these conditions. A good example of the role of alternative sigma factors is the master regulator Spo0A involved in the process of sporulation in *Bacillus subtilis*. Under growth-limiting conditions, Spo0A is a response regulator that is converted by phosphorelay proteins to SpoA-P. In its phosphorylated form, SpoA-P is an active DNA-binding protein. As a result, proteins for the initiation of sporulation are synthesized; this leads to an unequal cell division and septum formation (see Figure S15). This is the point where σ^F takes over. In growing cells, σ^F is bound to anti-σ^F and is therefore inactive. During septum formation, one of the chromosomes slowly migrates into the compartment that will become the forespore. In this transition phase, anti-σ^F is degraded by anti-anti-σ^F in the forespore compartment. The genetic information for making new anti-σ^F has not yet arrived in the forespore compartment because that part of the chromosome is still in the mother cell. Therefore, anti-σ^F can no longer be replenished, so σ^F is released. Its active form binds to RNA polymerase and allows forespore-specific genes to be transcribed.

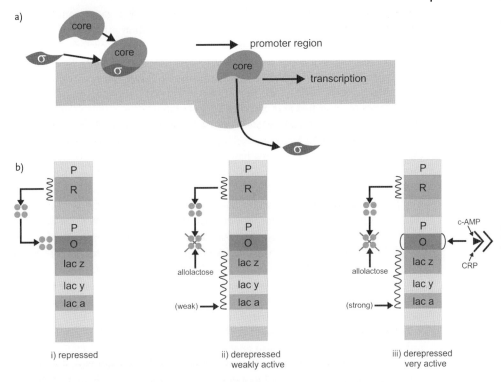

Figure S31 Regulation at the transcriptional level. (a) Binding of σ factor to the core enzyme gives active RNA-polymerase. (b) Regulation of the *lac* operon. i) the repressor gene is transcribed, the repressor R is synthesized and binds in the operator region, so there is no transcription. ii) if lactose is available, small amounts are transported into the cells (basic levels of LacZ are present). Allolactose formed in small amounts binds to the repressor that dissociates from the operator, allowing weak transcription of the *lac* operon. iii) if glucose is not available, the c-AMP level in the cells increases, and c-AMP binds to CRP that now functions as a transcription activator. *lacZ* codes for β-galactosidase, *lacY* for a lactose symporter driven by a proton gradient and *lacA* for a transacetylase.

After arrival of the gene for anti-σ^F in the forespore, anti-σ^F is synthesized and σ^F is again inactivated. In the meantime, however, forespore formation has already been initiated. In this case it is interesting to see how compartmental separation of a gene, the gene for anti-σ^F, is employed to allow σ^F to do its job.

Another plausible regulatory mechanism affecting transcription is represented by the lac operon of *E. coli*. It encodes genes whose products are required for the transport of lactose into the cells, for the cleavage of lactose by the well-studied enzyme beta-galactosidase, and for a transacetylase with a special function. In the absence of lactose, this operon is blocked by a repressor protein that binds in a special region, the operator region upstream of the lac genes. If lactose is present, however, this repressor loses its binding affinity to the DNA. Instead, the repressor binds to a derivative of lactose (allolactose) and simply diffuses away (Figure S31b).

This type of regulatory mechanism is called regulation by negative control. Actually, this system is a little more complicated. Transcription proceeds in the presence of lactose, but only when glucose is absent. Glucose is the preferred substrate of *E. coli*. As long as glucose is available, it is not necessary to make the enzymes for lactose utilization. If glucose is not available, a signal molecule is liberated: cyclic AMP. Cyclic AMP binds to the so-called c-AMP receptor protein (CRP), which binds as indicated and stimulates transcription.

The opposite of a repressor protein is an activator protein. Binding by the activator protein initiates or increases the transcription of a gene or the genes of an operon. Two-component systems have already been mentioned, systems such as the KdpD/KdpE system (Figure S16) and the ArcA/B system (Figure S20). More of these systems are listed in Table S3. Another regulatory principle is exemplified by FNR (fumarate-nitrate reductase regulator), which activates the synthesis of a number of proteins required for fermentative metabolism in *Escherichia coli*. This protein is oxygen sensitive because it contains ferrous iron (Fe^{2+} in the form of iron-sulfur clusters). It binds to the DNA as long as the partial pressure of oxygen is low. Pyruvate formate lyase is made under these conditions (see Figure S30), but genes for the aerobic respiratory chain are repressed. As soon as the pressure of oxygen increases, the FNR dimer dissociates and falls off the DNA. Then the genes for fermentative metabolism are no longer expressed.

Transcriptional control is also of great importance for anabolic metabolism. Not only energy metabolism but also anabolic metabolism has to be regulated. Let's take a look at protein synthesis. Proteins consist of 20 different amino acids, but these building blocks are not present in equimolar amounts. Glutamic and aspartic acids usually make up around 20 percent of the amino acids in proteins, whereas others such as histidine and lysine make up only 1 to 2 percent. Obviously, different amounts of amino acids have to be provided for the translational apparatus. Furthermore, when it comes to their synthesis, it has to be taken into account that 19 of the 20 amino acids are not synthesized independently of each other. There is for example the glutamate family of amino acids. The aromatic amino acids have common precursors as well as the branched-chain amino acids. Only histidine is synthesized by an independent pathway. The principle of repression is common in the regulation of amino acid biosynthetic pathways. Let's take tryptophan biosynthesis as an example. It proceeds along the trp operon as long as tryptophan is in demand. When tryptophan begins to accumulate, a repressor that requires tryptophan as corepressor is bound to the promoter/operator region of the trp operon, thereby preventing further tryptophan biosynthesis. However, an additional regulatory mechanism is required to harmonize the synthesis of proteins from 20 different precursors. This will be discussed in the next paragraph.

8.3 Translational regulation

The regulatory mechanism called attenuation combines transcriptional and translational elements. We shall take tryptophan synthesis as an example of this prin-

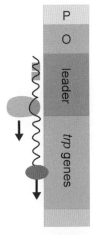
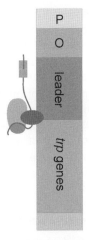

i) transcription proceeds   ii) transcription stops

Figure S32 Regulation of transcription of the *trp*-operon by attenuation. i) tryptophan is limiting, the two *trp* codons (ϒϒ) in the leader peptide cause movement of the ribosomes to be attenuated, but they finally synthesize the leader peptide, the *trp* operon is transcribed by the RNA polymerase (●), and the ribosomes proceed with the synthesis of the enzymes. ii) tryptophan is in excess, the RNA polymerase is followed by the ribosomes, and a secondary structure of the m-RNA is formed that blocks transcription and translation; large oval, ribosome; small oval, RNA polymerase.

ciple. Prior to the synthesis of Trp pathway proteins, a small leader peptide has to be synthesized at the ribosome (Figure S32). This leader peptide contains two adjacent codons for tryptophan. If tryptophan is limiting, the leader peptide can only be synthesized very slowly, so the movement of the ribosomes is attenuated. As a consequence, a secondary structure of the mRNA will form, allowing translation and tryptophan synthesis to proceed. If tryptophan is available, however, the leader peptide will be synthesized at the ribosomes, the mentioned structural change of the mRNA will not occur and translation is terminated.

These examples show that the cells try to avoid overproducing amino acids. Amino acid synthesis has to be finely regulated in order to balance the supply required for protein synthesis. We should take into account that proteins make up more than 50% of the constituents of microbial cells and that protein synthesis requires a lot of metabolic energy. The equivalent of the hydrolysis of four ATP is necessary for the formation of one peptide bond, two ATP for amino acid activation (see Table S1c), and two GTP for one ribosomal elongation cycle.

Important tools for translational control are referred to as small RNAs (sRNAs), between 50 and 250 nucleotides in length. The sRNAs are able to block the ribosomal binding site of certain genes. They have an antisense sequence, and only under certain conditions do they dissociate and thus allow translation. It was shown in recent years that a number of alternative sigma factors underlie translational control by small RNAs.

It should be mentioned that the physiological state known as the stationary phase of microbial cells also requires an arsenal of regulatory measures. In stationary-phase *Escherichia coli*, the concentration of the alternative sigma factor σ^{38} increases, with the result that the binding and function of ribosomes is disturbed and metabolism as a whole slows down.

Finally, when the provision of mRNA for the translationary machinery is discussed, one important aspect has to be mentioned, the half-life of mRNA. In the case of long-lived mRNA, it would take generations for its level to decrease significantly. However, the half-life of microbial mRNA is very short. In *E. coli* it is around two minutes. Thus, repression of expression has an almost immediate affect on the mRNA level in the cells.

8.4 The regulation of enzyme activity

In addition to the regulation of enzyme levels, microbial cells must also be able to instantly adjust the activity of enzymes to their metabolic requirements. If, for instance, more amino acids were to be synthesized than required for protein synthesis, it would not only be necessary to stop the synthesis of enzymes by repression, as already discussed, but also the activity of enzymes involved. One mechanism is known as feedback inhibition. If isoleucine is present in excess, for example, it will inhibit the first enzyme involved in its synthesis, threonine deaminase. Once its activity is reduced, the four subsequent enzymes involved in isoleucine biosynthesis run out of substrates, and isoleucine can no longer be synthesized. Let's look at a second example. A comparable inhibitory effect is observed in the case of the pyrimidine nucleotides, the uridine, thymidine, and cytidine phosphates. When CTP (cytidine 5′-triphosphate) accumulates in the cells, it inhibits the activity of the first enzyme of the pyrimidine nucleotide pathway, aspartate transcarbamoylase.

The activity of enzymes involved in energy metabolism is also under regulatory control. One example is the regulation of the activity of citrate synthase, the first enzyme in the tricarboxylic acid cycle. In *E. coli* and a number of other Gram-negative bacteria, enzyme activity is inhibited by elevated levels of NADH. In Gram-positive bacteria, there is a citrate synthase that is inhibited by elevated ATP levels. The function of both of these inhibitors makes sense because they indicate that there is no further need to produce NADH for energy conservation in the respiratory chain. In a microbial cell, there are hundreds of enzymes whose activity is tuned by certain metabolites that function as inhibitors or, in many cases, as activators. ADP, for instance, serves as an activator of phosphofructokinase. Elevated levels of ADP indicate a shortage of ATP, and this signal serves to accelerate glycolysis so that more ATP can be synthesized. There are a number of metabolites that control the energy charge: the concentration of ATP plus half the concentration of ADP divided by the sum of the concentrations of ATP, ADP, and AMP. When the energy charge is low, the activity of enzymes leading to increased ATP production is increased. A central metabolite in the cell is PEP (phosphoenolpyruvate), an intermediate in energy metabolism and glycolysis but also in several pathways leading to the synthesis of dicarboxylic acids or certain amino acids (phenylalanine, tyrosine, and tryptophan). Its level has to be maintained by anaplerotic reactions, as discussed in Section 7A.

Metabolites that act as inhibitors or activators of certain enzymes often are structurally unrelated to the actual substrates of these enzymes. CTP is chemically quite different from aspartate and carbamoyl phosphate, the substrates of aspartate transcarbamoylase. The substrates of citrate synthase, oxaloacetate and acetyl-CoA, are chemically quite different from the inhibitors NADH and ATP. In such cases we speak of allosteric effectors and of allosteric control.

The activity of certain enzymes can also be modified by covalent modification. One example in *Escherichia coli* is isocitrate dehydrogenase, which can be inhibited by phosphorylation. This plays a role when *E. coli* grows on acetate and has to perform the glyoxylate cycle. Under these conditions, isocitrate could serve as substrate for either isocitrate dehydrogenase or isocitrate lyase. The activity of isocitrate dehydrogenase needs to be reduced so that substrate is also available for the lyase reaction. The activity of glutamine synthetase in *E. coli* is regulated by covalent modification. The active form of the enzyme, which consists of 12 subunits, is unmodified. When conditions would allow too much ATP to flow into the glutamine synthetase reaction, the enzyme is inactivated by adenylylation of a tyrosine residue on each subunit. Following adenylylation, the OH group of tyrosine carries an AMP residue. This system is extremely complicated because the enzyme that adenylylates glutamine synthetase and the one that catalyzes the reverse reaction are also controlled by a modification system.

Finally, a model is presented to explain how the expression of the *nif* genes is regulated in *Klebsiella pneumoniae*. The *nif* gene cluster comprises about 20 genes. The conversion of atmospheric nitrogen into ammonia – the importance of this process was outlined in Chapter 11 – is from an energetic standpoint, a very costly process. For each molecule N_2 reduced, 16 molecules of ATP are hydrolyzed to 16 ADP and 16 P_i. For economic reasons, nitrogenase is only produced by the organisms when no other nitrogen sources are available. In addition, the nitrogenase is oxygen sensitive. Within these constraints, the expression of the *nif*-genes is under effective control. It is apparent from Figure S33 that expression requires the presence of free activator protein, NifA. In the presence of oxygen and ammonia, NifA is bound by NifL and is then inactive. Dissociation of this complex requires interaction with the modified nitrogen sensory protein, GlnK. Furthermore, NifL has to be bound to the cytoplasmic membrane, which is only possible when membrane components are "anaerobic" and reduced. This model is a good example for the complexity of regulatory mechanisms in microorganisms.

This section has shown that microbial cells can only proliferate and exist under widely differing physiological conditions if the proteome, the sum of all proteins present, is adjusted to the cellular needs. This requires the regulation of protein synthesis and fine tuning of the activities of certain key enzymes so that a well-functioning network is the result.

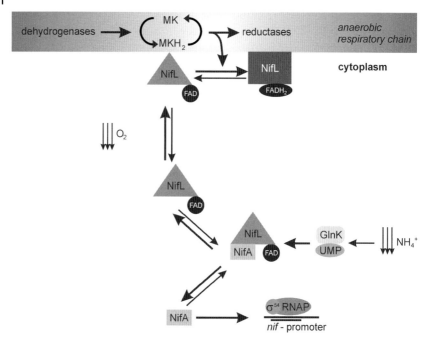

Figure S33 Expression of the *nif* genes in K. pneumoniae under nitrogen limitation and anaerobic conditions. NifL-NifA is the inhibitory complex, and NifA is the activator. Sequestration of the complex occurs by the modified nitrogen sensory protein GlnK (low ammonia concentration) and by reduction of the FAD-NifL to FADH$_2$-NifL (low oxygen concentration). FADH$_2$-NifL is bound at the cytoplasmic membrane; RNAP, RNA polymerase (R. Thummer *et al.*, J. Biol. Chem.; 282, 12517–12525; 2007, modified).

Questions

1) What could it mean if one organism has two ribosomal RNA gene clusters on its genome and another organism has ten?
2) What does σ^{32} do in E. coli?
3) Describe a system regulated by negative control.
4) What senses the FNR protein and which reactive groups are present in FNR?
5) Why is the regulation of amino acid synthesis so important, and how is overproduction prevented at the transcriptional and the translational levels?
6) What does regulation by antisense mean?
7) What are allosteric effectors?
8) Give an example of regulation by covalent modification.

Section 9
Genomes, genes, and gene transfer

9.1 Genomes

What is a genome? It is all of the genetic information present in the cell. In microbes it may consist of one chromosome plus, in some cases, one or more plasmids. If a cell contains one chromosome and two plasmids, we say that it contains three replicons. Plasmids are sometimes as large as chromosomes, then they are called megaplasmids. Plasmids differ from chromosomes by their mode of replication and usually by the lack of genes for ribosomal RNA clusters.

What have we learned from the more than one thousand microbial genomes that have been sequenced so far?

- Depending on the species, the size of microbial genomes varies dramatically. The largest genomes comprising almost 14 megabase pairs (Mbp) are present in myxobacteria such as *Sorangium cellulosum*. The smallest genomes are found in endosymbionts of insects such as *Carsonella ruddii*, which has a genome size of around 160 kilobase pairs. We recall that the genome size of *E. coli* is 4.6 Mbp, whereas that of *Clostridium tetani* is 2.8 Mbp. Microorganisms like the streptomyces that are capable of producing secondary metabolites such as antibiotics have genome sizes of more than 7 Mbp. We find the smallest genomes of free-living organisms among the thermophiles. One example is the archaeon *Picrophilus torridus* (Figure 12) with a genome size of around 1.5 Mbp. Extremely valuable information is revealed by studying the genome sequence and analyzing the number of genes present as well as the number of genes for ribosomal RNA clusters and for secondary metabolites. Data on *Clostridium tetani*, which is presented in Table S4, serves as an example

- The identification of genes in sequenced microbial genomes has shown that the microbial gene pool is the largest on our planet. Finding genes and identifying the encoded enzyme systems is still in its infancy. A convincing example is the Sargasso experiment mentioned in Chapter 31, which revealed a large number of genes of unknown function.

- As in the past, a continuous horizontal gene transfer is still taking place, not only within the two microbial domains but also between the domains.

Discover the World of Microbes: Bacteria, Archaea, and Viruses, First Edition. Gerhard Gottschalk.
© 2012 Wiley-VCH Verlag GmbH & Co. KGaA. Published 2012 by Wiley-VCH Verlag GmbH & Co. KGaA.

Table S4 Features of the genome of *Clostridium tetani*.

Site of chromosome	2 799 251 base pairs
Size of plasmid	74 082 base pairs
G + C content of chromosome	28.6%
G + C content of plasmid	24.5%
Orfs on chromosome	2372
Orfs on plasmid	61
Ribosomal RNA operons	6
tRNA genes	54
Prophages (incomplete)	3

For a circular presentation of the genome see Figure 70.

Examples include the bacterium *Thermotoga maritima*, which contains about 20 percent genes of archaeal origin, and the archaeon *Methanosarcina mazei* with nearly the same percentage of bacterial genes. An example of such gene transfer was given in Section 6, where it was pointed out that the oxidation of formaldehyde by methanotrophic bacteria involves the C_1 carrier tetrahydomethanopterin, which is a typical carrier in archaea.

- Microbial genomes are rich in inserts that often can be identified on the basis of their GC content. This term needs to be explained. Double-stranded DNA consists of GC and AT couples, so G and C as well as A and T have to be present in equimolar amounts because of base pairing. However, the G + C portion (and with it A + T) may vary considerably in microbes. Usually the G + C content of a genome in mole percent serves as a marker for a certain microbial species. If this content is 30 percent, then 30 percent of the base pairs in a particular DNA are represented by G + C and, logically, 70 percent by A + T. A few values for the GC contents of microbial genomes are given in Table S5. Now we come back to the inserts. If a 50-Kb fragment in a microbial genome has a G + C content of 35 percent but the average G + C content of the genome is perhaps 50 percent, this may indicate that the insert was acquired by gene transfer. Generally, we speak of genomic islands. The pathogenicity islands, which were discussed in Chapter 29, are one type of island. Prophages have often been identified on the basis of their GC content and the presence of certain genes. The genome of *Bacillus licheniformis*, for instance, contains four inserts of bacteriophage origin. The genome islands now bring us to gene transfer.

9.2 Gene transfer

Two important mechanisms for gene transfer have already been discussed in Chapter 20, gene transfer by transformation and by conjugation. A third mecha-

Table S5 G + C content of the chromosomes of some microbes.

Microorganism	G + C content (mol%)
Streptomyces coelicolor	72.1
Ralstonia eutropha	66.4
Mycobacterium tuberculosis	65.6
Sinorhizobium meliloti	62.7
Gluconobacter oxydans	60.8
Escherichia coli[a]	50.5
Bacillus licheniformis	46.2
Methanosarcina mazei	41.5
Prochlorococcus marinus[a]	35.1
Staphylococcus aureus	32.8
Rickettsia prowazekii	29.1
Clostridium tetani	28.6
Mycoplasma capricolum	23.8

a) GC content varies depending on strain.

nism for gene transfer is called transduction and involves bacteriophages. As an introduction, we should look at Figure 48, which depicts the transfer of genetic information into plant cells with the help of the Ti plasmid. Also refer to Figures 42 and 50. If a bacteriophage manages to inject its intact DNA, and this DNA gives rise to a new phage population or is incorporated into the chromosomal DNA and correctly excised again, there will be no gene transfer of interest. However, if host DNA is accidentally incorporated into the phage heads and these phage heads are still infectious, then the bacteriophage functions as a vector by which DNA is transferred from one cell to another. By the way, the "prophages" we see in many microbial genomes are most likely the remains of phage DNA that is no longer complete and difficult for the organism to remove.

Once gene transfer has taken place, the organisms can only take advantage of gene transfer when important post-requisites are met. When the incorporated plasmids replicate and thus provide the cells with resistance mechanisms against heavy metals or antibiotics, the utilization of extraneous genetic information is straightforward. However, any DNA fragments taken up by organisms have to be incorporated into the chromosome before they may contribute to microbial life. The mechanism by which this proceeds is called recombination. Two kinds of recombination are distinguished: I) site-specific recombination as discussed for lambda phage in connection with Figure 42 and II) generalized recombination that requires sequence homology between part of the DNA fragment and part of the chromosome. With the help of proteins such as RecA, the extraneous DNA fragment replaces the homologous fragment in the microbial chromosome.

Questions

1) Are the terms "genome" and "replicon" synonyms, and what is metagenomics?
2) How is it possible to identify in a particular organism genes that have been acquired by horizontal gene transfer?
3) What is the GC content of DNA and how does it vary among different microbial species?
4) Summarize the three mechanisms for gene transfer.
5) How is the DNA taken up by a particular organism incorporated into the genome?

Section 10
In-depth study of four special topics

The topics Antibiotics, Biotechnology, Pathogenetic Microorganisms and Viruses were extensively discussed in the corresponding chapters of this book. Additional information is provided here in the form of four tables and questions have been added.

10.1 Antibiotics

Table S6 Antibiotics.

Name	Source	Interference with
β-lactams		
Penicillin	Fungus *Penicillium*	Cell-wall synthesis
Ampicillin	Semisynthetic[a]	Cell-wall synthesis
Methicillin	Semisynthetic	Cell-wall synthesis
Amoxillin	Semisynthetic	Cell-wall synthesis
Cephalosporin	Fungus *Cephalosporium*	Cell-wall synthesis
Cefotaxime	Semisynthetic	Cell-wall synthesis
Bacitracin	*Bacillus* species	Cell-wall synthesis
Aminoglycosides		
Streptomycin	*Streptomyces griseus*	Protein synthesis
Gentamycin	*Micromonospora purpureochromogenes*	Protein synthesis
Neomycin	*Streptomyces fradiae*	Protein synthesis
Tetracyclines		
Tetracycline	*Streptomyces* species	Protein synthesis
Doxycycline	Semisynthetic	Protein synthesis
Macrolides		
Erythromycin D	*Saccharopolyspora erythraea*	Protein synthesis
Rifampicin	Semisynthetic	Transcription
Oxazolidinones		
Linezolid	Synthetic	Protein synthesis
Quinolones		
Ciprofloxacin	Synthetic	DNA replication
Metronidazole	Synthetic	DNA degradation in anaerobes

a) Semisynthetic, the basic structure of the antibiotic is chemically modified, see Figure 44.

Questions

1) Describe the key observation made by Alexander Fleming.
2) What is the definition of an antibiotic and who coined it?
3) Why do beta-lactam antibiotics preferentially act on Gram-positive bacteria?
4) Describe the experimental procedures you would use to isolate an antibiotic-producing microorganism.
5) How would you test the specificity and activity of the antibiotic in question 4?
6) Describe several targets for antibiotics.
7) Describe the features of a resistance plasmid.

10.2 Biotechnology

Table S7 Products of the bioindustries and their applications.

Alcohols

  Ethanol: 70 million t/year, beer, wine, biofuel, see Chapter 16

  Butanol: industrial production started again, see Chapter 15, biofuel, feedstock for chemical industry

  1,3-propanediol: first plant in operation, see Chapter 29, building block of polymers

  Future developments: ethanol from syngas ($CO + H_2$) by organisms such as *Clostridium ljungdahlii*

Acids

  Lactic and acetic acids: 200 000 and 150 000 t/year, see Chapter 18, food industries

  Citric acid: 1 million t/year, *Aspergillus niger*, soft drinks, preservative

  Gluconic acid: 100 000 t/year, *Aspergillus niger*, food additive

Biopolymers

  Poly-β-hydroxyalkanoates: see Chapter 29, biodegradable polymers

  Poly-amino acids: e.g., poly-glutamate, biodegradable materials

  Poly-lactic acid: medical equipment, e.g., stents

  Future developments: isoprene from glucose or syngas for rubber production

Amino acids

  L-glutamate: 1.5 million t/year, *Corynebactrium glutamicum*, see Chapter 29, taste intensifier

  L-lysine: 700 000 t/year, *Corynebacterium glutamicum*, animal feed

  L-phenylalanine, 15 000 t/year, recombinant *E. coli*, aspartame (artificial sweetener) production

  L-aspartate: 15 000 t/year, recombinant *E. coli*, aspartame production

Antibiotics

  50 000 t/year, see Chapter 22 and Table S6

Enzymes[a]

  For industrial use: amylases, proteases, lipases, cellulases, 100 000 t/year (estimated), see Chapter 28

  For analytical applications: alkaline phosphatases, luciferase, β-glucuronidase, peroxidase, creatinine deiminase

Recombinant alternatives to human proteins or hormones[a]

  Human albumin, insulin (see Chapter 28), α and β interferon, growth hormone, erythropoietine, hepatitis B surface antigen

Vitamins

  Ascorbic acid (vitamin C), 100 000 t/year, see Chapter 18 and Section 7B, vitamin B12, 10 t/year, see Chapter 19, vitamin B2, 2500 t/year

a) Especially under enzymes and recombinant alternatives, only a few representative products are mentioned.

Questions

1) Describe the features of a microorganism capable of producing insulin.
2) Which parameters have an important effect on glutamate production by *Corynebacterium glutamicum*?
3) Why is biotechnology of importance for the laundry?
4) Which biodegradable polymers may be able to replace plastic materials?
5) Which contribution would you expect of bioethanol and butanol toward fulfilling our energy demand?

10.3 Pathogenic microorganisms

Table S8 Pathogenic bacteria.

Microorganism	Infectious disease	Source of infection
Bacillus anthracis	Anthrax	Contact with infected animals/spores (bioterrorism)
Bordetella pertussis	Whooping cough	Humans
Borrelia burgdorferi	Lyme disease	Bites from infected ticks
Chlamydia trachomatis	Conjunctivitis	Humans or contaminated water
Clostridium botulinum	Botulism	Soil-contaminated meat
Clostridium difficile[a]	Diarrhea	Regularly present in the human gut
Clostridium histolyticum	Gas gangrene	Soil
Clostridium tetani	Tetanus	Soil
Corynebacterium diphtheriae	Diphtheria	Humans
Escherichia coli		
Enterotoxinogenic strains (ETEC)	Food poisoning	Contaminated water or food
Enterohemorrhagic strains (EHEC)	Bloody diarrhea	Contaminated water or food
Uropathogenic strains (UPEC)	Bladder infection	Humans
Helicobacter pylori[a]	Stomach ulcers, chronic gastritis	Almost omnipresent in humans
Legionella pneumophila	Pneumonia	Warm water/air-conditioning systems
Listeria monocytogenes	Listeriosis	Cheese and other food
Mycobacterium leprae	Leprosy	Various contacts in tropical environment
Mycobacterium tuberculosis	Tuberculosis	Humans
Neisseria gonorrhoeae	Gonorrhea	Sexual contact/humans
Neisseria meningitidis	Meningitis	Nasopharynx
Pseudomonas aeruginosa	Pneumonia	Nosocomial/cystic fibrosis
Rickettsia prowazekii	Epidemic typhus	Humans, lice
Salmonella typhi	Typhoid fever	Fecal contamination of water or food
Salmonella typhimurium	Gastroenteritis	Contaminated food
Shigella dysenteriae	Diarrhea	Contaminated water and food
Staphylococcus aureus[a]	Skin infections, abscesses	Present on skin/humans
Streptococcus pneumonia[a]	Meningitis, pneumonia	Present in nasopharynx
Treponema pallidum	Syphilis	Sexual contact/humans
Vibrio cholerae	Cholera	Contaminated water
Yersinia pestis	Bubonic plague	Fleas

a) Regularly present, develops a pathogenic lifestyle under certain conditions.

Questions

1) Which microorganisms like to settle in the respiratory tract?
2) What is the potential hazard of soil?
3) Which effects do the toxins of *Vibrio cholerae*, *Salmonella*, or enterotoxigenic *E. coli* strains have in the gastrointestinal tract?
4) Why is it so difficult to combat *Mycobacterium tuberculosis*?
5) What is the infection strategy of *Pseudomonas syringae*?
6) Which compound serves as a growth substrate for microorganisms on leaves?

10.4 Viruses

Table S9 Human viruses.

Virus	Disease	Genome	Shape
Picornaviridae			
Poliovirus (Chapter 30)	Poliomyelitis	+ssRNA	Icosahedral nucleocapsid
Rhinovirus	Common cold	+ssRNA	Icosahedral nucleocapsid
Hepatitis A virus	Hepatitis A	+ssRNA	Icosahedral nucleocapsid
Flaviviridae			
Yellow fever virus	Yellow fever	+ssRNA	Envelope around spherical nucleocapsid
Dengue virus	Dengue fever	+ssRNA	Envelope around spherical nucleocapsid
Hepatitis C virus	Hepatitis C	+ssRNA	Envelope around spherical nucleocapsid
Coronaviridae			
Coronavirus	Common cold	+ssRNA	Envelope around helical nucleocapsid
SARS	Respiratory tract infection	+ssRNA	Envelope around helical Nucleocapsid
Filoviridae			
Marburg and Ebola viruses	Hemorrhagic fever	−ssRNA	Filamentous envelope around a helical nucleocapsid
Paramyxoviridae			
Measles virus	Measles	−ssRNA	Envelope around helical nucleocapsid
Mumps virus	Mumps	−ssRNA	Envelope around helical nucleocapsid
Orthomyxoviridae			
Influenza viruses A and B (Chapter 30)	Influenza, flu	−ssRNA	Envelope around spherical nucleocapsid

(continued)

Table S9 (continued)

Virus	Disease	Genome	Shape
Arenaviridae			
Arenavirus	Lymphocytic choriomeningitis	$-$ssRNA	Envelope with two nucleocapsids
Lassa virus	Hemorrhagic fever	$-$ssRNA	Envelope with nucleocapsid
Reoviridae			
Rotavirus	Diarrhea	dsRNA	Icosahedral nucleocapsid
Retroviridae			
HIV (Chapter 30)	AIDS	$+$ssRNA	Envelope around icosahedral nucleocapsid
Hepaduaviridae			
Hepatitis B virus	Hepatitis B	dsDNA with a single-stranded gap	Envelope with nucleocapsid
Papovaviridae			
Papilloma virus	Genital carcinoma	dsDNA	Icosahedral nucleocapsid
Herpesviridae			
Herpes simplex (Chapter 30)	Watery blisters on the skin	dsDNA	Envelope around icosahedral nucleocapsid
Varicella-zoster virus	Shingles	dsDNA	Envelope around icosahedral nucleocapsid
Epstein-Barr virus	Infectious mononucleosis	dsDNA	Envelope around icosahedral nucleocapsid

Questions

1) Interpret the statement of David Baltimore.
2) Why are viruses not considered to be living organisms?
3) Interpret the importance of the chemical synthesis of the poliovirus.
4) What is the basis for referring to influenza viruses as H1N1 or a similar term?
5) Why is replication accuracy a problem in RNA viruses?
6) Define a retrovirus.
7) How are viruses employed in gene therapy?
8) Give reasons why DNA may have evolved in viruses, from which it subsequently entered the microbial world.

Appendix A Selected literature

General Microbiology and Bacterial Metabolism (Chapters 1, 2, 8, 9, 11 and 12)

Atlas, R.M., and Bartha, R. (1997) *Microbial Ecology*, Addison-Wesley, Boston.
Dawkins, R. (1976) *The Selfish Gene*, Oxford University Press.
de Kruif, P. (1926) *Microbe Hunters*, Harcourt Brace, New York.
Dixon, B. (2009) *Animalcules*, ASM Press, Washington, DC.
Gest, H. (2003) *Microbes*, ASM Press, Washington DC.
Gottschalk, G. (1986) *Bacterial Metabolism*, Springer Verlag, New York–Berlin–Heidelberg–Tokyo.
Harwood, C., and Buckley, M. (2008) Uncharted Microbial World: Microbes and their Activities in the Environment. American Academy of Microbiology Colloquia Reports.
Ingraham, J., *et al.* (1983) *Growth of the Bacterial Cell*, Sinauer Ass., Inc, Sunderland, MA.
Ingram, J.L. (2010) *March of the Microbes*, The Belknap Press of Harvard University Press, Cambridge, Massachusetts, London, England.
Kartal, B., *et al.* (2010) Sewage treatment with anammox. *Science*, **328**, 702–703.
Madigan, M.T., *et al.* (2009) Brock Biology of Microorganisms, Pearson Benjamin Cummings, San Francisco, Boston, New York, Cape Town, Hong Kong, London, Madrid, Mexico City, Montreal, Munich, Paris, Singapore, Sydney, Tokyo, Toronto.
Pace, N.R. (2006) Time for a change. *Nature*, **441**, 289.
Postgate, J. (2000) *Microbes and Man*, Cambridge University Press, Cambridge, New York, Port Chester, Melbourne, Sydney.
Ruben, S., and Kamen, M.D. (1941) Long-lived radioactive carbon: C^{14}. *Phys. Rev.*, **59**, 349–354.
Slonczewski, J.L., and Foster, J.W. (2009) *Microbiology – An Evolving Science*, W.W. Norton, New York, London.
Staley, J.T., *et al.* (2007) *Microbial Life*, Sinauer Ass., Inc. Publishers, Sunderland, MA.
Thimann, K.V. (1963) *The Life of Bacteria*. The Macmillan Company, New York, Collier Macmillan Limited, London.
Whitman, W.B., *et al.* (1998) Prokaryotes: the unseen majority. *Proc. Natl. Acad. Sci.*, **95**, 6578–6583.

Evolution (Chapters 3–5)

Eigen, M., and Schuster, P. (1979) *The Hypercycle, A Principle of Natural Self-Organization*, Springer Verlag, Berlin–Heidelberg–New York.
Lane, N. (2003) *Oxygen, The Molecule That Made the World*, Oxford University Press, Oxford.
Lane, N., *et al.* (2010) How did LUCA make a living? Chemiosmosis in the origin of life. *BioEssays*, **32**, 271–280.
Lincoln, T.A., and Joyce, G.F. (2009) Self-sustained replication of an RNA Enzyme. *Science*, **323**, 1229–1232.
Lunine, J.I. (1999) *Earth, Evolution of A Habitable World*, Cambridge University Press, Cambridge.

Miller, S.L., and Urey, H.C. (1959) Organic compound synthesis on the primitive earth. *Science*, **130**, 245–251.

Woese, C.R. (2000) Interpreting the universal phylogenetic tree. *Proc. Natl. Acad. Sci.*, **97**, 8392–8396.

Woese, C.R., and Fox, G.E. (1977) The phylogenetic structure of the prokaryotic domain: the primary kingdom. *Proc. Natl. Acad. Sci.*, **74**, 5088–5090.

Zimmer, C. (2006) *Evolution, The Triumph of An Idea*, Harper Perennial, New York–London–Toronto–Sydney.

Life under extreme Conditions, Archaea (Chapters 6, 7 and 13)

Cavanaugh, C., et al. (1981) Prokaryotic cells in the hydrothermal vent tube worm *Riftia pachyptila* Jones: possible chemoautotrophic symbionts. *Science*, **213**, 340–342.

Cavicchioli, R. (2006) *Archaea*, ASM Press, Washington DC.

Deppenmeier, U, et al. (1996) Pathways of energy conservation in methanogenic archaea. *Arch. Microbiol.*, **165**, 149–153.

Ferry, J.G. (ed.) (1993) *Methanogenesis*, Chapman and Hall, New York–London.

Horikoshi, K., et al. (2010) *Extremophiles Handbook*, Springer Japan KK, Tokyo.

Jannasch, H.W. (1995) Life at the sea floor. *Nature*, **374**, 676–677.

Miroshnichenko, M.L., and Bonch-Osmolovskaya, E.A. (2006) Recent developments in the thermophilic microbiology of deep-sea hydrothermal vents. *Extremophiles*, **10**, 85–96.

Mitchell, L., et al. (2006) Microbial diversity in the deep sea and the underexplored "rare biosphere". *Proc. Natl. Acad. Sci.*, **103**, 12115–12120.

Sloan Jr, E.D. (2003) Fundamental principles and applications of natural gas hydrates. *Nature*, **426**, 353–359.

Stetter, K.O., et al. (1990) Hyperthermophilic microorganisms. *FEMS Microbiol. Rev.*, **75**, 117–124.

Thauer, R.K., et al. (2008) methanogenic archaea: ecologically relevant differences in energy conservation. *Nat. Rev.*, **6**, 579–591.

Vieille, C., and Zeikus, J.G. (2001) Hypertermophilic enzymes: sources, uses, and molecular mechanisms for thermo-stability. *Microbiol. Mol. Biol. Rev.*, **65**, 1–43.

Wilkansky, B. (1936) Life in the dead sea. *Nature*, **Sept 12**, 467.

Climate and Energy (Chapters 14–17)

Nordrhein-Westfälische Akademie der Wissenschaften: Die Energieversorgung sichern. Denkschrift, 2006.

Buckley, M., and Wall, J. (2006) Microbial Energy Conversion. American Academy of Microbiology Colloquia Reports.

Gu, L., et al. (2003) Response of a deciduous forest to the Mount Pinatubo eruption: enhanced photosynthesis. *Science*, **299**, 2035–2038.

Houghton, R.A. (2007) Balancing the global carbon budget. *Ann. Rev. Earth Planet Sci.*, **35**, 313–347.

Intergovernmental Panel on Climate Change (IPCC).

Khan Khalil, M.A., et al. (2007) Atmospheric methane: trends and cycles of sources and sinks. *Environ. Sci. Technol.*, **41**, 2131–2137.

Olah, GA, et al. (2006) *Beyond Oil and Gas: The Methanol Economy*, Wiley-VCH Verlag GmbH & Co. KGaA, Weinheim.

Reeburgh, W.S. (2007) Oceanic Methane Biogeochemistry. *Chem. Rev.*, **107**, 486–513.

Román-Leshkov, Y., et al. (2007) Production of dimethylfuran for liquid fuels from biomass-derived carbohydrates. *Nature*, **447**, 982–986.

Bacteria as Commensals, Parasites, or Endosymbionts (Chapters 10, 26 and 27)

Adler, J. (1987) How motile bacteria are attracted and repelled by chemicals: an approach to neurobiology. *Biol. Chem. Hoppe-Seyler*, **368**, 163–174.

Andersson, S.G.E., et al. (1998) The genome sequence of *Rickettsia prowazekii* and the origin of mitochondria. *Nature*, **396**, 133–143.

Bourret, R.B., and Stock, A.M. (2002) Molecular information processing: lessons from bacterial chemotaxis. *J. Biol. Chem.*, **277**, 9625–9628.

Comstock, L.E. (2007) The inside story. *Nature*, **448**, 542–544.

Dethlefsen, L., et al. (2007) An ecological and evolutionary perspective on human–microbe mutualism and disease. *Nature*, **449**, 811–818.

Eckburg, P.B., et al. (2005) Diversity of the human intestinal microbial flora. *Science*, **308**, 1635–1638.

Gill, S.R., et al. (2006) Metagenomic analysis of the human distal gut microbiome. *Science*, **312**, 1355–1359.

Kemmling, A., et al. (2004) Biofilms and extracellular matrices on geomaterials. *Environ. Geol.*, **46**, 429–435.

Kutschera, U., and Niklas, K.J. (2005) Endosymbiosis, cell evolution, and speciation. *Theory Biosci.*, **124**, 1–24.

Margulis, L. (1998) *Symbiotic Planet: A New Look at Evolution*, Perseus Books Group, Amhurst, MA.

Palmer, C., et al. (2007) Development of the Human Infant Intestinal Microbiota. *PLoS Biol.*, **5**, 1556–1573.

Tamas, I., et al. (2002) 50 million years of genomic stasis in endosymbiotic bacteria. *Science*, **296**, 2376–2379.

Whiteley, M., et al. (1999) Identification of genes controlled by quorum sensing in *Pseudomonas aeruginosa*. *Proc. Natl. Acad. Sci.*, **96**, 13904–13909.

Wilson, M. (2005) *Microbial Inhabitants of Humans*, Cambridge University Press, Cambridge.

Antibiotics, Plasmids, and Biotechnology (Chapters 18, 22, 23, 24 and 28)

Antibiotic Resistance: An Ecological Perspective on an Old Problem. American Academy of Microbiology Colloquia Reports, 2009.

Antimicrobial Resistance: An Ecological Perspective. American Academy of Microbiology Colloquia Reports, 2002.

Dröge, M., et al. (2000) Phenotypic and molecular characterization of conjugative antibiotic resistance plasmids isolated from bacterial communities of activated sludge. *Mol. Gen. Genet.*, **263**, 471–482.

Dürre, P., and Bahl, H. (2001) *Clostridia, Biotechnology and Medical Application*, Wiley-VCH Verlag GmbH & Co. KGaA, Weinheim.

Gottfried, T. (1997) *Alexander Fleming, Discoverer of Penicillin*, Franklin Watts, A Division of Grolier Publishing, New York, London, Hong Kong, Sydney, Danbury (CT).

Holden, M.T.G., et al. (2004) Complete genomes of two clinical *Staphylococcus aureus* strains: evidence for the rapid evolution of virulence and drug resistance. *Proc. Natl. Acad. Sci.*, **101**, 9786–9791.

Hopwood, D.A. (2007) *Streptomyces in Nature and Medicine. The Antibiotic Makers*, Oxford University Press.

Ingram, L.O. (1999) Enteric bacterial catalysts for fuel ethanol production. *Biotechnol. Prog.*, **15**, 855–866.

Jördening, H.J., and Winter, J. (eds) (2004) *Environmental Biotechnology – Concepts and Applications*, Wiley-VCH.

Kado, C.I. (1998) Origin and evolution of plasmids. *Antonie van Leeuwenhoek*, **73**, 117–126.

Koepke, M, et al. (2010) *Clostridium ljungdahlii* represents a microbial production platform based on syngas. *Proc. Natl. Acad. Sci.*, **107**, 13087–13092.

Levine, S.E., et al. (2010) A mechanistic model of the enzymatic hydrolysis of cellulose. *Biotechnol. Bioeng.*, **107**, 37–51.

Maurer, K.-H. (2004) Detergent protease. *Curr. Opin. Biotechnol.*, **15**, 330–334.

Nakamura, C.E., and Whited, G.M. (2003) Metabolic engineering for the microbial production of 1,3-propanediol. *Curr. Opin. Biotechnol.*, **14**, 454–459.

Nester, E, et al. 100 Years of *Bacillus thuringiensis*: A Critical Scientific Assessment. American Academy of Microbiology Colloquia Reports, 2002.

Schlüter, A., et al (2003) The 64 508 bp IncP-1β antibiotic multiresistance plasmid pB10 isolated from a waste-water treatment plant provides evidence for recombination between members of different branches of the IncP-1β group. *Microbiology*, **149**, 3139–3153.

Steinbüchel, A., and Rhee, S.K. (eds) (2005) *Biopolymers in the Food Industry. Properties, Patents and Production*, Wiley-VCH, Weinheim.

Wang, J., et al. (2006) Platensimycin is a selective FabF inhibitor with potent antibiotic properties. *Nature*, **441**, 358–361.

Wink, M. (ed.) (2006) *An Introduction to Molecular Biotechnology*, Wiley-VCH.

Bioelements (Chapter 19)

Levi, P. (1996) The Periodic Table. Random House.

Wackett, L.P., et al. (2004) Microbial genomics and the periodic table. *Appl. Environ. Microbiol.*, **70**, 647–655.

Wolfe-Simon, F., et al. (2010) A bacterium that can grow by using arsenic instead of phosphorus. *Sciencexpress*, December 2010 online.

Molecular Microbiology (Chapters 20, 21 and 25)

Alberts, B., et al. (2002) *Molecular Biology of the Cell*, Garland Science, New York.

Falkow, S., et al. (1961) Episomic transfer between *Salmonella typhosa* and *Serratia marcescens*. *Genetics*, **46**, 703–706.

Rabinow, P. (1997) *Making PCR*, The University of Chicago Press, Chicago and London.

Pathogenic Bacteria (Chapter 29)

Brock, T.D. (1988) *Robert Koch. A Life in Medicine and Biotechnolgy*, Springer, Heidelberg–New York.

Büttner, D, and Bonas, U (2003) Common infection strategies of plant and animal pathogenic bacteria. *Curr. Opin. Plant Biol.*, **6**, 312–319.

Cangelosi, G.A., et al. (2005) From Outside to Inside: Environmental Microorganisms as Human Pathogens. American Academy of Microbiology Colloquia Report.

Hacker, J., and Dobrindt, U. (eds) (2006) *Pathogenomics*, Wiley-VCH Verlag & Co. KGaA, Weinheim.

Kaufmann, S.H.E., et al. (2010) *The Immune Response to Infection*, ASM.

Medzhitov, R. (2007) Recognition of microorganisms and activation of the immune response. *Nature*, **449**, 819–826.

Pallen, M.J., and Wren, B.W. (2007) Bacterial pathogenomics. *Nature*, **449**, 835–842.

Rose, J.B., et al. (2001) Health, Climate and Infectious Disease: A Global Perspective. American Academy of Microbiology Colloquia Reports.

Ryan, F. (1992) *The Forgotten Plague*, Little, Brown and Comp, Boston–New York–Toronto–London.

Wilson, B., et al. (2010) *Bacterial Pathogenesis: A Molecular Approach*, 3rd edn, ASM.

Viruses (Chapter 30)

Belser, J.A., and Tumpey, T.M. (2010) What we learned from reconstructing the 1918 influenza pandemic virus. *Microbe*, **5**, 477–483.

Flint, S., et al. (2004) *Principles of Virology: Molecular Biology, Pathogenesis and Control of Animal Viruses*, ASM Press, Washington, DC.

Gallant, J. (2010) *100 Questions and Answers about HIV and AIDS*, Jones and Bartlett Learning, Sudbury, MA.

Sax, P.E., et al. (2010) *HIV Essentials*, Jones and Bartlett Publishers, Sudbury, MA.

Stevens, J., et al. (2006) Structure and receptor specificity of the hemagglutinin from an H5N1 influenza virus. *Science*, **312**, 404–410.

Villareal, L.P., and Witzany, G. (2010) Viruses are essential agents within the roots and stem of the tree of life. *J. Theor. Biol.*, **262**, 698–710.

Wimmer, E. (2006) The Test-Tube Synthesis of a Chemical Called Poliovirus. EMBO reports 7, Special Issue.

Omics Technologies (Chapter 31)

Béjà, O., et al. (2000) Bacterial rhodopsin: evidence for a new type of phototrophy in the sea. *Science*, **289**, 1902–1906.

Daniel, R. (2005) The metagenomics of soil. *Nature Rev. Microbiol.*, **3**, 470–478.

Frigaard, N.-U., et al. (2006) Proteorhodopsin lateral gene transfer between marine planktonic Bacteria and Archaea. Nature, 439, 847–850.

Gibson, D.G., et al. (2008) Complete chemical synthesis, assembly, and cloning of a Mycoplasma genitalium genome. Science, 319, 1215–1220.

Gill, S.R., et al. (2006) Metagenomic analysis of the human distal gut microbiome. Science, 312, 1355–1359.

Giovannoni, S.J., et al. (2005) Genome streamlining in a cosmopolitan oceanic bacterium. Science, 309, 1242–1245.

Jungblut, P.R., and Hecker, M. (eds) (2007) Proteomics of Microbial Pathogens, Wiley-VCH Verlag GmbH & Co. KGaA, Weinheim.

Schuster, S., and Gottschalk, G. (2005) Microbial genomics in its second decade. Curr. Opin. Microbiol., 8, 561–563; and subsequent articles in this issue.

Large-Scale Sequencing: The Future of Genomic Sciences? American Academy of Microbiology Colloquia Reports, 2009.

Stahl, DA, and Tiedje, J (2002) Microbial Ecology and Genomics: A Crossroads of Opportunity. American Academy of Microbiology Colloquia Reports.

Streit, W.R., and Daniel, R. (2010) Metagenomics. Methods and Protocols, Humana Press, Springer–New York–Dordrecht–Heidelberg–London.

Turnbaugh, PJ, et al. (2007) The human microbiome project. Nature, 449, 804–810.

Venter, J.C., et al. (2004) Environmental genome shotgun sequencing of the Sargasso Sea. Science, 304, 66–74.

Incredible Microbes (Chapter 32)

Cohen-Bazire, G., et al. (1969) Comparative study of the structure of gas vacuoles. J. Bacteriol., 100, 1049–1061.

Daly, M.J., et al. (2007) Protein oxidation implicated as the primary determinant of bacterial radioresistance. PLoS Biol., 5, 0769–0779.

Ha, S.W., et al. (2007) Interaction of potassium cyanide with the Ni-4Fe-5S active site cluster of the CO dehydrogenase of Carboxydothermus hydrogenoformans. J. Biol. Chem., 282, 10639–10646.

Harris, H.W., et al. (2010) Electrokinesis is a microbial behavior that requires extracellular electron transport. Proc. Natl. Acad. Sci., 107, 326–331.

Horikoshi, K. (2006) Alkaliphiles, Kodausha Springer, Tokyo–Berlin–Heidelberg–New York.

McInerney, M.J. (2009) Syntrophy in anaerobic global carbon cycles. Curr. Opin. Biotechnol., 20, 623–632.

Nealson, K., et al. (2001) Geobiology: Exploring the Interface Between the Biosphere and the Geosphere. American Academy of Microbiology Colloquia Reports.

Paper, W., et al. (2007) Ignicoccus hospitalis sp. nov., the host of "Nanoarchaeum equitans". Int. J. Syst. Evol. Microbiol, 57, 803–808.

Schüler, D. (2006) Magnetoreception and Magnetosomes in Bacteria, Springer Verlag, Berlin–Heidelberg.

Schulz, H.M., and Jørgensen, B.B. (2001) Big bacteria. Ann. Rev. Microbiol., 55, 105–137.

Ting, C.S., et al. (2002) Cyanobacterial photosynthesis in the oceans: the origins and significance of divergent light-harvesting strategies. Trends Microbiol., 10, 134–142.

White, O., et al. (1999) Genome sequence of the radioresistant bacterium Deinococcus radiodurans R1. Science, 286, 1571–1577.

Wu, M., et al. (2005) Life in hot carbon monoxide: the complete genome sequence of Carboxydothermus hydrogenoformans Z-2901. PLoS Genet., 1, 0563–0574.

Appendix B Glossary

ABC: ATP-binding cassette of transporters, see Figure S16

Acetogenesis: synthesis of acetate from H_2 plus CO_2 or from organic compounds (3 acetates from 1 glucose), see Figure S27

Active transport: uphill transport of solutes from the outside into the interior of cells, it is energy-dependent

Adhesine: surface structures that provoke attachment of cells to solid substances

Aerobe: organisms growing in the presence of O_2. Usually O_2 is utilized, if not we speak of aerotolerant organisms

Agar: a polymer of algae generally used for surface growth of microorganisms on plates, see Figure 4

Akne: inflammation of skin involving *Propionibacterium acnes*, see Figure 23

Alkaliphile: organisms growing at pH values above pH 9

Ames test: examination of the carcinogenicity of compounds, see Figure 9

Anabolic metabolism: the sum of the biochemical reactions which lead to the synthesis of cell constituents from the growth substrates

Anaerobe: organisms growing in the absence of O_2; obligate anaerobes, for example, methanogens or certain clostridia are even killed by O_2

Anaerobic respiration: O_2 is replaced by other electron acceptors in the respiratory chain such as nitrate, see Figure 26

Anammox: a microbial process in which ammonia is oxidized with nitrite, see Figure 26

Discover the World of Microbes: Bacteria, Archaea, and Viruses, First Edition. Gerhard Gottschalk.
© 2012 Wiley-VCH Verlag GmbH & Co. KGaA. Published 2012 by Wiley-VCH Verlag GmbH & Co. KGaA.

Annotation: identification of genes on DNA sequences and prediction of functions of the assigned proteins

Anoxygenic photosynthesis: microbial energy metabolism in which light is used for energy conservation, O_2 is not produced

Antibiotics: compounds produced by microorganisms that kill or inhibit other microorganisms, see Table S6

Antibiotic resistance: inactivation of an antibiotic; mechanisms: enzymatic degradation, modification of the target, see Figure 47

Antibody: a protein that specifically binds to an antigen, it is produced by B cells

Anticodon: a base triplet in t-RNAs that pairs with the codon on messenger RNA

Antigenic drift: minor changes in the composition of viral proteins, for example, of proteins of the influenza virus

Antigenic shift: major changes in viral proteins

Antiparallel: refers to double stranded DNA in which the two strands run in opposite directions, from 5' to 3' and from 3' to 5'

Archaea: one of the three domains of life, see Figure S11

Attenuation: a regulatory mechanism to control the synthesis of full-length messenger RNA molecules, see Figure S32

Autotrophy: growth of organisms on CO_2 as sole carbon source

B cells: lymphocytes which in contact with an antigen produce antibodies. They may differentiate into plasma B cells, which are the actual antibody producers, and in memory B cells which live for a long time

Bacillus: an important genus of aerobic spore-forming bacteria

Bacteria: they represent the second domain of the phylogenetic tree, see Figure S14

Bacteriochlorophyll: pigments in phototrophic bacteria, see Figures 20 and 21

Bacteroids: they develop from *Rhizobium* cells and form nodules capable of nitrogen fixation, see Figure 25

Bacteriophages: viruses of prokaryotic organisms

Bacteriorhodopsin: a light-driven proton pump present in membranes of certain halophilic archaea. It is a protein containing retinal

Batch culture: a microbial culture growing in a fixed volume

β-lactam antibiotics: they contain a β-lactam ring like the penicillins, see Figure 44

Biofilm: mats of microbial cells held together by polysaccharides and attached to surfaces, see Figure 55

Bioluminescence: production of visible light by bacteria, see Figure 54

Black smokers: hydrothermal vents in the deep sea emitting water (250 °C–400 °C) and minerals

Borreliosis: see Lyme disease

Botulism: food poisoning caused by a toxin produced by *Clostridium botulinum*

Calvin-Benson-Bassham cycle: a series of reactions by which most autotrophic bacteria convert CO_2 into organic material. The key enzyme is rubisco, see Figure S26

Capsid: the protein shell around the viral genome

Catabolic reactions: they are performed by microbes in order to connect breakdown of their substrates with the generation of metabolic energy

Catabolite repression: the preferred substrate, for example, glucose represses the induction of pathways for the utilization of other substrates present, for example, lactose, see Figure S31

CD4 cells: these are T helper cells which function as targets for HIV

Chemiosmosis: proton or sodium ion gradients are generated at the cytoplasmic membrane and are used for ATP synthesis

Chemolithotrophy: energy generation by oxidation of inorganic compounds, for example, of nitrogen or sulfur compounds, see Figures 26 and S12

Chemoorganotrophy: generation of energy by catabolism of organic compounds that are oxidized or undergo fermentations

Chemostat: a continuous culture in which the cell density remains constant. This is achieved by a growth-limiting nutrient in the medium supply

Chemotaxis: movement toward or away from a chemical, see Figures 53 and S19

Chemotherapy: the use of antibiotics or of synthetic compounds to treat infectious diseases

Chitin: a polymer of N-acetylglucosamine, see Table S2

Chloroplast: the organelle harboring the photosynthetic apparatus in phototrophic eukaryotes

***Chromatium okenii*:** representative sulfur-oxidizing purple bacterium, see Figure 8

Chromosome: the in most cases circular DNA molecule which harbors the essential genetic information of a prokaryote

Citric acid cycle, also tricarboxylic acid cycle (TCA cycle): a series of reactions by which the acetate residue of acetyl-CoA is oxidized to CO_2, the reducing equivalents are transferred to NAD^+ (3 sites in the cycle) and to FAD (1 site)

Cloning: isolation of a DNA fragment containing a certain gene, incorporation of this fragment into a vector which can be replicated and expressed in a host, see Figure 15

Clostridium: an important genus of anaerobic spore-forming bacteria

Codon usage: the way organisms employ different codons for the same amino acid, see Figure S8

Compatible solutes: compounds, for example, potassium ions, betaine or others to balance the osmotic pressure inside and outside the microbial cells

Complementary: a nucleic acid sequence that can pair with a second sequence, see Figures S5 and S6

Conjugation: transfer of DNA from one prokaryotic cell to another, see Figure 39

Corals: marine organisms, most of which harbor dinoflagellates (Zooxanthellae) that perform photosynthesis and from which nutrients are derived. Corals are reef-builders, see Figure 56

Coral bleeching: whitening of the corals due to the dying off of Zooxanthellae.

Crenarchaeota: a phylum of archaea, see Figure S11

CTX: cholera toxin

Cyanobacteria: oxygenic phototrophs, previously called blue-green algae

Cyclic AMP: a nucleotide involved in catabolite repression, see Figure S31

Cyst: a resting stage of a number of bacteria

Cytochromes: ion-containing porphyrins involved in electron transport

Cytoplasm: the fluid of the cells' interior

Cytoplasmic membrane: the membrane surrounding the cytoplasm, see Figures 2 and S9

Denitrification: microbial reduction of nitrate to nitrogen gas under anaerobic conditions; dinitrogen oxide (laughing gas) is a byproduct, see Figure 26

Deoxyribonucleic acid: DNA in short, see Figure S6

Deinococcus radiodurans: a bacterium highly resistant to x-rays

Dipicolinic acid: a compound present in the spores of bacilli and chlostridia as well as in other sporeformers; it largely contributes to refractibility and heat-resistance

Dismutation: conversion of a compound into two compounds, an oxidized and a reduced one, example: superoxide dismutase

Dissimilation: degradation of compounds and release of the products to the environment; example: dissimilatory sulfate reduction, see Figure S12

DNA: deoxyribonucleic acid, the informational macromolecule in the microbial cell

DNA polymerase: it catalyzes the synthesis of a new strand of DNA along a DNA template. The direction of synthesis is 5' to 3' and the new strand runs antiparallel to the DNA template strand

Doubling time: the time needed for a microbial population to double its cell mass

Ecosystem: a defined environment with all the organisms living there

EET: extracellular electron transport

EHEC: enterohemorrhagic *E. coli*

Electron acceptor: a compound accepting electrons, for example, oxygen in the respiratory chain, see Figure S18

Electron donor: a compound such as NADH that donates electrons to components of the respiratory chain or to other systems, see Figure S18

Electron transport phosphorylation: another term to describe chemiosmotic ATP synthesis; electron transport gives rise to the generation of a protonmotive force which is used for ATP synthesis

Embden-Meyerhof-Parnas (EMP) pathway: degradation of glucose to two molecules of pyruvate via fructose 1,6-bisphosphate, see Figure S28

Enantiomer: one of the two forms of a chiral compound, for example, D- or L-lactic acid

Endergonic reactions: they require an input of energy, see Table S1

Endocytosis: uptake of a virus particle by an animal or plant cell, see Figure 67

Endoplasmatic reticulum: internal membrane systems in eukaryotic cells, see Figure 2

Endospores: the spores of bacilli, clostridia and related genera; these spores are extremely resistant to heat, see Figures 31 and S15

Endosymbiosis: uptake of one cell type by another cell type and development of a stable association of the two types, for example, mitochondria in eukaryotic cells and chloroplasts in plant cells

Enterotoxin: the lipopolysaccharides of the outer membrane of certain Gram-negative bacteria

Entner-Doudoroff pathway: degradation of glucose to two molecules of pyruvate via 2-keto-3-deoxy-6-phosphogluconate, see Figure S28

Epidemic: a disease that spreads rapidly

Escherichia coli: see EHEC, ETEC, and UPEC

ETEC: enterotoxinogenic *E. coli* (travelers' diarrhea)

Eukarya: a domain of the phylogenetic tree, see Figure 6

Euryarchaeota: a phylum of the archaea, see Figure S11

Exergonic reactions: they proceed with the liberation of energy, the free energy change is negative, see *Thermoproteus tenax* in Section 4c

Exoenzymes: enzymes excreted by microorganisms, see Table S2

Exponential growth phase: also called log phase, see Figure S1

Extremophiles: organisms that grow at high temperature, at high salt concentrations or at high or low pH values

F⁻cell: cell receiving DNA during conjugation, see Figure 39

Facultative anaerobe: this indicates that a microorganism, for example, *Escherichia coli*, is able to switch from aerobic growth to anaerobic growth

Feedback inhibition: a regulatory mechanism by which the product of a biochemical pathway, for example, tryptophan accumulation in the cell inhibits the activity of the first enzyme of its biosynthetic pathway

Fermentation: classically, the conversion of substrates to products under anaerobic conditions, for example, alcohol fermentation or lactic acid fermentation, see Figures S28 and S17. These fermentations are coupled to the generation of ATP for growth; in a broader sense, the conversion of certain feedstocks into useful products in a biotechnological process, see Table S7

Ferredoxins: these are proteins containing so-called iron-sulfur clusters, they serve as electron carriers

Fimbria: filamentous structures on bacterial cells; they do not contribute to the motility of the cells, but play a role in adherence to surfaces, see Figure 64

Firmicutes: phylum of Gram-positive bacteria with a low GC content

Frameshift: a mutation in which one or two bases of the DNA are deleted. As a result, the orfs (open reading frames) are shifted because they rest upon three base frames

Free energy: see exergonic reactions

Fruiting body: macroscopic structures developed by certain myxobacteria, see Figure 74

FtsZ: a ring-like protein involved in the division of microbial cells

Fungi: eukaryotic microorganisms including yeasts

GC content: mol percent of G + C in the DNA, a characteristic for a given microbial species, see Table S5

Gas vesicles: gas-filled protein structures in some microorganisms, that contribute to the regulation of the buoancy of the cells

Gel electrophoresis: agarose gels are used to separate nucleic acid molecules in an electric current and polyacrylamide gels to separate proteins, see Figure 51

Gene therapy: replacement of a defective gene by a functioning copy of this gene

Generation time: the time required for doubling of the cell number, see Figure S1

Genetic code: base triplets of the DNA that contain the information for one amino acid in a protein, see Figure S8

Genome: the total genetic information in organisms; in microorganisms it comprises the chromosome, and if present, also the plasmids, see Figure 70

Genomics: this discipline recently emerged; it involves sequencing genomes and interpreting the sequence data

Genus: species related to each other are grouped together into a genus, for example, into the genus *Bacillus* or *Pseudomonas*

Glycosydic bonds: they link sugar molecules to form polymers such as starch or cellulose

Glyoxylate cycle: a series of reactions required by aerobic microorganisms to grow on acetate or compounds which are degraded via acetyl-CoA, see Figure S22

Gram-negative cell walls: they contain a thin peptidoglycan (murein) layer, and in addition an outer membrane consisting of lipopolysaccharides. The space between the two layers is called periplasmic space. Organisms with Gram-negative cell walls are, for instance, the proteobacteria, see Figure S14

Gram-positive cell walls: they consist of a rather thick layer of peptidoglycan (murein). Representatives are the bacilli and the clostridia

Gram stain: a staining technique introduced by the Danish bacteriologist Christian Gram, see Chapter 22

Green sulfur bacteria: they belong to the anoxygenic phototrophs, for example, the genus *Chlorobium*

Group translocation: transport of sugars by phosphotransferase systems, see Figure S17

Growth rate μ (hours^{-1}): describes the increase of cell mass of a certain microorganism per unit of time, see Section 1

HAART: abbreviation of highly active antiretroviral therapy, a combination of drugs used to treat HIV-infections

Haber-Bosch process: industrial process for the production of ammonia from N_2 and H_2

Habitat: the place where an organism lives

Halobacteriales: extremely halophilic archaea

Halophiles: organisms able to live in the presence of high salt concentrations

Hemolysins: bacterial toxins that lyse red blood cells

Hemorrhagic fever: induced by viruses or by EHEC bacteria, it is characterized by bleeding disorders, also resulting in kidney failure, see Table S9

Heptatitis: liver inflammation caused by several types of viruses, see Table S9

Herpes simplex: a virus containing double-stranded DNA, see Figure 66

Heterocysts: these are cyanobacterial cells specialized on nitrogen fixation; they occur in filamentous algae such as *Nostoc* or *Anabaena* species. Heterocysts do not operate photosystem II and do not evolve oxygen

Heterofermentative: lactic acid bacteria which, in addition to lactic acid, produce carbon dioxide, ethanol or acetate are called heterofermentative

Heterotrophy: the source of carbon is organic material

Hfr: abbreviation for high frequency of recombinants. Cells of this type have the F-plasmid integrated into the chromosome, see Figure 39

Homoacetogens: see acetogens

Homofermentative: bacteria that produce two molecules of lactic acid from one molecule of glucose, see Figure S17

Horizontal gene transfer: also called lateral gene transfer: mechanism by which genetic material is transferred from one organism to another organism, see Figures 38 and 39

Humoral immunity: immunity caused by antibodies

Hybridization: joining of two complementary strands of DNA

Hydrogen bond: a bond between a covalently bound hydrogen atom and an electronegative atom such as oxygen or nitrogen. The development of hydrogen bonds is the basis of base pairing in DNA and RNA

Hydrogenases: these enzymes either take up H_2 or evolve H_2

Hydrothermal vents: hot water emitting springs on the sea floor, see Figure 14

Hyperthermophiles: organisms growing at temperatures higher than 80 °C

Hypervariable regions: they occur in immunoglobulins and in mitochondrial DNA, see Figure 57

Icosahedron: a platonic body bounded by 20 equilateral triangles, a very common shape of virus particles, see Figure 66

Immunity: an immune organism is in the possession of systems to resist infections

Immunodeficiency: lack of a functional immune system, for example, caused by HIV infections

In vitro: reactions outside of living organisms

In vivo: reactions in a living organism

Induction: for example, induction of enzyme synthesis in response to a specific signal, see Figures S31 to S33

Informational macromolecules: DNA, RNA, and proteins, information is stored in the sequence of these macromolecules

Innate immunity: the non-inducible ability of organisms to destroy invaders

Interspecies hydrogen transfer: consumption of H_2 by one organism allows H_2 production of the other partner organism; an example is given in Chapter 32, *Syntrophus aciditrophicus* and *Methanospirillum hungatei*

Isotopes: chemical elements containing the same number of protons and electrons, but differing in the number of neutrons and therefore in atomic weight. The atomic weight of carbon is 12. The discovery of carbon 14 and its importance is discussed in Chapter 8

Joule: a unit of energy, where one joule equals 10^7 ergs and 0.2390 calories

Kb: a fragment of nucleic acid, 1000 bases long

Koch's postulates or Henle-Koch postulates: they have to be fulfilled in order to prove that a certain microorganism is responsible for a certain disease

Korarchaeota: phylum of the archaea, see Figure S11

Lag phase: the phase after inoculating a culture, in which there is little apparent activity, see Figure S1

Lagging strand: the DNA strand which has to be synthesized in the opposite direction of the process of replication, see Figures 3 and S6

Lateral gene transfer: see horizontal gene transfer

Leading strand: the strand of DNA synthesized in the direction of DNA replication

Leghemoglobin: a protein in the root nodules that binds oxygen, see Figure 25

Lethal Factor (LF): component of anthrax toxin

Lipopolysaccharide (LPS): major component of the outer membrane of Gram-negative bacteria. LPS of many bacteria is toxic, it is the endotoxin

Lyme disease: disease caused by *Borrelia burgdorferi*

Lysogeny: the phage DNA is integrated into the bacterial genome and is replicated as prophage

Lysozyme: an enzyme degrading peptidoglycan, the major component of microbial cell walls

Lytic pathway: following infection, virus particles are produced and the host cell undergoes lysis

Macromolecules: see informational macromolecules; polysaccharides such as starch or cellulose are non-informational macromolecules

Magnetosome: a pearl of particles consisting of iron oxide, see Figure 73

Megabase (Mb): 1 million of bases in nucleic acids

Membrane: see cytoplasmic membrane

Membrane potential: component of the protonmotive force

Memory B cells: they are long-lived and specifically bind their antigen

Meningitis: this is an inflammation of brain tissue caused by several bacterial species including *Neisseria meningitidis* and *Listeria monocytogenes*

Mesophiles: organisms that live at temperatures between 15 °C and 40 °C

Messenger RNA (mRNA): genes at the DNA are transcribed into messenger RNA. From there the information is used at the ribosomes for protein synthesis, see Figure 3

Metagenome: the total of the genetic information present in all cells of a particular habitat

Methanogens: methane-producing archaea

Methanogenesis: the process leading to the formation of methane

Methanotrophy: the process of oxidizing methane to CO_2

Methylotrophy: the process of oxidizing methyl group-containing compounds such as methanol, methylamine or methane, so methanotrophs are also methylotrophs; compounds containing carbon-carbon-bonds do not belong here

Methyl transferases: enzyme systems transferring methyl groups, one functioning as Na^+ pump is involved in methanogenesis, see Figure S13

Microbial leaching: growth of acid-producing microorganisms such as the thiobacilli on sulfide-containing ores results in acid production and leaching of metals such as copper

Miller-Urey experiment: production of organic compounds from gases, see Figure 7

Mitochondria: organelles of microbial origin in eukaryotic cells, involved in the generation of ATP, see Figures 2 and 57

Mutation: a change in the base sequence of the genome, see Figure 72

Nanoarchaeota: phylum of the archaeota, see Figure S11

Negative RNA strand: it is complementary to the positive viral mRNA strand, see Figure 67

Nickel-tetrapyrrole: a cofactor involved in methanogenesis, see Figure 28

Nitrification: microbial oxidation of ammonia to nitrate, see Figure 26

Nitrogen fixation: reduction of N_2 to ammonia as catalyzed by nitrogenase, see Figure 26

Nodules: root tubercles filled with nitrogen-fixing bacteroids, see Figure 25

Non-coding RNA: it is different from messenger RNA and is not translated

Nosocomial infection: an infection acquired in hospitals

Nucleoside: a base connected to ribose

Nucleotide: a base connected to ribose and to phosphate

Obligate: it refers to a distinct quality, for example, an obligate chemolitotroph will not grow as organotroph

Oligotrophy: ability to live under very poor nutrient conditions

Open reading frame (orf): a DNA fragment with all the information for transcription and translation into a protein

Operator: a region of the DNA where a repressor may bind to prevent transcription, see Figure S31

Operon: genes transcribed starting from a single site, see Figure S31

Opportunistic pathogen: expression of a pathogenic lifestyle under certain conditions

Outer membrane: part of the cell wall of Gram-negative bacteria. It consists of lipopolysaccharides and proteins

Oxidation: process in which electrons are removed from a chemical element or a compound, for example, Fe^{2+} to Fe^{3+}

Oxidative phosphorylation: see electron transport phosphorylation

Oxygenases: enzyme systems by which oxygen from O_2 is incorporated into compounds, see Figure S23

Oxygenic photosynthesis: it leads to O_2 evolution and is performed by cyanobacteria, see Figure 21

Pasteurization: significant reduction of the number of spoilage- or disease-producing bacteria by short-term heating, for example, of milk

Pathogenicity islands: defined regions on the bacterial chromsome with genes encoding virulence factors, see Figure 64

Penicillin: antibiotic belonging to the β-lactam antibiotics, see Figure 44

Peptidoglycan: also called murein, major component of the bacterial cell walls

Peritrichous flagellation: flagellae are distributed all over the bacterial surface

Pertussis: whooping cough caused by *Bordetella pertussis*

Phosphodiester bond: a covalent bond between nucleotides

Phosphotransferase system (PTS): a sugar transport system, see Figure S17

Photoautotrophy: ability to use light as energy source and CO_2 as carbon source; cyanobacteria are photoautotrophs

Photosynthesis: the use of light for ATP synthesis and for generation of reducing equivalents to convert CO_2 into cell material

Phylogeny: the history and the relationship of organisms

Phylum: major lineage of species and genera within the domains of life

Phytanyl: a hydrocarbon that is a major component of the cytoplasmic membrane of archaea, see Figure S9

Phytopathogens: microorganisms causing plant diseases

***Picrophilus torridus*:** the most robust archaeon, see Figure 12

Pilus: a structure formed by Hfr or F^+ cells during conjugation, see Figure 39

Plaques: clear zones of lysis on a lawn of cells caused by bacteriophages, see Figure 40

Plasma cell: short-lived B lymphocytes which produce antibodies

Plus RNA strand: it serves as mRNA in viruses

Point mutations: just one base pair is altered

Polar flagellation: flagellae are attached at one or at both ends of the cell, see Figure 8

Poly- β-hydroxybutyrate (PHB): storage material of many bacteria, see Figure 60

Polymerase chain reaction (PCR): millionfold amplification of a DNA sequence, see Figures 51 and 52

Polynucleotide: a polymer of nucleotides linked by phosphodiester bonds, see Figures 5 and S6

Polypeptide: it is made of several amino acids held together by peptide bonds; polypeptides are also called proteins, see Figure 58

Porins: channels in the outer membrane of Gram-negative bacteria allowing the passage of small molecules from the outside into the periplasmic space

Primary producers: organisms that make biomass from CO_2

Primer: a short DNA fragment required for the initiation of the DNA polymerase reaction, see Figure 52

Prion: an infectious protein

Probiotic bacteria: living microorganisms, mostly lactic acid bacteria taken for the benefit of the patient

Prokaryotes: all members of the bacterial and archaeal domain; they lack a nucleus, see Figure 6

Promoter: the region on the DNA at which transcription starts by binding RNA polymerase, see Figure S31

Prophage: they reside in the bacterial genome and are replicated together with the genome of the host, see Figure 42

Proteobacteria: a phylum of the bacteria including Gram-negative bacteria

Proteome: all the proteins present in a particular organism, see Figure 71

Proteorhodopsin: a retinal-containing protein present in certain marine bacteria, it is related to bacteriorhodopsin

Proton motive force: it consists of a proton gradient and the membrane potential across the cytoplasmic membrane

Protoplast: cell without cell wall

Psychrophiles: organisms that grow at low temperatures, usually below 10 °C

Pure culture: it represents a clone originating from one cell, see Figure 4

Purines: compounds such as adenine and guanine, two of the four bases in DNA

Purple nonsulfur bacteria: they use H_2 or organic compounds as H donors

Purple sulfur bacteria: they use H_2S as H donor

Pyrimidine: chemical structures such as cytosine and thymine present in DNA, and uracil present in RNA

Pyrite: the iron ore FeS_2

Pyrodictium occultum: one of the most "heat-loving" archaea, see Figure 11

Pyrogenic: fever-inducing

Quorum sensing: a regulatory mechanism depending on the population size; above a certain cell density expression of certain genes is activated, see Figures 54 and S21

Reaction center: complex of pigments in the photosynthetic apparatus in which the electrons are released which drive the generation of the protonmotive force during photosynthesis, see Figures 20 and 21

Recombinant DNA: DNA fragments in an organism originating from other sources, see Figure 50

Recombination: a process by which DNA fragments are incorporated into a chromosome

Redox potential: E_{01}, the tendency of a compound to become reduced or oxidized. The redox potential of the hydrogen electrode under standard conditions is 0 mV, of $NAD^+/NADH + H^+$ −320 mV, and of $\frac{1}{2}O_2/O_2$ +810 mV

Reduction: a reaction in which a chemical element or a compound accepts electrons, for example, $Fe^{3+} \rightarrow Fe^{2+}$ or $O \rightarrow O^{2-}$

Replication: synthesis of DNA by DNA-dependent DNA polymerase, see Figure S6

Replication fork: the site on the DNA at which replication takes place, see Figure S6

Repression: repression of protein synthesis by inhibition of transcription or translation

Repressor: a protein that blocks transcription, see Figure S31

Respiration: generation of a proton motive force by redox reactions and use of this protonmotive force for ATP synthesis, see Figure S18

Response regulator: one member of two-component-systems, see Figure S20

Restriction enzymes: enzymes that cleave DNA at specific sequences, see Figure 49

Retinal: cofactor of bacteriorhodopsin, see Figure 18

Retrovirus: a virus containing RNA which is transcribed into DNA for replication, a prominent example being HIV, see Figure 68

Reverse transcriptase: a viral enzyme that synthesizes DNA using RNA as template, see Figure 68

Rhizosphere: the region around plant roots

Ribonucleic acid (RNA): an informational molecule consisting of a base and ribose connected by phosphate bridges

Ribosomal RNA: 5S, 16S, and 23S RNA are constituents of ribosomes, for 16S RNA see Figure 5

Ribosomes: the protein-synthesizing factories of the cell, in prokaryotes they consist of 30 S and 55 S subunits; they contain three types of ribosomal RNAs and 50 proteins

Ribozyme: an RNA molecule with catalytic properties

Rickettsia prowazekii: obligate intercellular bacterium causing typhus

RNA polymerase: an enzyme synthesizing RNA using DNA as template, see Figure S31

RNA world: an ancient world in which RNA was the only informational macromolecule; this RNA also exhibited catalytic activities

Rubisco: abbreviation for ribulose 1,5-bisphosphate carboxylase

Rumen: one of the stomachs of ruminant animals (cows, goats, and sheep)

Secondary metabolites: compounds like antibiotics produced by a number of microorganisms in the stationary phase of growth

Sensor kinase: the component of two-component systems which responds to external signals; it develops autophosphorylation activity and transfers the phosphoryl moiety to the second component, see Figure S20

Sequencing: the determination of the order of the four types of nucleotides in DNA or RNA molecules, see Figure 69

Shotgun sequencing: the whole genome is converted randomly into small fragments which are cloned and sequenced. By applying computational methods, the whole genome sequence is reconstructed

Signal sequence: a stretch of about 20 amino acids at the N-terminus of a protein, it represents the so-called signal peptide indicating that this protein should be exported

16S rRNA: the differences in the sequence of this molecule in microbial species has been the basis for the construction of phylogenetic trees, see Figures 5, 6, S11 and S14

Solfatara: hot and acidic springs which are the habitat of hyperthermophilic archaea, for example, the Solfatara north of Naples (Italy) or in Japan, see Figure 12

Species: taxon below the genus level, for example, the genus *Clostridium* comprises about 100 species such as *Clostridium botulinum* (highly pathogenic) or *Clostridium pasteurianum* (harmless soil bacterium). These species have properties in common (e.g. anaerobic metabolism, spore formation etc.), at the species level they are characterized by the sequence of their 16S-rRNA and by other properties

Spheroplast: a bacterium that has lost most of its cell wall structures

Spirillum: spiral-shaped microorganisms, for example, *Magnetospirillum*, see Figure 73

Spirochete: tightly coiled bacteria such as *Treponema pallidum* (syphilis) or *Borrelia burgdorferi* (Lyme disease)

Spores: resistant resting structures of organisms; of special interest are the spores of the genera *Bacillus* and *Clostridium* often called endospores; they are extremely resistant to heat and to dessication

Stalk: extracellular structure of organisms which may serve for attachment to surfaces

Start codon: it signals the beginning of translation; usually it is AUG

Stationary phase: a phase which follows the growth phase of a culture, see Figure S1

Sterilization: removal of all living organisms from an environment, it can be accomplished by heating or by ultrafiltration

Stickland reaction: pairwise fermentation of amino acids, for example, glycine and alanine. It is common among amino acid fermenting clostridia

Stop codon: signal to terminate translation; stop codons are UAG, UAA and UGA, see Figure S8

Stromatolites: microbial mats consisting of cyanobacteria, often sulfur-oxidizing bacteria and others; they were fossilized and represent the earliest records of life on earth, see Figure 10

Substrate level phosphorylation: from compounds containing high-energy phosphate bonds, phosphate is transferred to ADP to give ATP, such compounds are phosphoenol pyruvate, 1,3-bisphosphoglycerate and acetyl phosphate, see Figures S17 and S28 to S30

Sulfate-reducing bacteria and archaea: they convert sulfate into hydrogen sulfide, see Figure S12

Symbiosis: relationship between two organisms beneficial to both

T3S: type III secretion systems, see Figure 63

Taxonomy: identification and classification of organisms

Teichoic acids: constituents of the cell walls of Gram-positive bacteria

Temperate phages: phages coexisting with their hosts, their DNA is integrated in the microbial genome, see Figure 42

Tetanus: disease caused by *Clostridium tetani*, see Figure 61

Tetracycline: a group of antibiotics, see Table S6

Thermophiles: organisms growing optimally between 45 °C and 80 °C

Thermosome: a heat-shock protein complex in archaea

Thiomargarita namibiensis: the largest bacterium, see Figure 75

Ti-plasmid: it is present in *Agrobacterium tumefaciens* and is involved in gene transfer to plants, see Figure 48

Transcription: synthesis of RNA by RNA-polymerase with DNA as template, see Figures 3 and S31

Transcriptome: the total of mRNAs produced by an organism

Transduction: transfer of host genes by a virus from one cell to another

Transfection: transfer of genes cloned into the genome of a virus

Transfer RNA (tRNA): it carries its amino acid to the ribosome

Transformation: transport of free DNA into microbial cells, see Figure 38

Transgenic organisms: they contain foreign DNA in their genomes

Translation: protein synthesis at the ribosomes, see Figure 3

TT dimers: they are formed in DNA by UV light

Two-component systems: they consist of a sensor kinase and a respone regulator, see Figure S20

Typhus: a disease caused by *Rickettsia prowazekii*

Universal phylogenetic tree: a tree including all domains of life, see Figure 6

UPEC: uropathogenic *E. coli*

Vaccination: introduction of an inactivated or weakened pathogen or toxin to stimulate immune response

Vacuole: a compartment inside the cells containing a fluid or gas

Vector: a self-replicating DNA molecule, for example, a plasmid carrying cloned genes

Virulence: degree of pathogenicity expressed by an organism

Virus: a particle containing either RNA or DNA and capable of penetrating host cells in which it proliferates

Vital counts: number of living cells in a microbial population

Vitamin C: L-ascorbic acid synthesis, see Figure S25

Wild-type: a strain of a particular species isolated from nature

Winogradsky columns: glass columns filled with mud and water, also containing calcium carbonate and an organic compound. They make it possible to follow the diverse microbial activities in such a habitat over weeks and months

Yeast: single celled growth-form of some fungi

Appendix C Subject index of figures and tables

Antibiotics
Antibiotic susceptibility	Figure 45
Antibiotics used in medicine	Table S6
Discovery of penicillin	Figure 43
Structure of penicillin	Figure 44

Archaea, micrographs and electron micrographs
Ignicoccus hospitalis and *Nanoarchaeum equitans*	Figure 76
Methanoarchaea	Figure 15
Picrophilus torridus	Figure 12
Pyrodictium occultum	Figure 11

Bacteria, micrographs and electron micrographs
Chromatium okenii	Figure 8
Escherichia coli strain 536	Figure 64
Magnetospirillum gryphiswaldense	Figure 73
Mycobacterium tuberculosis	Figure 62
Ralstonia eutropha	Figure 1
Ralstonia eutropha accumulating PHB	Figure 60
Sorangium cellulosum and *Chondromyces robustus*	Figure 74
Sporomusa spec	Figure 31
Sulfate-reducing bacteria	Figure 15
Thiomargarita namibiensis	Figure 75

Bacteriophages
Bacteriophage T$_4$	Figure 41
Integration of phage DNA	Figure 42
Plaques on an *E. coli* lawn	Figure 40

Biotechnology
Biofuel	Figure 32
Exoenzymes	Table S2

Making goat cheese — Figure 33
1,3-propanediol — Figure 59
Poly-β-hydroxybutyric acid — Figure 60
Products of bioindustries — Table S7
Synthesis of ascorbate — Figure S25
A vinegar plant — Figure 34

Cycles
Carbon cycle — Figure 16
Carbon cycle — Figure 29
Nitrogen cycle — Figure 26
Sulfur cycle — Figure S12

DNA
Base pairing — Figure S5
Cloning of a gene — Figure 50
Features of the *C. tetani* genome — Table S4
G + C content of microorganisms — Table S5
Genomics, see also DNA transfer and plasmids — Figure 70
Human mitochondrial DNA — Figure S7
PCR products — Figure 51
Principle of PCR — Figure 52
Repair of UV-damaged DNA — Figure 72
Replication — Figure S6
Replication, transcription, translation — Figure 3
Sequencing of DNA fragments — Figure 69

DNA transfer
Conjugation — Figure 39
Integration of phage DNA — Figure 42
Transformation — Figure 38

Evolution
Evolution of azoreductases — Figure 37
The Miller-Urey apparatus — Figure 7
Proterozoic stromatolites — Figure 10
See also Phylogenetic trees

Growth
Bacterial colonies on agar — Figure 4
Continuous culture — Figure S2
Growth curve — Figure S1

Models
Anoxygenic photosynthesis — Figure 20
ATP-consuming reactions — Table S1

Bacteriorhodopsin proton pump   Figure 18
Composition of ATP synthase   Figure 19
The eukaryotic and the prokaryotic cell   Figure 2
Nitrogenase   Figure 25
Oxygenic photosynthesis   Figure 21

Organisms en Masse
Anaerobic methane oxidation   Figure 15
Bacterial imprint on soybean leaf   Figure 65
Giant tube worms   Figure 14
Phototrophic bacteria   Figure 8
Salt works on Lanzarote   Figure 17
Subaerial and subaquatic biofilms   Figure 55
Symbiotic nitrogen fixation   Figure 25

Pathogenic bacteria
Compilation of pathogenic bacteria   Table S8
Man suffering from tetanus   Figure 61
Microbes and men   Figure 22
Mycobacterium tuberculosis   Figure 62
Uropathogenic *E. coli*   Figure 64

Pathways
Acetone-butanol fermentation   Figure S29
Alcohol fermentation via Embden-Meyerhof-Parnas and Entner-Doudoroff pathways   Figure S28
Calvin-Benson-Bassham cycle   Figure S26
Degradation of long-chain hydrocarbons   Figure S24
Glyoxylate cycle   Figure S22
Lactic acid fermentation   Figure S17
Methane oxidation   Figure S23
Methanogenesis   Figure S13
Mixed acid fermentation   Figure S30
Synthesis of L-ascorbate   Figure S25
Wood-Ljungdahl pathway   Figure S27

Phylogenetic trees
Archaeal and bacterial 16S-rRNA sequences   Figure 5
Tree of archaea   Figure S11
Tree of bacteria   Figure S14
Tree of the three domains   Figure 6

Plasmids
Electron micrograph of a plasmid   Figure 46
Plasmid of enterotoxinogenic *E. coli*   Figure 47
Ti-plasmid   Figure 48

Processes at the cytoplasmic membrane

Active transport	Figure S16
The aerobic respiratory chain	Figure S18
Capillary test for demonstration of chemotaxis	Figure 53
Chemotaxis	Figure S19
Two-component systems	Figure S20
See also Models	

Proteins

Composition of insulin	Figure 58
Proteomics	Figure 71
The genetic code	Figure S8

Regulation

Attenuation	Figure S32
Bioluminescent bacteria	Figure 54
Quorum sensing	Figure S21
Regulation of *nif* genes	Figure S33
Regulation of transcription	Figure S31

RNA

Replication, transcription, translation	Figure 3
Sequence of 16S rRNA	Figure 5
See also Regulation	

Special figures

Ames test	Figure 9
"Burning air"	Figure 27
Coral bleaching	Figure 56
Madaba map	Figure 14
Table of bioelements	Figure 36

Sporulation

Sporomusa	Figure 31
Steps in spore formation	Figure S15

Structures

The 20 amino acids	Figure S7
Structure of ATP	Figure S10
Building blocks of DNA	Figure S2
Characteristics of RNA	Figure S4
Coenzyme B_{12}	Figure 35
Composition of the cytoplasmic membrane and of murein	Figure S9
Nickel-tetrapyrrole	Figure 28

Viruses

Composition of viruses	Figure 66
Human viruses	Table S9
Infection cycle of influenza viruses	Figure 67
Infection cycle of HIV	Figure 68

Credits

Permissions were kindly granted for following chapter openings (poems and quotes):

p.23: ASM Press/Carl R. Woese: The Archae: An Invitation to Evolution, in: Ricardo Cavicchioli: Archae – *Cellular and Molecular Biology*, ASM Press, Washington, p. 11, 2007.

p. 65: Springer/ C.B. Niel: On the morphology and physiology of the purple and green sulphur bacteria, *Archives of Microbiology* **3**, 1–112, 1932.

p. 73: "New Year Greeting", ©1969 by W.H. Auden, from *Collected Poems of W. H. Auden* by W.H. Auden. Used by permission of Random House, Inc. In the UK: "A New Year Greeting" from *Collected Poems*, © The Estate of W.H. Auden 1976, 1991, All rights reserved.

p. 81: The Literary Trustees of Walter de la Mare and the Society of Authors/Walter de la Mare: *Peacock Pie: A Book of Rhymes*, Faber Kids, 2001.

p. 105: The Jewish Publication Society/Chaim Weizmann: *Trial and Error – The Autobiography of Chaim Weizmann*, JPS, 1949.

p. 221: Macmillan Publishers Ltd/David Baltimore: Our genome unveiled, *Nature* **409**, 814–816, 2001.

index

Index

a

ABC transporter 297
acetate 311f.
acetic acid bacteria 124
Acetobacter xylinum 124
Acetobacterium woodii 292, 318
acetone 105ff., 321
acetone-butanol fermentation 321f.
acetyl-CoA 313
Acidianus ambivalens 282ff.
Acinetobacter
– *baumannii* 215
– *calcoaceticus* 313
– species 249
acyclovir (Zovirax) 227
adenine 268
adenosine 5'-diphosphate (ADP) 9, 60, 330
adenosine 5'-triphosphate (ATP) 9, 60, 274f., 301, 317, 330
– synthesis in microbes 301
Agrobacterium tumefaciens 165ff., 291
AIDS (acquired immunodeficiency syndrome) 222ff.
D-alanine 123
L-alanine 123
Alcanivorax
– *borkumensis* 313
– species 249
alcohol dehydrogenase 111ff.
alcohol fermentation 318ff.
– via the EMP and the ED pathways 321
Ames test 33
amino acid 271
– fermentation 321
ammonia 87ff.
ammonia monooxygenase 313
amphotericin 156
amylase 300

Anabaena 289
– *azollae* 85
anabolic metabolism 288
anammox 89
animal
– bacteria 203ff.
anthrax 209f.
antibiotic 151ff., 338f.
– resistance gene 160ff.
antimicrobial 150
Aquifex aeolicus 289
Arc system 307ff.
archaea 7, 17ff., 53, 278, 281ff.
– metabolism 283
– methane 93
– methanogenic 292
– phylogenetic tree 282
Archaeoglobus fulgidus 284
Arthrobacter simplex 313
ascorbic acid (vitamin C) 124f.
Aspergillus species 215
ATP, *see* adenosine 5'-triphosphate
ATP synthase 60, 66, 303
autotrophic growth 316
Azoarcus 85, 185
Azolla 85
Azospirillum 85, 185
Azotobacter vinelandii 293

b

Bacillus 156, 194ff.
– *anthracis* 149, 186, 209, 290ff.
– *licheniformis* 194, 290ff., 334f.
– *subtilis* 194, 240, 274, 293, 325f.
– *thuringensis* (Bt) 167, 293
bacteria 53f., 73f.
– animal 203ff.
– climate 99f.
– competence 139

Discover the World of Microbes: Bacteria, Archaea, and Viruses, First Edition. Gerhard Gottschalk.
© 2012 Wiley-VCH Verlag GmbH & Co. KGaA. Published 2012 by Wiley-VCH Verlag GmbH & Co. KGaA.

– deep sea 47f.
– diversity 289ff.
– ethanol-producing 111
– food resource 203ff.
– Gram-negative 152f., 274
– Gram-positive 152f., 274
– human 203ff.
– methanotrophic 314
– nickel-resistant 131
– nitrifying 313
– phylogenetic tree 289f.
– plant 217f.
– production factory 191ff.
– resistance 160ff.
– sulfate-reducing 50
– symbiotic 48
bacterial fermentation 318
bacterial food 203
bacterial genome 239
bacterial growth 4ff., 261ff.
bacterial sex life 133ff.
bacterial spore 108
bacterial sulfur cycle 284
bacteriophage 145ff.
bacteriorhodopsin 58f., 283, 304
Bacteroides fragilis 77
base pairing 269
beer 115
Beggiatoa 252
– *alba* 285
beta (β)-lactam antibiotic 153f.
beta (β)-lactamase 135
Bifidobacterium 79
bile acid 78
biodiesel 118
bioelement 127
bioenergy 118
bioethanol 114f.
biofilm 182
biofuel 102, 109, 111ff.
biogas 93ff., 117f.
bioinformatics 235
biological warfare 209f.
bioluminescent bacteria 178
biomass 101, 109, 114, 117
biotechnology 191, 339f.
Bordetella pertussis 204
Borrelia burgdorferi 206
borreliosis 205
bronchitis 209
bubonic plague 205
Buchnera
– *aphidicola* 187
butanol 105ff., 117, 321
γ-butyrolactone 309

c

c-AMP receptor protein (CRP) 328
Calvin-Benson-Bassham (CBB) cycle 62, 316f.
cancer 215ff.
candicidin 156
Candida
– *albicans* 76, 215
– *lipolytica* 313
– species 249
capsid 221
capsomer 221
carbohydrate 311
carbon cycle 56, 101
carbon dioxide 100ff., 111
– fixation 317ff.
carbon dioxide cycle 101
carbon metabolism 311ff.
carbon monoxide 246
Carboxydothermus hydrogenoformans 246, 277
carcinogenic compound 33f.
Carsonella
– *ruddii* 333
– species 187
cell membrane 273
cell wall 273f.
cellulase 113f., 300
cellulose 114
Cenarchaeum symbiosum 281
Cephalosporium 154ff.
– *roseum* 313
Che protein 306
cheese 121ff.
chemiosmotic mechanism 301
chemolithotrophic way of life 49
chemostat 263
chemosynthetic symbiosis 48
chemotaxis 306
chitinase 300
Chlamydia pneumoniae 209
Chlorella 61f.
Chlorobium limicola 291
Chloroflexus
– *aurantiacus* 289
chloroplast 187
cholera toxin (CTX) 211
Chondromyces robustus 294
Chromatium okenii 179, 285, 291
Citrobacter species 196f.
Clamydia trachomatis 76
cloning 173
Clostridium
– *aceticum* 292, 318, 325
– *acetobutylicum* 105ff., 290ff., 321f., 325

– *acidiurici* 325
– *botulinum* 290ff., 323
– *difficile* 77, 211
– *histolyticum* 204, 323
– *ljungdahlii* 277, 292, 318
– *pasteurianum* 325
– *perfringens* 204
– species 196
– *sporogenes* 323
– *sticklandii* 323
– *tetani* 204f., 238, 290ff., 326, 333
– *thermoaceticum* 130
cobalt 127
coenzyme B 95
coenzyme B_{12} 128
coenzyme F_{430} 94, 127
coenzyme M 95
cold seep 50
cold vent 50
combustible air 91ff.
competence
– bacteria 139
complementary DNA (cDNA) 224
conjugation 140ff.
conjugative plasmid 142
coral bleaching 186
Corynebacterium glutamicum 179, 192
Crenarchaeota 281
CTP (cytidine 5′-triphosphate) 330
cyanobacteria 31ff., 69f.
cyclic AMP (cAMP) 328
cytoplasmic membrane 273
cytosine 268

d

deep-sea vent 48
Deinococcus radiodurans 245f., 289
denitrification 304
Desulfobacterium autotrophicum 285, 304, 318
Desulfovibrio vulgaris 179, 285
Desulfurococcales 281
Desulfuromonas acetoxidans 285
diabetes 194f.
diarrhea 210
dicarboxylic azobenzene (DCAB) 134
dihydroxyacetone phosphate (DAP) 316
DNA (deoxyribonucleic acid) 7ff., 267
– base 268
– complementary (cDNA) 224
– methylation 171, 326
– mitochondrial (mtDNA) 187f.
– regulation 326
– replication 10, 270
– restriction enzyme 169ff.
– sequence 236
– transfer 136ff.
DNA polymerase 174, 228
DNA virus 225
Dunaliella 45f.

e

Eco R1 169
EET (extracellular electron transfer) 248
EHEC (enterohemorrhagic *E. coli*) 211f.
Embden-Meyerhof-Parnas (EMP) pathway 302, 311, 321ff.
endosymbiont 185ff.
endosymbiosis 70f.
energy 297ff.
energy conservation 117ff.
Enterococcus
– *aecium* 215
– *faecalis* 215
Entner-Doudoroff (ED) pathway 311ff.
enzyme 9, 269
– anaerobic 321
enzyme activity
– regulation 330
epidemics 221
Epstein-Barr virus (EBV) 227, 344
Erwinia uredovora 167
erythromycin 155
Escherichia coli 76f., 94, 145ff., 179, 196, 213ff., 291, 326ff.
ETEC (enterotoxigenic *E. coli*) 211f.
ethanol 111
Eubacterium rectale 77
eukaryotic cell 13
Euryarchaeota 281
exoenzyme 300
extremozyme 200

f

F-factor 159
fermentation
– acetone-butanol 105ff., 318ff., 321f.
– alcohol 318ff.
– amino acid 321
– butyrate 321
– homoacetate 319, 321
– lactic acid 318
– mixed acid 321ff.
– propionate 123
ferredoxin 119
Firmicutes 293
flagella 179f., 306
flavin adenine dinucleotide (FAD) 303
food chain 291

Frankia 85, 185
furanosyl borate ester 309

g
GC content 335
gene 333
gene cloning 173
gene technology 171
gene therapy 232
gene transfer 137, 142, 333f.
– horizontal 333, 137
genetic code 272
genetic engineer 165ff.
genetic engineering 196
genome 333
genomics 235ff.
Geobacter metallireducens 304
glacier ice 244
global warming potential (GWP) 103f.
Gloeocapsa 290
Gluconobacter oxydans (*Gluconobacter oxidans*) 124, 316
glucose 313
glucose isomerase enzyme 192
glyceraldehyde 3-phosphate (GAP) 316f.
glyoxylic acid (glyoxylate) cycle 311f.
GMO (genetically modified microorganism) 219
Gram-negative bacteria 152f., 274
Gram-positive bacteria 152f., 274
green manuring 84
green sulfur bacteria 67
growth
– aerobic heterotrophic 311
– microbial 4ff., 261ff.
guanine 268

h
HAART (highly active antiretroviral therapy) 231
habitat 283
Haemophilus influenzae 76, 235ff., 305
Haloarcula marismortui 46, 283
Halobacteriales 283
Haloferax volcanii 46
Haloraptus utahensis 42
Halorubrum sodomense 47
Helicobacter pylori 74, 216, 291
Heliobacterium 290ff.
hemagglutinin (HA, H) 225f.
hemicellulose 114
Herpes 222

Herpes simplex virus 227
Herpes virus 8 (HHV-8) 231
heterodisulfide 95
HIV (human immunodeficiency virus) 228ff.
homoacetate fermentation 321
homoserine lactone 309
horizontal gene transfer 137, 333
human
– bacteria 73ff., 203ff.
hydrocarbon 250
– degradation 315
hydrochlorofluorocarbon (HCFC) 99
hydrogen sulfide 49
Hydrogenobacter thermophilus 318
hyper cycle 27
hyperthermophiles 281

i
Ignicoccus hospitalis 255, 281
immunotherapy 149
infection 204
– in hospital 215
influenza virus 222ff.
– infection cycle 226
informational macromolecule 267
insulin 195
interbacterial relationship 177ff.
intestinal microflora 78
isoamyl alcohol 105

k
Klebsiella 197f.
– *aerogenes* 325
– *pneumoniae* 331
Korachaeota 281

l
β-lactam antibiotic 153f.
β-lactamase 135
lactic acid bacteria 121
lactic acid fermentation 318
Lactobacillus
– *bulgaricus* 121, 318
– *helveticus* 318
– *lactis* 318
– species 196
lambda (λ) bacteriophage 147f.
Legionella pneumophila 206ff.
Legionnaire's disease 206
Leuconostoc mesenteroides 318
lignocellulose 113
lipase 300
Listeria monocytogenes 214

listeriosis   214
LUCA (last universal common ancestor)   15ff., 277
Lyme disease   205

m

macromolecule
– informational   267
Magnetospirillum gryphiswaldense   247
malaria   205
MCP (methyl-accepting chemotaxis protein)   306f.
membrane   297ff.
metabolism
– anabolic   288
– archaea   283
– carbon   311ff.
– microbial   325
metagenomic library   242
metagenomics   241f.
metal needle effect   94
methane   91ff., 103, 117, 313f.
– aerobic oxidation   314
– anaerobic oxidation   50
methane monooxygenase   313
methane seep   50
methane-oxidizing archaea   51
methanoarchaea   50, 93f., 118
Methanobrevibacter smithii   282
Methanocaldococcus jannaschii   282, 292
Methanococcus   53
methanogen   318
methanogenesis   94, 285
methanogenic archaea   292
methanogenic pathway   286f.
Methanohalobium evestigatum   282
methanol   313
Methanopyrus kandleri   282
Methanosaeta concilli   282
Methanosarcina   305
– *acetivorans*   282
– *barkeri*   282
– *mazei*   282, 292, 326, 334
Methanosarcinales   285
Methanosphera stadtmanae   282
Methanospirillum hungatei   253
Methanothermobacter
– *marburgensis*   282
– *thermoautotrophicus*   282
methanotroph   313
methanotrophic bacteria   314
methicillin-resistant *Staphylococcus aureus* (MRSA)   215
methyl-coenzyme M   95

Methylobacterium   104, 218
Methylomonas   104
methylotrophy   217f.
microbe
– ATP synthesis   301
– climate maker   99ff.
– incredible   245ff.
– resistance to radiation   245
microbial growth   261ff.
– batch culture   261
– condition   263
– continuous culture   263
– logarithmic   261
microbial metabolism
– regulation   325ff.
microbial shape   264
microbial size   264
microorganism
– aerobic   264
– anaerobic   264
– chemotrophic   264
– dead sea   45ff.
– facultative anaerobic   264
Miller-Urey soup   24
mitochondria   187
mitochondrial DNA (mtDNA)   187f.
mixed acid fermentation   321ff.
molybdenum   83
Moorella thermoacetica   130, 318
motility   305
MRSA (methicillin-resistant *Staphylococcus aureus*, multiresistant *Staphylococcus aureus*)   161, 215
Mycobacterium
– *smegmatis*   313
– species   249
– *tuberculosis*   186, 206ff., 290
Mycoderma aceti   123ff.
Mycoplasma
– *genitalium*   238
– *mycoides*   256
– *pneumoniae*   209
Myxobacteria   293

n

NADH   302ff., 330
Nanoarchaeota   281
Nanoarchaeum equitans   255, 281
Natronomonas pharaonis   42, 254, 283   (connected on p. 254)
Neisseria meningitides   76
neuraminidase (NA, N)   225f.
nickel   94, 127ff.
nickel-resistant bacteria   131

nif gene 331f.
nitrate 88f.
nitrifying bacteria 313
nitrite 88f.
Nitrobacter winogradskyi 291
nitrogen 81ff.
nitrogen cycle 88, 304
nitrogen fixation
– symbiotic 84
– regulation 331
nitrogenase 82f.
Nitrosomonas europaea 291, 313
Nocardia petroleophila 313

o

Oculina patagonica 186
Oligotropha carboxydovorans 291
"omics" era 235
oncogene 232
Oscillatoria 289
oxidation
– aerobic 314
– D-glucose 316
– incomplete 315
oxygen 34ff., 278

p

P700 305
P840 305
P870 system 304
Paracoccus denitrificans 325, 326
pathogenic microorganism 341
pathogenicity island (PAI) 212f.
PCR (polymerase chain reaction) 172ff.
Pelagibacter ubique 249, 304
penicillin 150ff.
Penicillium 150ff.
phage 145, 335
phosphoenolpyruvate (PEP) 298, 312, 330
phosphotransferase system (PTS) 298
Photobacterium phosphoreum 180
photosynthesis 29ff., 59, 65ff.
– aerobic (oxygenic) 31, 68f.
– anaerobic (anoxygenic) 31, 69, 291
– first type 65
– second type 59
photosystem
– PSI 305
– PSII 305
phototrophic bacteria 30
phototrophic microorganism 263
phylogenetic tree
– archaea 282
– bacteria 289f.

Picrophilus torridus 41, 255, 283, 333
plague 211
Planctomyces 291
plant
– bacteria 217f.
plasmid 159ff., 165ff.
Plasmodium falciparum 205
pneumonia 209
Pocillopora damicornis 186
polio virus 222ff.
poly-*beta*-hydroxy fatty acid (PHF) 198ff.
poly-*beta*-hydroxybutyric acid (PHB) 198ff.
Prochlorococcus marinus 248f., 289
prokaryote 278
prokaryotic cell 13
prontosil 150
1,3-propanediol 196
propanediol dehydrogenase 197
prophage 335
Propionibacterium acnes 73ff., 290
propionic acid bacteria 123
Propionigenium modestum 252, 305
protease 193, 300
protein 269
proteomics 241ff.
proton motive force 59, 302ff.
proton translocation
– light-driven 304
Pseudomonas
– *aeruginosa* 76, 181, 209ff., 291, 309
– *fluorescens* 313
– *lindneri*, see *Zymomonas mobilis*
– *syringae* 218, 291
pullulanase 300
purple bacteria 66
Pyrococcus 201
– *furiosus* 283
Pyrodictium occultum 39f., 255, 282
Pyrolobus fumarii 282
pyruvate decarboxylase 111ff.

q

quorum sensing 309

r

radical 35
Ralstonia eutropha 199f., 291, 325
Reclinomonas americana 189
reducing equivalent 31
renewable resources 117
repair system 294
replication 10, 270
resistance
– bacteria 160ff.

resistome 163
respiratory chain
– aerobic 302f.
restriction enzyme 169ff.
retrovirus 228ff.
Rhizobium radiobacter 165
rhizosphere 96
Rhodospirillum rubrum 291
ribonucleotide reductase 233
ribozyme 26ff.
ribulose 1,5-bisphosphate 62, 316
ribulose 1,5-bisphosphate carboxylase (rubisco) 63
Rickettsia
– *prowazekii* 186ff.
– species 291
RNA (ribonucleic acid) 7ff., 222ff.
– 16S-rRNA 17ff., 55
– messenger (mRNA) 267, 326ff.
– ribosomal (rRNA) 267
– small (sRNA) 329
– synthesis 12
– transfer (tRNA) 269
– translation 272
– triplet 272
RNA polymerase 228
RNA virus 225
RNA world 277
ROS (reactive oxygen species) 35, 278
Roseobacter 304

s

Saccharomyces cerevisiae 111, 197, 318
Salmonella
– species 204ff.
– *typhi* 211ff., 291
– *typhimurium* 33, 179
saltpeter 87
salvarsan 150
selenocysteine 130
sensor kinase 307ff.
sequencing
– DNA 236
Shewanella oneidensis 248, 304
shingles 227
SHV-1 β-lactamase 135
Sinorhizobium meliloti 84
sodium glutamate 191
sodium ion pump 305
Sorangium cellulosum 333
Spanish influenza virus 224
Sporomusa 109, 293
sporulation 294

Staphylococcus
– *aureus* 179, 204ff., 274, 290, 309
– *epidermidis* 73, 215
Streptococcus 76
– *cremoris* 121
– infection 150
– *pneumonia* 209
– *salivarius* 76
– *thermophilus* 121
Streptomyces 154ff., 193, 290
– *griseus* 151ff.
streptomycin 151ff.
stromatolite 35f.
substrate-level phosphorylation 301
sugar 111
sulfate-reducing bacteria 50
Sulfolobales 281
Sulfolobus 252, 282
– *acidocaldarius* 281ff.
– *solfataricus* 246, 291
sulfur cycle 284
survival strategy 292
symbiont 186
symbiosis 185
symbiotic bacteria 48
symbiotic nitrogen fixation 84
symporter 297
Syntrophus aciditrophus 252f.

t

T-DNA region 165
T4 bacteriophage 147f.
Taq polymerase 174, 201
TCA, *see* tricarboxylic acid
TEM-1 β-lactamase 135f.
tetanus 204ff.
tetracycline 155
Thermobacterium mobile, see *Zymomonas mobilis*
Thermococcus 201
thermophiles 281
Thermoplasma acidophilum 283
Thermoproteales 281
Thermoproteus 282
– *tenax* 179, 255, 277, 281, 318
Thermotoga maritima 289, 334
thermozyme 42
Thermus
– *aquaticus* 39, 228, 289
– *thermophilus* 289
Thiobacillus 252
– *ferrooxidans* 255
– *thiooxidans* 255, 285, 291
thioester 26

Thiomargarita namibiensis 250f.
Thioploca 252
Thiospirillum jenense 180, 291
thymine 268
Ti plasmid 165
tobacco mosaic virus (TMV) 28
transcription 12, 326
– regulation 327ff.
transformation 139
translation 12, 272, 326
– regulation 328f.
transport 297
– active 298
– primary 297
– secondary 297
tricarboxylic acid (TCA) cycle 302, 312
Trichodesmium 289
tuberculosis 206ff.
tungsten 129
two-component system 307f.
typhoid fever 211

u
UPEC (uropathogenic *E. coli*) 212
uracil 269
urease 94

v
vanadium 83
vancomycin 215
vancomycin-resistant enterococci (VRE) 215
Varicella zoster 227
Vibrio
– *cholerae* 210ff., 291
– *fischeri* 180, 309
– *harveyi* 309
– *shiloi* 185f.
vinegar 123ff.
viral hypothesis 234
virus 7, 145, 221ff., 343f.
vitamin B_{12} 127
vitamin C synthesis 124
VRSA 215

w
wine 115
Wood-Ljungdahl pathway 318f.

x
Xanthomonas campestris 218
xylanase 300

y
Yersinia pestis 205ff.
yogurt 121

z
Zooxanthellae 185f.
zovirax 227
Zymomonas mobilis 112f., 291, 318